Formal Logic

Propositional Logic, Predicate Logic, Modal Logic, and Non-Classical Logics

Lucien Sina

13.5.2026

Contents

Preface

Formal logic is one of the foundations of scientific thinking. It provides a precise language for analyzing statements, concepts, inferences, and proofs, thereby creating tools that are of central importance in mathematics, computer science, philosophy, linguistics, and many other disciplines. Anyone who understands logical relationships gains not only a deeper understanding of formal systems, but also of the structure of arguments and the limits of inference itself.

This book aims to offer an accessible yet systematic introduction to formal logic. It is intended for readers who wish to engage with logical methods at either an introductory or a more advanced level, and it accompanies them from the first concepts of propositional logic through predicate logic to modal logics and selected non-classical logics. Special emphasis is placed on presenting the individual topics not only in a formal way, but also in a manner that makes their interconnections visible.

The focus is not merely on learning definitions and

theorems, but on developing a robust understanding of logical methods. For this reason, central concepts such as syntax, semantics, model, inference, equivalence, and normal form are introduced with care and supplemented by examples, exercises, and solutions. The path leads from the foundations of formal reasoning through algorithmic procedures such as resolution and SAT solvers to more advanced topics such as Herbrand theory, description logic, epistemic logic, fuzzy logic, default logic, and answer set programming.

This book follows a didactic approach that combines formal precision with gradual development. Many sections are designed not only to present individual results, but also to make their motivation, significance, and application apparent. In this way, logic is not meant to appear as a collection of isolated rules, but as a coherent field with a clear internal structure and considerable practical reach.

My thanks go to all those who have contributed to the creation of this book through suggestions, feedback, and critical comments. I am equally grateful to the readers who embark on the sometimes demanding, yet rewarding journey through formal logic. Feedback, corrections, and suggestions for improvement are always welcome, as they help refine and strengthen this book further.

I wish you an insightful and stimulating read.

Lucien Sina

0.1 Introduction and Overview

Logic is the discipline that systematically studies inference and reasoning. At its core are formal languages (syntax), which specify the rules for forming well-structured expressions, and semantics, which give those expressions meaning. With these tools, questions of truth, consequence, and the representation of knowledge can be formulated and analyzed precisely. A logical system therefore provides both an account of which expressions are valid and a framework for interpreting statements with respect to structures – so-called models.

0.1.1 Motivation and Relevance

The study of logic is fundamental for several reasons. On the one hand, logic forms the formal foundation of mathematics and of many areas of computer science. Concepts such as algorithms, computability, and complexity are historically closely connected with formal logic; their development and proof techniques are essential for understanding what can be solved by automated means. On the other hand, logic has direct practical applications: in the specification and verification of software and hardware, in knowledge representation in artificial intelligence, in database querying, in automated reasoning within theorem provers, and in the formal modeling used across many scientific disciplines.

0.1.2 Historical Context

Modern formal logic has its roots in the philosophy and mathematics of the 19th and 20th centuries. Key milestones include the work of Frege, Russell, and Hilbert, who advanced formal systems and axiomatic presentations, as well as Turing, Church, and Gödel, whose results clarified the relationship between computability and formal provability. These developments gave rise to central concepts that still shape both the technical and the theoretical study of logic today.

0.1.3 Aims of This Book

This book pursues three complementary aims:

1. To provide a precise and self-contained introduction to the most important logical systems (propositional logic, predicate logic, modal logics, selected non-classical logics, and answer set programming).

2. To build a bridge to computer science by covering algorithmic aspects, questions of decidability and complexity, as well as concrete methods (e.g. SAT solvers, model checking, and answer set programming).

3. To develop methodological skills: precise formalization, proofs by induction (both ordinary and structural), and a confident command of syntax and semantics.

0.1.4 Approach and Structure

Throughout this book, we follow a recurring pattern:

- **Syntax**: We define the formal language in terms of its signature, terms, and formulas; we specify formation rules and notation.

- **Semantics**: We explain how formulas are interpreted in structures and define notions such as truth, satisfiability, and validity.

- **Proof and Decision Procedures**: For each logic, we consider common proof systems (axiomatic systems, natural deduction, resolution, and related methods) as well as algorithmic procedures and their limits.

- **Examples and Applications**: At the end of each central chapter, we discuss typical applications and provide exercises.

0.1.5 Who Is This Book For?

This book is intended for advanced undergraduate and master's students in computer science, mathematics, and related fields, as well as for practitioners seeking a solid and mathematically grounded understanding of logic. Basic familiarity with elementary set theory, discrete mathematics, and proof techniques (in particular mathematical induction) is assumed; where necessary, we briefly review these foundations.

What This Book Does Not Cover

This work is not a collection of extensive programming exercises for specialized logic tools, nor does it aim to provide a complete historical account of all schools of logic. Instead, it focuses on the formal concepts, typical algorithms, and the connection to practical computer science. More specialized aspects (e.g. detailed implementations of SMT or SAT solvers) are discussed only insofar as they help illustrate theoretical results.

0.1.6 Learning Goals and How to Use This Book

After studying the first six chapters, readers should be able to:

- define formal languages and describe their semantics in structures;
- apply typical proof techniques in logic (mathematical and structural induction, proof by contradiction, and direct construction);
- understand the basic principles of algorithmic methods such as SAT solving and model checking, and carry out simple cases by hand;
- recognize the limits of formal systems (undecidability and complexity classifications) and draw conclusions from them for computer science.

Notes on Notation

We aim to use notation consistently; the main conventions are:

- $\mathbb{N}, \mathbb{Z}, \mathbb{R}$ denote the usual number sets;
- formula variables are usually written as p, q, r, φ, ψ, and individuals as x, y, z;
- for set operations we use $\cup, \cap, \setminus, \times$;
- quantifiers appear in the form $\forall x$ and $\exists x$.

Next Steps

We begin with a concise yet complete introduction to notation and review the necessary basics of set theory and induction. After that, we turn to propositional logic and then proceed step by step to the more advanced topics.

What follows are brief descriptions of each chapter, intended to provide orientation regarding the topics covered and the skills developed in the respective sections.

0.1.7 Chapter Summaries

In the following short sections, we outline the topics addressed in each chapter and indicate which competencies the reader is expected to gain from them.

Chapter 0 – Introduction and Foundations. This chapter establishes the central notational and conceptual foundations: set and relation theory, functions, principles of induction, the structure of formal languages, and the basic notions of semantics (models, satisfiability, validity, and logical consequence). It serves as a reference point to which the following chapters repeatedly return.

Chapter 1 – Propositional Logic. Propositional logic provides the entry point into formal reasoning: we introduce propositional variables and connectives ("and," "or," "not," "implies"), define semantic evaluation by means of truth tables, and discuss syntactic normal forms (CNF and DNF). Important topics include proof methods (resolution-based procedures and natural deduction), satisfiability (SAT), and basic complexity results.

Chapter 2 – Predicate Logic. Predicate logic extends propositional logic by adding quantifiers and predicates, thereby making it possible to express relations between individuals. We discuss signatures, structures, the interpretation of terms and formulas, as well as central metatheoretical results (completeness, incompleteness, Löwenheim–Skolem, and questions of decidability), and we show their consequences for computer science (e.g. limits of automatability).

Chapter 3 – Modal Logics (including Temporal Logic). Modal logics make it possible to qualify statements by means of operators such as "possible," "necessary," "always," and "eventually." We introduce Kripke semantics, discuss standard modal systems (K, T, S4, S5), and describe special instances such as temporal logics (LTL, CTL) together with their applications in model checking and in the specification of systems.

Chapter 4 – Further Non-Classical Logics and Answer Set Programming. In this chapter, we consider a selection of non-classical logics that represent typical extensions of, or alternatives to, classical logic: intuitionistic logic, paraconsistent logic, many-valued and fuzzy logic, default logic, and probabilistic logics. We also introduce Answer Set Programming (ASP) as a practical technique for knowledge representation and non-monotonic inference.

Chapter 5 – Summary and Outlook. The concluding chapter brings together the main results, highlights the connections between the areas discussed, and offers a brief outlook on automated reasoning, current research topics, and further applications in computer science, artificial intelligence, and mathematics.

Important concepts that recur throughout this book include:

- **Formula** – a syntactically well-formed expression of a logical language.

- **Structure or model** – an interpretation that assigns values to the symbols of the language.
- **Satisfiability** – the existence of a model that makes a formula true.
- **Logical consequence** – the relation in which a formula follows logically from a set of formulas.
- **Model checking** – the algorithmic verification of whether a structure is a model of a formula.

This book aims to provide a precise and application-oriented introduction to the central systems of logic. We place particular emphasis on formal precision, motivating examples, and algorithmic aspects such as computability and complexity.

0.2 Foundations

0.2.1 Notation and Set Theory

We use the standard notation of set theory. If X is a set and x is an object, then $x \in X$ means that x is an element of X. For two sets X and Y, we write $X \subseteq Y$ if X is a subset of Y.

The sets of natural numbers, integers, and real numbers are denoted by $\mathbb{N}$, $\mathbb{Z}$, and $\mathbb{R}$, respectively.

The Cartesian product of two sets X and Y is denoted by $X \times Y$. The power set of a set X is written as $\mathcal{P}(X)$ or 2^X.

Sets, Operations, and Relations

Sets can be specified implicitly by defining properties. For example,

$$D = \left\{ x \in \mathbb{N} \;\middle|\; \begin{array}{l} \exists y \in \mathbb{N} : y \mid x \\ \qquad\qquad \text{and } 1 < y < x \end{array} \right\}$$

is the set of all composite natural numbers.

Important set operations are union $\cup$, intersection $\cap$, and difference $\setminus$. For two sets X, X', we have:

$$\begin{aligned} X \cup X' &= \{\, x \mid x \in X \text{ or } x \in X' \,\}, \\ X \cap X' &= \{\, x \mid x \in X \text{ and } x \in X' \,\}, \\ X \setminus X' &= \{\, x \mid x \in X \text{ and } x \notin X' \,\}. \end{aligned}$$

The Cartesian product of X and X' is

$$X \times X' = \{\, (x, x') \mid x \in X,\ x' \in X' \,\}.$$

More generally, we write $X^n = X \times \cdots \times X$ (n times) for the n-fold Cartesian product.

A (binary) relation R between sets X and X' is a subset $R \subseteq X \times X'$. Instead of writing $(x, x') \in R$, we sometimes also write xRx'. An n-ary relation on X is a subset of X^n.

If $R \subseteq X \times X$ (that is, $X' = X$), we define:

- R is *reflexive* iff $(x, x) \in R$ holds for all $x \in X$.
- R is *symmetric* iff $(x, y) \in R$ implies $(y, x) \in R$.

- R is *transitive* iff $(x, y) \in R$ and $(y, z) \in R$ imply $(x, z) \in R$.

If R is reflexive, symmetric, and transitive, then R is called an *equivalence relation.*

A function f is a mapping $f : X \to Y$. For $x \in X$, we write $f(x)$ for the image of x. Any function can also be viewed as a special relation $f \subseteq X \times Y$ satisfying the functionality property: if $(x, y) \in f$, then there is no $y' \in Y$ with $y' \neq y$ and $(x, y') \in f$.

Let $f : X \to Y$.

- f is *injective* (one-to-one) iff for all $x, x' \in X$, $f(x) = f(x')$ always implies $x = x'$.
- f is *surjective* iff for every $y \in Y$, there exists an $x \in X$ such that $f(x) = y$.
- f is *bijective* iff it is both injective and surjective.

The inverse map f^{-1} is generally defined as a mapping into the power set:

$$f^{-1} : Y \to \mathcal{P}(X), \qquad f^{-1}(y) = \{\, x \in X \mid f(x) = y \,\}.$$

If f is a bijection, then f^{-1} is a mapping $f^{-1} : Y \to X$, and for each $y \in Y$, $f^{-1}(y)$ is the unique $x \in X$ such that $f(x) = y$.

If $f : X \to Y$ and $g : Y \to Z$ are mappings, then $g \circ f : X \to Z$ denotes their composition, defined by $(g \circ f)(x) = g(f(x))$ for all $x \in X$.

Asymptotic Growth Notation. For functions $f, g : \mathbb{N} \to \mathbb{N}$, one writes $g \in O(f)$ iff there exist constants $c > 0$ and $n_0 \in \mathbb{N}$ such that

$$\forall n \geq n_0 : \quad g(n) \leq c \cdot f(n).$$

Intuitively, $g \in O(f)$ means that asymptotically g does not grow faster than f.

0.2.2 Structures and Semantics

A structure $\mathcal{A}$ for a logical signature consists of an underlying set A together with an interpretation of the symbols of the signature. A formula φ is true in a structure $\mathcal{A}$ if $\mathcal{A} \models \varphi$ holds.

Important notions include:

- Satisfiability: There exists a structure $\mathcal{A}$ such that $\mathcal{A} \models \varphi$
- Validity: For all structures $\mathcal{A}$, we have $\mathcal{A} \models \varphi$
- Logical consequence: $\Phi \models \psi$

0.2.3 Principles of Induction

Inductive proofs are among the central proof techniques in logic and theoretical computer science. Here we distinguish between two common forms: *mathematical induction* over the natural numbers and *structural induction* over syntactically defined objects (terms, formulas, trees). Both methods are based on the principle of well-ordered or well-founded structures, but they differ in the way the induction step is formulated.

Mathematical Induction

Mathematical induction (also called simple or strong induction, depending on the formulation) is used for statements about the natural numbers. To prove a statement $A(n)$ for all $n \in \mathbb{N}$, one shows:

1. (*Base case*) $A(1)$ holds (or $A(0)$, depending on the convention).

2. (*Induction step*) For an arbitrary $n > 1$, one shows that from the assumption that $A(k)$ holds for all $k < n$, it follows that $A(n)$ also holds.

Mathematical induction then follows from the fact that no smallest set of counterexamples can exist.

Example 0.1. *(Gauss's summation formula) For all $n \in \mathbb{N}$, we have*

$$\sum_{i=1}^{n} i = \frac{n(n+1)}{2}.$$

Proof. Base case: For $n = 1$, we have $\sum_{i=1}^{1} i = 1$ and $\frac{1\cdot(1+1)}{2} = 1$. Hence the equation holds for $n = 1$.

Induction step: Let $n > 1$ and assume that the formula holds for $n - 1$, that is,

$$\sum_{i=1}^{n-1} i = \frac{(n-1)n}{2}.$$

Then for n we obtain:

$$\begin{aligned}\sum_{i=1}^{n} i &= \Big(\sum_{i=1}^{n-1} i\Big) + n \\ &= \frac{(n-1)n}{2} + n \\ &= \frac{n(n-1)+2n}{2} \\ &= \frac{n(n+1)}{2}.\end{aligned}$$

This proves the induction step, and therefore the statement holds for all n. □

Structural Induction

Structural induction is well suited for statements about recursively defined and tree-like objects such as terms or formulas. The recursive definition of the object type serves as a guide: one proves the desired property for the elementary (atomic) objects and then shows that the property is preserved when more complex objects are built from them according to the formation rules.

Formally, let $\mathcal{F}$ be the set of formulas of a language, defined by the rules

$$\varphi ::= p \mid \neg\varphi \mid (\varphi \wedge \psi) \mid (\varphi \vee \psi)$$

with atomic formulas p. To prove a property $P(\varphi)$ for all $\varphi \in \mathcal{F}$, one shows:

- $P(p)$ for every atomic p.

- If $P(\varphi)$ holds, then $P(\neg\varphi)$ holds.
- If $P(\varphi)$ and $P(\psi)$ hold, then $P((\varphi \wedge \psi))$ holds (and analogously for other binary operators).

Example 0.2. *Every formula $\varphi \in \mathcal{F}$ has only finitely many subformulas.*

Proof. We prove the statement *"φ has finitely many subformulas"* for all $\varphi \in \mathcal{F}$ by structural induction.

Base case: Let $\varphi = p$ be atomic. Then the set of subformulas is $\{ p \}$, and is therefore finite.

Inductive step: Assume that for φ and ψ, the sets of their subformulas are finite. Then the subformulas of $\neg\varphi$ are given by $\{\neg\varphi\} \cup \{\text{subformulas of } \varphi\}$, which is a finite union of finite sets and is therefore finite. Similarly, for the compound formula $(\varphi \wedge \psi)$, the set of subformulas is

$$\{ (\varphi \wedge \psi) \} \cup \{\text{subformulas of } \varphi\} \cup \{\text{subformulas of } \psi\}$$

which is again a finite union of finite sets and hence finite.

Therefore, by structural induction, every $\varphi \in \mathcal{F}$ has only finitely many subformulas. □

Comparison, Remarks, and Typical Applications

- **Which form of induction should be used?** Use ordinary mathematical induction when the statement is formulated over the natural numbers (or directly over $\mathbb{N}$), for example in combinatorics, sums, or recursively defined sequences. Use structural induction when the statement concerns recursively defined, tree-like objects such as terms, formulas, proof trees, or abstract syntax trees.

- **Form of the inductive step.** In ordinary induction, the inductive step typically refers to the immediate predecessor $n-1$ or to all $k < n$ in the case of strong induction. In structural induction, one shows that the property is preserved by the formation or construction rules.

- **Common underlying idea.** Both methods are applications of the principle that there are no infinite strictly descending chains in a well-founded relation. Formally, both can be viewed as special cases of well-founded induction.

- **Practical advice.** Formulate inductive claims in such a way that the induction hypothesis contains exactly the information needed for the proof. In structural induction, it is often helpful to work with a stronger induction claim (for example, a statement about the class of all substructures), since this makes composite cases easier to handle.

0.3 Structure and Style of Mathematical Writing

Mathematical texts in this book follow a clear and recognizable pattern: definitions and notational conventions establish precise concepts and formal usage, examples illustrate the underlying intuition, results (lemmas, propositions, theorems, and corollaries) state the central claims, and proofs justify them rigorously. This structure helps the reader distinguish quickly between formal statements, arguments, and illustrative material.

Definitions and Notation

- **Definitions.** Definitions formalize new concepts. They should be concise, complete, and phrased in such a way that the relevant properties can be checked directly.

- **Notation.** Notation introduces abbreviated forms of writing (for example, $\mathcal{P}(X)$, $X \times Y$, or $O(\cdot)$). Notational conventions should be used consistently and collected in a prominent place, such as at the beginning of a chapter or in a notation box.

Examples and Illustrations

Examples serve to build intuition and to show typical or contrasting cases. They should be concise, demonstrate the use of newly introduced definitions, and, whenever possible, point ahead to later exercises or applications.

Results: Lemma, Proposition, Theorem, Corollary

- **Lemma.** An auxiliary result used primarily in the proof of another result.
- **Proposition.** A useful result, usually of lesser central importance.
- **Theorem.** A central and important result.
- **Corollary.** A statement that follows directly from a previous result.

All results should include a precise formulation, assumptions where appropriate, and a clear proof.

Proofs

Proofs are part of standard mathematical argumentation. Whenever possible, they should be clearly structured:

1. A brief opening statement indicating what is to be proved.
2. If necessary, a division into cases or subsidiary lemmas.
3. A clear concluding mark at the end (usually written as $\square$).

Remarks, Exercises, and References

- **Remarks** often contain additional information, intuition, or generalizations. They are typically shorter and less formal.
- **Exercises** at the end of a chapter reinforce understanding.
- **References** point to further sources and should be cited precisely.

A Short Illustrative Example

Definition 0.3. *A function $f : X \to Y$ is called* injective *if from $f(x) = f(x')$ it follows that $x = x'$.*

Theorem 0.4. *Let $f : X \to Y$ be injective and let $g : Y \to Z$ be arbitrary. Then $g \circ f$ is injective if and only if g is injective on the image of f.*

Proof. The claim follows by directly checking the definition: if $g \circ f$ is injective and $g(f(x)) = g(f(x'))$, then $f(x) = f(x')$, and by the injectivity of f it follows that $x = x'$. The converse direction is equally direct. □

Chapter 1

Propositional Logic

1.1 Introduction and Overview

Propositional logic is the most basic formal system for describing simple statements and their logical combinations with precision. Its core elements are atomic propositions, each of which can take exactly one of two truth values: true or false. By combining these atomic building blocks with connectives such as *and*, *or*, and *not*, we obtain compound formulas that can express more complex logical relationships.

Despite its elementary syntax, propositional logic has a wide range of applications. It provides an essential foundation for many further areas of logic and also plays a central role in computer science, for example in automated theorem proving, the analysis of logical circuits, verification, and methods in artificial intelligence.

One of its great strengths is that it is fully formalized: both the formation of formulas and their meaning can be specified exactly.

This chapter first introduces the syntactic and semantic foundations of propositional logic. We begin with signatures, formulas, and interpretations, and then clarify when a formula is true under a given interpretation and what it means for a set of formulas to have a model. We then turn to the notion of logical consequence and ask when one formula follows from another or from a set of formulas. Building on this, we examine logical equivalence between formulas and present important transformation rules and normal forms that are especially relevant for algorithmic processing.

In the remainder of the chapter, we study methods that not only describe logical reasoning semantically but also carry it out syntactically. These include general proof systems as well as the resolution calculus, which is one of the central tools of automated deduction. Finally, we look more closely at the algorithmic side of propositional logic: in particular, we investigate the satisfiability problem, explain why it is theoretically difficult, and introduce Tseitin encoding and the DPLL algorithm as two important methods that often allow satisfiability to be checked very efficiently in practice.

The chapter is organized as follows: Subsection 1.1 introduces the syntax and semantics of propositional logic, beginning with signatures and then moving to formulas, interpretations, and models. Subsection 1.2 discusses properties of models, the notion of consequence, and the independence of the signature. Subsection 1.3 is devoted to equivalence, including both the definition of logical equivalence and important equivalence-preserving transformations, as well as the representation of statements in normal forms. Subsection 1.4 presents several calculi for automated inference, with particular emphasis on the resolution calculus. Finally, Subsection 1.5 explains the reduction to satisfiability problems, Tseitin encoding, and the DPLL algorithm as central methods for solving satisfiability problems.

In this way, the chapter connects the conceptual foundations of propositional logic with its central algorithmic methods. It thereby provides the basis for everything that follows concerning formal reasoning, normal forms, and automated proof procedures.

1.2 Syntax and Semantics

1.2.1 Introduction

Like every logical system, propositional logic is defined by two components:

- *Syntax:* the description of the well-formed expressions (formulas), that is, which strings of symbols count as valid objects, and
- *Semantics:* the assignment of meanings (truth values) to these expressions.

In the following, we first introduce signatures and formulas and then define their model-theoretic semantics.

1.2.2 Signatures

Definition 1.1. *A* propositional signature Σ *is a (finite) set of propositional variables (also called atoms or propositions).*

Example 1.2. *To model a smart-home alarm system, one may choose, for example,*

$$\Sigma_{\mathit{Alarm}} = \{\mathit{Smoke},\ \mathit{Heat},\ \mathit{Motion},\ \mathit{Alarm},\ \mathit{Sprinkler}\},$$

where the intended meanings of the variables are roughly as follows:

- *Smoke: smoke is detected*
- *Heat: high temperature is detected*
- *Motion: movement in the house is registered*
- *Alarm: the alarm is triggered*
- *Sprinkler: the sprinklers are activated*

A typical formula in $L(\Sigma_{\mathit{Alarm}})$ *would be, for example,*

$$(\mathit{Smoke} \vee \mathit{Heat}) \to \mathit{Alarm},$$

which means: "If smoke or high temperature is detected, then the alarm is triggered." ■

1.2.3 Formulas

Definition 1.3. *Let Σ be a propositional signature. The set $L(\Sigma)$ of* formulas *over Σ is defined inductively as follows:*

1. *every propositional variable $p \in \Sigma$ is a formula (an atomic formula);*
2. *if φ is a formula, then $\neg\varphi$ is also a formula;*
3. *if φ and ψ are formulas, then $(\varphi \wedge \psi)$ and $(\varphi \vee \psi)$ are formulas as well.*

Alternatively, in BNF:

$$\varphi ::= p \mid \neg\varphi \mid (\varphi \wedge \varphi) \mid (\varphi \vee \varphi) \qquad (p \in \Sigma).$$

Notation 1.4. *We use the following abbreviations and conventions:*

- *Connectives: $\neg$ (negation), $\wedge$ (conjunction), $\vee$ (disjunction).*
- *Implication and equivalence are defined as $\varphi \to \psi := \neg\varphi \vee \psi$ and $\varphi \leftrightarrow \psi := (\varphi \to \psi) \wedge (\psi \to \varphi)$.*
- *Parenthesis convention: $\neg$ binds more strongly than $\wedge$, which binds more strongly than $\vee$, which binds more strongly than $\to$, which in turn binds more strongly than $\leftrightarrow$.*
- *A* literal *is either an atom p or its negation $\neg p$.*

Example 1.5. *Using Σ_{Alarm} from the example above, the following formulas, for instance, belong to $L(\Sigma_{Alarm})$:*

$$
\begin{aligned}
&(Smoke \vee Heat) \rightarrow Alarm,\\
&\neg Smoke \wedge (Alarm \vee Sprinkler),\\
&Motion \rightarrow Alarm.
\end{aligned}
$$

The set of literals is

$$
\begin{aligned}
\mathrm{Lit}(\Sigma_{Alarm}) = \{&Smoke, \neg Smoke,\ Heat, \neg Heat,\\
&Motion, \neg Motion,\ Alarm, \neg Alarm,\\
&Sprinkler, \neg Sprinkler\}.
\end{aligned}
$$

■

1.2.4 Semantics: Interpretations, Satisfaction, and Models

Definition 1.6. *Let Σ be a signature. An* interpretation *(or valuation) ω over Σ is a function $\omega : \Sigma \rightarrow \{true, false\}$. We denote the set of all interpretations by $\Omega(\Sigma)$.*

Definition 1.7. *The satisfaction relation $\models_{\mathrm{AL}} \subseteq \Omega(\Sigma) \times L(\Sigma)$ is defined inductively:*

$$
\begin{aligned}
\omega \models_{\mathrm{AL}} p \quad &\textit{iff} \quad \omega(p) = true,\\
\omega \models_{\mathrm{AL}} \neg\varphi \quad &\textit{iff} \quad \textit{not } (\omega \models_{\mathrm{AL}} \varphi),\\
\omega \models_{\mathrm{AL}} \varphi \wedge \psi \quad &\textit{iff} \quad \omega \models_{\mathrm{AL}} \varphi \textit{ and } \omega \models_{\mathrm{AL}} \psi,\\
\omega \models_{\mathrm{AL}} \varphi \vee \psi \quad &\textit{iff} \quad \omega \models_{\mathrm{AL}} \varphi \textit{ or } \omega \models_{\mathrm{AL}} \psi.
\end{aligned}
$$

If $\omega \models_{\mathrm{AL}} \varphi$, then ω is called a model *of φ.*

Definition 1.8. *For a formula φ, we define*

$$\mathrm{Mod}_{\mathrm{AL}}(\varphi) = \{\omega \in \Omega(\Sigma) \mid \omega \models_{\mathrm{AL}} \varphi\},$$

the set of all models of φ. Furthermore, φ is satisfiable *iff* $\mathrm{Mod}_{\mathrm{AL}}(\varphi) \neq \varnothing$, *and φ is* valid *(a tautology) iff* $\mathrm{Mod}_{\mathrm{AL}}(\varphi) = \Omega(\Sigma)$.

Explanation. If φ is a formula and ω is a model of φ ($\omega \in \mathrm{Mod}_{\mathrm{AL}}(\varphi)$), then ω can be understood intuitively as a *possible world* or as a *situation* in which the statement φ is in fact true. The interpretation ω determines, for each atomic statement $p \in \Sigma$, whether p is true or false in that world; in this way, ω gives a *complete* explanation of why φ is true: namely through the assignment of truth values to the atoms from which φ is built.

The term *model* highlights precisely this explanatory role: a model is a concrete witness showing that a formula is satisfiable. Accordingly, the set $\Omega(\Sigma)$ is the set of all possible worlds (all possible valuations), and $\mathrm{Mod}_{\mathrm{AL}}(\varphi)$ consists of those worlds that make φ true.

Two short but useful notions:

- *Satisfiability:* The formula φ is satisfiable if and only if $\mathrm{Mod}_{\mathrm{AL}}(\varphi) \neq \varnothing$. A specific $\omega \in \mathrm{Mod}_{\mathrm{AL}}(\varphi)$ is also called a *model* or a *satisfying assignment* of φ.

- *Countermodel:* To show that a formula φ is *not* valid (not tautological), it suffices to give some

$\omega \in \Omega(\Sigma)$ such that $\omega \notin \mathrm{Mod}_{\mathrm{AL}}(\varphi)$; such an ω is called a *countermodel*.

Remark. In applications (e.g. model checking), finding a model often amounts to showing that a desired system behavior can occur, whereas providing a countermodel shows that a desired property is violated under certain valuations.

Example 1.9. *Consider Σ_{Alarm} and the interpretation ω given by $\omega(Smoke) = true$, $\omega(Heat) = false$, $\omega(Motion) = true$, $\omega(Alarm) = true$, $\omega(Sprinkler) = false$. Then, for example, $\omega \models_{\mathrm{AL}} Smoke$, $\omega \models_{\mathrm{AL}} \neg Heat$, $\omega \models_{\mathrm{AL}} (Smoke \vee Heat) \to Alarm$ (because $\omega(Smoke) = true$ and $\omega(Alarm) = true$), $\omega \models_{\mathrm{AL}} Motion \to Alarm$ (since $\omega(Motion) = true$ and $\omega(Alarm) = true$) and $\omega \models_{\mathrm{AL}} \neg Sprinkler$.* ■

1.2.5 Truth Tables

The connectives can be described by truth tables; here is the compact representation (1 = true, 0 = false):

φ	ψ	$\varphi \wedge \psi$	$\varphi \vee \psi$	$\neg\varphi$
0	0	0	0	1
0	1	0	1	1
1	0	0	1	0
1	1	1	1	0

1.2.6 Remarks

- If $|\Sigma| = n$, then $|\Omega(\Sigma)| = 2^n$ interpretations.

- The inductive definition of semantics (satisfaction) is equivalent to directly specifying a truth function $[\![\cdot]\!]_\omega$.
- Model-theoretic semantics provides the foundation for notions such as *logical consequence* $\Phi \models \psi$ (every interpretation that satisfies all formulas in Φ also satisfies ψ).

1.2.7 Truth Functions

Definition 1.10. *Let Σ be a propositional signature and $\omega \in \Omega(\Sigma)$ an interpretation. The* truth function $[\![\cdot]\!]_\omega : L(\Sigma) \to \{\text{true}, \text{false}\}$ *is defined inductively as follows:*

$$[\![p]\!]_\omega = \omega(p) \quad (p \in \Sigma),$$

$$[\![\neg\varphi]\!]_\omega = \begin{cases} \text{true}, & \textit{if } [\![\varphi]\!]_\omega = \text{false}, \\ \text{false}, & \textit{otherwise}, \end{cases}$$

$$[\![\varphi \land \psi]\!]_\omega = \begin{cases} \text{true}, & \textit{if } [\![\varphi]\!]_\omega = \text{true } \textit{and } [\![\psi]\!]_\omega = \text{true}, \\ \text{false}, & \textit{otherwise}, \end{cases}$$

$$[\![\varphi \lor \psi]\!]_\omega = \begin{cases} \text{false}, & \textit{if } [\![\varphi]\!]_\omega = \text{false } \textit{and } [\![\psi]\!]_\omega = \text{false}, \\ \text{true}, & \textit{otherwise}. \end{cases}$$

Proposition 1.11. *Let Σ be a signature, $\omega \in \Omega(\Sigma)$ an interpretation, and $\varphi \in L(\Sigma)$. Then*

$$\omega \models_{\text{AL}} \varphi \quad \textit{iff} \quad [\![\varphi]\!]_\omega = \text{true}.$$

Proof. Let Σ be a propositional signature, $\omega \in \Omega(\Sigma)$ an interpretation, and $\varphi \in L(\Sigma)$ a formula. We prove the equivalence

$$\omega \models_{\mathrm{AL}} \varphi \quad \text{iff} \quad [\![\varphi]\!]_\omega = \text{true}$$

by structural induction on the formula φ.

Base case. Let $\varphi = p$ with $p \in \Sigma$ be an atom. By definition of the satisfaction relation, $\omega \models_{\mathrm{AL}} p$ holds exactly if $\omega(p) = \text{true}$. By definition of the truth function, $[\![p]\!]_\omega = \omega(p)$. Hence it follows immediately that $\omega \models_{\mathrm{PL}} p \iff [\![p]\!]_\omega = \text{true}$.

Inductive step. Now let φ be a formula, and assume that the statement already holds for all proper subformulas of φ (induction hypothesis). We consider the different construction cases.

- $\varphi = \neg\psi$.
  $\Rightarrow$: Assume that $\omega \models_{\mathrm{PL}} \neg\psi$. Then, by definition, $\omega \not\models_{\mathrm{PL}} \psi$. By the induction hypothesis, it follows that $[\![\psi]\!]_\omega \neq \text{true}$, hence $[\![\psi]\!]_\omega = \text{false}$. By definition of the truth function, this implies $[\![\neg\psi]\!]_\omega = \text{true}$.
  $\Leftarrow$: Assume that $[\![\neg\psi]\!]_\omega = \text{true}$. Then $[\![\psi]\!]_\omega = \text{false}$. By the induction hypothesis, $\omega \not\models_{\mathrm{PL}} \psi$, and therefore $\omega \models_{\mathrm{PL}} \neg\psi$.

- $\varphi = \psi \wedge \xi$.
  $\Rightarrow$: Assume that $\omega \models_{\mathrm{PL}} \psi \wedge \xi$. Then $\omega \models_{\mathrm{PL}} \psi$ and $\omega \models_{\mathrm{AL}} \xi$. By the induction hypothesis, therefore, $[\![\psi]\!]_\omega = \text{true}$ and $[\![\xi]\!]_\omega = \text{true}$. Therefore, $[\![\psi \wedge \xi]\!]_\omega =$

true.
$\Leftarrow$: Assume that $[\![\psi \wedge \xi]\!]_\omega = \text{true}$. Then $[\![\psi]\!]_\omega = \text{true}$ and $[\![\xi]\!]_\omega = \text{true}$. By the induction hypothesis, it follows that $\omega \models_{\text{PL}} \psi$ and $\omega \models_{\text{PL}} \xi$, hence $\omega \models_{\text{PL}} \psi \wedge \xi$.

- $\varphi = \psi \vee \xi$.
  $\Rightarrow$: Assume that $\omega \models_{\text{PL}} \psi \vee \xi$. Then $\omega \models_{\text{PL}} \psi$ or $\omega \models_{\text{PL}} \xi$. Without loss of generality, assume $\omega \models_{\text{PL}} \psi$. By the induction hypothesis, $[\![\psi]\!]_\omega = \text{true}$, so $[\![\psi \vee \xi]\!]_\omega = \text{true}$. (The other case is analogous.)
  $\Leftarrow$: Assume that $[\![\psi \vee \xi]\!]_\omega = \text{true}$. Then $[\![\psi]\!]_\omega = \text{true}$ or $[\![\xi]\!]_\omega = \text{true}$. By the induction hypothesis, it follows that $\omega \models_{\text{PL}} \psi$ or $\omega \models_{\text{PL}} \xi$, and hence $\omega \models_{\text{PL}} \psi \vee \xi$.

Since all construction cases have been covered, the claim follows for every formula φ. □

The logical connectives can also be presented compactly by means of truth tables. We begin with the version using the labels true/false:

φ	ψ	$\varphi \wedge \psi$	$\varphi \vee \psi$	$\neg\varphi$
false	false	false	false	true
false	true	false	true	true
true	false	false	true	false
true	true	true	true	false

Here is the compact 0/1 representation (0 = false, 1 = true):

φ	ψ	$\varphi \wedge \psi$	$\varphi \vee \psi$	$\neg\varphi$
0	0	0	0	1
0	1	0	1	1
1	0	0	1	0
1	1	1	1	0

Remark. The above definition of the truth function is equivalent to the inductive definition of the satisfaction relation $\models_{\text{PL}}$. In proofs, one or the other formulation may be more convenient depending on the context: the satisfaction relation emphasizes the model-theoretic perspective ($\omega \models \varphi$), whereas the truth function explicitly specifies the value of the formula under a given interpretation ($[\![\varphi]\!]_\omega \in \{\text{true}, \text{false}\}$).

Outlook

In the next section, we examine syntactic normal forms, concrete transformation rules (e.g. CNF and DNF), and initial proof methods (resolution and natural deduction), all of which are relevant both theoretically and practically (for example, in the context of SAT solvers).

1.3 Exercises with Solutions

Exercise 1.12. *Let the propositional signature be* $\Sigma = \{P, Q, R, S, T, U\}$. *Consider the expressions*

$$\varphi_1 = S \land \neg(P \leftrightarrow \neg R),$$

$$\omega_2 : P \mapsto \mathit{true},\ Q \mapsto \mathit{false},\ R \mapsto \mathit{true},$$
$$S \mapsto \mathit{false},\ T \mapsto \mathit{true},\ U \mapsto \mathit{false}.$$

and

$$\varphi_3 = \neg T \land \neg(\neg T \lor R).$$

Decide for each of the following statements whether it is ***true*** *or* ***false*** *(briefly justify your answer in each case):*

1. $\neg P \lor R \in L(\Sigma)$.

2. *The given formula*

   $$\overline{\varphi_1} = \neg S \lor (\neg P \lor \neg R) \land (R \lor P) \lor \neg Q$$

   is the complement (the negation) of φ_1.

3. $[\![\neg T \lor (\neg R \leftrightarrow P)]\!]_{\omega_2} = \mathit{false}$.

4. $|\,\mathrm{Mod}_{\mathrm{PL}}(\varphi_3)| = 0$.

Solution 1.13. *1.* ***True.*** $\neg P \lor R$ *is a well-formed expression, built from atoms in* Σ *using the connectives* $\neg$ *and* $\lor$, *and is therefore a formula in* $L(\Sigma)$.

2. ***False.*** *The claimed formula is not the negation of* φ_1. *First, it contains the atom* Q, *which does not occur in* φ_1; *second, by De Morgan's laws, the syntactic structure of a negation of* $\varphi_1 = S \land \neg(P \leftrightarrow \neg R)$ *is equivalent to*

$$\neg S \lor \neg\neg(P \leftrightarrow \neg R) \;=\; \neg S \lor (P \leftrightarrow \neg R),$$

not to the given complicated formula with $\neg Q$ *and the additional term* $(\neg P \lor \neg R) \land (R \lor P)$.

3. ***True.*** *We evaluate the formula* $\neg T \lor (\neg R \leftrightarrow P)$ *under* ω_2. *Under* ω_2, *we have* $T = \textit{true}$, *so* $\neg T = \textit{false}$. *Also,* $R = \textit{true}$, *hence* $\neg R = \textit{false}$. *Since* $P = \textit{true}$, $\neg R \leftrightarrow P$ *is the equivalence between false and true, and is therefore false. Thus,* $\neg T \lor (\neg R \leftrightarrow P)$ *equals* $\textit{false} \lor \textit{false} = \textit{false}$. *Hence the statement is correct.*

4. ***True.*** *We simplify* $\varphi_3 = \neg T \land \neg(\neg T \lor R)$. *The subformula* $\neg(\neg T \lor R)$ *is equivalent to* $\neg\neg T \land \neg R$, *that is, to* $T \land \neg R$. *Therefore,* $\varphi_3 \equiv \neg T \land (T \land \neg R)$. *This formula is clearly unsatisfiable, since* $\neg T$ *and* T *would have to hold simultaneously. Hence* φ_3 *has* no *models and therefore* $|\mathrm{Mod}_{\mathrm{PL}}(\varphi_3)| = 0$.

Feedback (Brief):

- *(1) The formula belongs to the language over* Σ *— correct.*

- *(2) The given negation formula is incorrect — it contains an extraneous atom and does not have the right structure (it is not equivalent to the negation of φ_1).*
- *(3) True — the evaluation yields false.*
- *(4) True — φ_3 is contradictory and therefore unsatisfiable.*

Exercise 1.14. *Vary the signature Σ (for example, use only three atoms) and determine, for all $2^{|\Sigma|}$ interpretations, which formulas from Exercise 1 are true. Compare the number of models of the respective formulas.*

Solution 1.15. *We choose $\Sigma' = \{P, Q, R\}$ and define the formulas analogous to those in the exercise (adapting the original task to the smaller signature):*

$$\begin{aligned}
\varphi_1 &:= Q \wedge \neg(P \leftrightarrow \neg R),\\
\alpha &:= \neg P \vee R,\\
\beta &:= \neg Q \vee ((\neg P \vee \neg R) \wedge (R \vee P)) \quad \textit{(candidate for } \neg\varphi_1\textit{)},\\
\gamma &:= \neg R \vee (\neg P \leftrightarrow Q),\\
\delta &:= \neg P \wedge \neg(\neg P \vee R).
\end{aligned}$$

The formula $(\neg P \vee \neg R) \wedge (R \vee P)$ is equivalent to $(P \leftrightarrow \neg R)$, which means that β is precisely $\neg Q \vee (P \leftrightarrow \neg R)$ — formally, the negation of $\varphi_1 = Q \wedge \neg(P \leftrightarrow \neg R)$.

The complete truth table (T = true, F = false):

P	Q	R	φ_1	α	β	γ	δ
F	F	F	F	T	T	T	F
F	F	T	F	T	T	F	F
F	T	F	T	T	F	T	F
F	T	T	F	T	T	T	F
T	F	F	F	F	T	T	F
T	F	T	F	T	T	T	F
T	T	F	F	F	T	T	F
T	T	T	T	T	F	F	F

From the table, we can read off the number of models (the number of interpretations for which the formula is true):

- $|\text{Mod}(\varphi_1)| = 2$ *(true for the assignments* $(P, Q, R) = (F, T, F)$ *and* (T, T, T)*)*
- $|\text{Mod}(\alpha)| = 6$
- $|\text{Mod}(\beta)| = 6$
- $|\text{Mod}(\gamma)| = 6$
- $|\text{Mod}(\delta)| = 0$ *(unsatisfiable)*

Interpretation and conclusions

- $\alpha = \neg P \lor R$ *is not a tautology (it is not true in every interpretation), but it is satisfied in 6 out of 8 cases — so it is common, but not universally valid.*

- *β is, in this concrete form, logically equivalent to $\neg\varphi_1$. This can be seen algebraically:*

$$\begin{aligned}\neg\varphi_1 &= \neg(Q \wedge \neg(P \leftrightarrow \neg R)) \\ &= \neg Q \vee \neg\neg(P \leftrightarrow \neg R) \\ &= \neg Q \vee (P \leftrightarrow \neg R)\end{aligned}$$

  and $P \leftrightarrow \neg R$ is equivalent to $(\neg P \vee \neg R) \wedge (R \vee P)$. Therefore, β and $\neg\varphi_1$ have the same models (6 interpretations).

- *γ is also true in 6 interpretations; its set of models differs from that of α and β in its exact composition, but the cardinalities are the same (6).*

- *δ is contradictory: $\delta = \neg P \wedge (\ P \wedge \neg R\ )$ (after simplification) requires both P and $\neg P$ at the same time — hence it has no models.*

Note on the Procedure (Generalization) For arbitrary signatures Σ with n atoms, one can proceed in the same way by listing all 2^n interpretations (or generating them programmatically) and evaluating each formula automatically. The number of models of a formula is exactly the number of rows in the truth table or evaluation table in which the formula takes the value true. For manual exercises with $n = 3$ or $n = 4$, this is still quite manageable; for larger n, however, it is more practical to use a small script (or a SAT solver) to count the models.

1.4 Logical Consequence

In practice, we use logical formulas not only to describe individual states of affairs, but above all to derive new information from what is already known. Formally, *logical consequence* asks the following question: which statements must hold in every world in which our assumptions are true?

This chapter introduces the semantic notion of consequence, presents simple calculation rules for sets of models, and relates the concepts of "model," "theory," and "deductive closure" to one another.

1.4.1 Models as Descriptions of Situations

Recall: for a signature Σ, $\Omega(\Sigma)$ denotes the set of all possible assignments of truth values to the atomic variables. Associated with a formula φ is its set of models

$$\mathcal{M}(\varphi) = \{\omega \in \Omega(\Sigma) \mid \omega \models_{\mathrm{AL}} \varphi\},$$

that is, the set of all assignments that make φ true.

Intuition. One may think of $\omega \in \mathcal{M}(\varphi)$ as a concrete "situation" in which the description φ holds. In this way, a model fully explains why φ is true: namely through the truth values assigned to its atomic constituents.

1.4.2 Operations on Sets of Models

The classical connectives correspond to simple set-theoretic operations on sets of models. These relationships are useful because they translate semantic arguments into the language of set operations.

Proposition 1.16. *For all formulas $\varphi, \psi \in L(\Sigma)$, we have*

$$\mathcal{M}(\varphi \wedge \psi) = \mathcal{M}(\varphi) \cap \mathcal{M}(\psi), \qquad \mathcal{M}(\varphi \vee \psi) = \mathcal{M}(\varphi) \cup \mathcal{M}(\psi),$$

$$\mathcal{M}(\neg\varphi) = \Omega(\Sigma) \setminus \mathcal{M}(\varphi).$$

Proof. Let Σ be a propositional signature, and let $\varphi, \psi \in L(\Sigma)$ be formulas. We prove the three identities one by one, in each case by showing both inclusions.

(1) $\mathcal{M}(\varphi \wedge \psi) = \mathcal{M}(\varphi) \cap \mathcal{M}(\psi)$.
"$\subseteq$": Let $\omega \in \mathcal{M}(\varphi \wedge \psi)$. Then $\omega \models_{\mathrm{PL}} \varphi \wedge \psi$. By the semantic definition of conjunction, this means that both $\omega \models_{\mathrm{PL}} \varphi$ and $\omega \models_{\mathrm{PL}} \psi$ hold. Hence $\omega \in \mathcal{M}(\varphi)$ and $\omega \in \mathcal{M}(\psi)$, so $\omega \in \mathcal{M}(\varphi) \cap \mathcal{M}(\psi)$.

"$\supseteq$": Let $\omega \in \mathcal{M}(\varphi) \cap \mathcal{M}(\psi)$. Then $\omega \models_{\mathrm{PL}} \varphi$ and $\omega \models_{\mathrm{PL}} \psi$. By the semantic definition of conjunction, it follows that $\omega \models_{\mathrm{PL}} \varphi \wedge \psi$, hence $\omega \in \mathcal{M}(\varphi \wedge \psi)$.

Thus the desired equality follows.

(2) $\mathcal{M}(\varphi \vee \psi) = \mathcal{M}(\varphi) \cup \mathcal{M}(\psi)$.
"$\subseteq$": Let $\omega \in \mathcal{M}(\varphi \vee \psi)$. Then $\omega \models_{\mathrm{PL}} \varphi \vee \psi$, i.e. $\omega \models_{\mathrm{PL}} \varphi$ or $\omega \models_{\mathrm{PL}} \psi$. In the first case, $\omega \in \mathcal{M}(\varphi)$; in the second case, $\omega \in \mathcal{M}(\psi)$. In either case, $\omega \in \mathcal{M}(\varphi) \cup \mathcal{M}(\psi)$.

"$\supseteq$": Let $\omega \in \mathcal{M}(\varphi) \cup \mathcal{M}(\psi)$. Then either $\omega \models_{\mathrm{PL}} \varphi$ or $\omega \models_{\mathrm{PL}} \psi$. In either case, it follows immediately that $\omega \models_{\mathrm{PL}} \varphi \vee \psi$, and hence $\omega \in \mathcal{M}(\varphi \vee \psi)$.

This proves the second equality as well.

(3) $\mathcal{M}(\neg\varphi) = \Omega(\Sigma) \setminus \mathcal{M}(\varphi)$.
We prove the two statements $\Omega(\Sigma) = \mathcal{M}(\varphi) \cup \mathcal{M}(\neg\varphi)$ and $\mathcal{M}(\varphi) \cap \mathcal{M}(\neg\varphi) = \varnothing$, from which the claim follows.

First, let $\omega \in \Omega(\Sigma)$ be arbitrary. Either $\omega \models_{\mathrm{PL}} \varphi$ or not. In the first case, $\omega \in \mathcal{M}(\varphi)$. In the second case, $\omega \models_{\mathrm{PL}} \neg\varphi$, and therefore $\omega \in \mathcal{M}(\neg\varphi)$. Thus, $\Omega(\Sigma) \subseteq \mathcal{M}(\varphi) \cup \mathcal{M}(\neg\varphi)$; the reverse inclusion is trivial, so equality holds.

Next, suppose for contradiction that $\omega \in \mathcal{M}(\varphi) \cap \mathcal{M}(\neg\varphi)$. Then $\omega \models_{\mathrm{PL}} \varphi$ and $\omega \models_{\mathrm{PL}} \neg\varphi$. But this is impossible, since $\omega \models_{\mathrm{PL}} \neg\varphi$ holds exactly when $\omega \not\models_{\mathrm{PL}} \varphi$. Therefore, the intersection is empty.

From $\Omega(\Sigma) = \mathcal{M}(\varphi) \cup \mathcal{M}(\neg\varphi)$ and $\mathcal{M}(\varphi) \cap \mathcal{M}(\neg\varphi) = \varnothing$, it follows immediately that $\mathcal{M}(\neg\varphi) = \Omega(\Sigma) \setminus \mathcal{M}(\varphi)$.

This proves all three identities. □

Remark 1.17. *Conjunctions correspond to intersections, disjunctions to unions, and negation to the complement of the set of models.*

Notation 1.18. *Let Σ be a propositional signature, let $\omega \in \Omega(\Sigma)$ be a propositional interpretation, and let $\Phi \subseteq L(\Sigma)$ be a set of propositional formulas. If $\omega \models_{\mathrm{PL}} \varphi$ for all $\varphi \in \Phi$, we write $\omega \models_{\mathrm{PL}} \Phi$ for short and call ω a* model of Φ. *Furthermore, we set*

$$\mathrm{Mod}_{\mathrm{PL}}(\Phi) = \{\omega \in \Omega(\Sigma) \mid \omega \models_{\mathrm{PL}} \Phi\}$$

for the set of all models of Φ.

Proposition 1.19. *Let* Σ *be a propositional signature, and let* $\Phi = \{\varphi_1, \ldots, \varphi_n\} \subseteq L(\Sigma)$ *be a finite set of formulas. Then the following hold:*

1. $\mathrm{Mod}_{\mathrm{PL}}(\Phi) = \bigcap_{i=1}^{n} \mathrm{Mod}_{\mathrm{PL}}(\varphi_i)$.

2. $\mathrm{Mod}_{\mathrm{PL}}(\Phi) = \mathrm{Mod}_{\mathrm{PL}}(\varphi_1 \wedge (\varphi_2 \wedge (\cdots \wedge \varphi_n) \ldots))$.

Proof. (1) We prove both inclusions.

"$\subseteq$": Let $\omega \in \mathrm{Mod}_{\mathrm{PL}}(\Phi)$. By definition, $\omega \models_{\mathrm{PL}} \varphi_i$ for every $i = 1, \ldots, n$. Hence ω belongs to each set $\mathrm{Mod}_{\mathrm{PL}}(\varphi_i)$, and therefore to their intersection. Thus $\mathrm{Mod}_{\mathrm{PL}}(\Phi) \subseteq \bigcap_{i=1}^{n} \mathrm{Mod}_{\mathrm{PL}}(\varphi_i)$.

"$\supseteq$": Let $\omega \in \bigcap_{i=1}^{n} \mathrm{Mod}_{\mathrm{PL}}(\varphi_i)$. Then ω satisfies every φ_i, so $\omega \models_{\mathrm{PL}} \Phi$. Therefore $\omega \in \mathrm{Mod}_{\mathrm{PL}}(\Phi)$, which proves the reverse inclusion. This completes the proof of (1).

(2) In view of (1), it suffices to show that

$$\bigcap_{i=1}^{n} \mathrm{Mod}_{\mathrm{PL}}(\varphi_i) = \mathrm{Mod}_{\mathrm{PL}}(\varphi_1 \wedge (\varphi_2 \wedge (\cdots \wedge \varphi_n) \ldots)).$$

This follows directly by repeated application of the identity $\mathrm{Mod}_{\mathrm{PL}}(\alpha \wedge \beta) = \mathrm{Mod}_{\mathrm{PL}}(\alpha) \cap \mathrm{Mod}_{\mathrm{PL}}(\beta)$ (cf. Proposition 1.12). By induction, the set of models of the nested conjunction is exactly the intersection of the individual sets of models, which establishes (2). □

Remark 1.20. *Both statements extend immediately to infinite sets* Φ*, provided that the right-hand side of (1) is understood as* $\bigcap_{\varphi \in \Phi} \mathrm{Mod}_{\mathrm{PL}}(\varphi)$*. The representation in (2) as a finite nested conjunction is only directly meaningful for finite* Φ*. For infinite* Φ*, one may speak formally of the (algebraic) conjunction of all* $\varphi \in \Phi$*, which in general cannot be expressed as a finite formula term, but is still represented semantically by the intersection of the corresponding sets of models.*

1.4.3 Satisfiability Classes of Formulas

Many questions in logic revolve around whether a formula φ can ever be true at all, whether it is always true, or whether it becomes false in some cases. To this end, we introduce the following notions.

Definition 1.21. *Let* Σ *be a propositional signature and* $\varphi \in L(\Sigma)$*. We write* $\Omega(\Sigma)$ *for the set of all interpretations over* Σ *and* $\mathcal{M}(\varphi) = \{\omega \in \Omega(\Sigma) \mid \omega \models_{\mathrm{PL}} \varphi\}$*. Then* φ *is called*

1. satisfiable *if* $\mathcal{M}(\varphi) \neq \varnothing$*.*

2. unsatisfiable *if* $\mathcal{M}(\varphi) = \varnothing$*.*

3. valid *(tautological) if* $\mathcal{M}(\varphi) = \Omega(\Sigma)$*.*

4. falsifiable *if* $\mathcal{M}(\varphi) \neq \Omega(\Sigma)$ *(i.e. there exists an interpretation that does not satisfy* φ*).*

Remark 1.22. *For every signature Σ, we have $\Omega(\Sigma) \neq \varnothing$: there is always at least one assignment of the atomic variables (for example, every variable may be set to* false*). The following simple relationships are based on this observation.*

Proposition 1.23. *Let Σ be a propositional signature and $\varphi \in L(\Sigma)$. Then the following hold:*

1. *φ is satisfiable if and only if φ is not unsatisfiable.*
2. *φ is valid if and only if φ is not falsifiable.*
3. *If φ is valid, then φ is satisfiable.*
4. *If φ is unsatisfiable, then φ is falsifiable.*
5. *φ is valid if and only if $\neg\varphi$ is unsatisfiable.*
6. *φ is satisfiable if and only if $\neg\varphi$ is falsifiable.*

Proof. We prove the statements one by one.

(1) By definition, φ is satisfiable if and only if $\mathcal{M}(\varphi) \neq \varnothing$, and φ is unsatisfiable if and only if $\mathcal{M}(\varphi) = \varnothing$. The first claim follows immediately from the logical negation of this set-theoretic condition.

(2) φ is valid if and only if $\mathcal{M}(\varphi) = \Omega(\Sigma)$. The formula φ is falsifiable exactly when $\mathcal{M}(\varphi) \neq \Omega(\Sigma)$. Thus the two properties are the negation of one another, and the stated equivalence follows.

(3) If φ is valid, then $\mathcal{M}(\varphi) = \Omega(\Sigma)$. Since $\Omega(\Sigma)$ is nonempty, it follows that $\mathcal{M}(\varphi) \neq \varnothing$, hence φ is satisfiable.

(4) If φ is unsatisfiable, then $\mathcal{M}(\varphi) = \varnothing$. Because $\Omega(\Sigma) \neq \varnothing$, we obtain $\mathcal{M}(\varphi) \neq \Omega(\Sigma)$, so φ is falsifiable.

(5) We use the identity $\mathcal{M}(\neg\varphi) = \Omega(\Sigma) \setminus \mathcal{M}(\varphi)$. Then we have

$$\begin{aligned} \neg\varphi \text{ is unsatisfiable} &\iff \mathcal{M}(\neg\varphi) = \varnothing \\ &\iff \Omega(\Sigma) \setminus \mathcal{M}(\varphi) = \varnothing \\ &\iff \mathcal{M}(\varphi) = \Omega(\Sigma). \end{aligned}$$

Hence φ is valid. The converse follows analogously.

(6) Analogously to (5), we obtain

$$\begin{aligned} \neg\varphi \text{ is falsifiable} &\iff \mathcal{M}(\neg\varphi) \neq \Omega(\Sigma) \\ &\iff \Omega(\Sigma) \setminus \mathcal{M}(\varphi) \neq \Omega(\Sigma) \\ &\iff \mathcal{M}(\varphi) \neq \varnothing. \end{aligned}$$

from which the equivalence follows: φ is satisfiable if and only if $\neg\varphi$ is falsifiable.

This proves all six statements. □

Example 1.24. *Consider the signature $\Sigma = \{P, Q\}$. An interpretation $\omega \in \Omega(\Sigma)$ assigns a truth value to each atom. Altogether, there are four possible interpretations:*

ω	P	Q
ω_1	false	false
ω_2	false	true
ω_3	true	false
ω_4	true	true

We now consider several formulas and classify them according to the categories introduced above.

1. ***Satisfiable Formula.***

   The formula

   $$\varphi_1 := P \wedge Q$$

   is satisfiable, since it is satisfied, for example, under the interpretation ω_4.

2. ***Unsatisfiable Formula.***

   The formula

   $$\varphi_2 := P \wedge \neg P$$

   is unsatisfiable, because no interpretation can satisfy both conditions simultaneously.

3. ***Valid Formula.***

   The formula

   $$\varphi_3 := P \vee \neg P$$

   is true under every interpretation. Hence,

   $$\mathrm{Mod}_{\mathrm{AL}}(\varphi_3) = \Omega(\Sigma),$$

   which means that φ_3 *is valid.*

4. ***Falsifiable Formula.***

   The formula

   $$\varphi_4 := P$$

   is falsifiable, because it is false, for instance, under the interpretations ω_1 *and* ω_2.

This simple example shows how the different classes of propositional formulas can be distinguished by means of their sets of models.

1.4.4 Semantic Consequence

We now want to formalize when a formula ψ follows necessarily from another formula φ.

Definition 1.25 (Semantic Consequence / Entailment). *For $\varphi, \psi \in L(\Sigma)$, we write*

$$\varphi \models \psi \qquad \textit{iff} \qquad \mathcal{M}(\varphi) \subseteq \mathcal{M}(\psi).$$

This means: whenever φ holds, ψ must hold as well. For sets $\Phi, \Psi \subseteq L(\Sigma)$, we define analogously $\Phi \models \Psi$ by $\mathcal{M}(\Phi) \subseteq \mathcal{M}(\Psi)$.

Examples (intuitive).

- From Smoke $\wedge$ SprinklerOn it follows that Smoke.
- From BatteryDead it does not necessarily follow that PowerOff (there are models in which the device continues to run on a backup battery).

Proposition 1.26. *The relation $\models$ is reflexive and transitive; for sets, monotonicity holds: $\Phi \subseteq \Phi'$ implies* $\mathrm{Cn}(\Phi) \subseteq \mathrm{Cn}(\Phi')$.

Proof. Reflexivity: $\mathcal{M}(\varphi) \subseteq \mathcal{M}(\varphi)$ is trivial. Transitivity: from $\mathcal{M}(\varphi) \subseteq \mathcal{M}(\psi)$ and $\mathcal{M}(\psi) \subseteq \mathcal{M}(\chi)$, it follows immediately that $\mathcal{M}(\varphi) \subseteq \mathcal{M}(\chi)$. Monotonicity follows from the definition for sets of formulas and the natural inclusion relation. □

1.4.5 Deductive Closure (What Follows from the Assumptions?)

Definition 1.27. *For a set of assumptions* $\Phi \subseteq L(\Sigma)$*, the* deductive closure *is*

$$\mathrm{Cn}(\Phi) = \{\psi \in L(\Sigma) \mid \Phi \models \psi\}.$$

It contains exactly those formulas that follow semantically from Φ.

Essential Properties.

- $\Phi \subseteq \mathrm{Cn}(\Phi)$ (preservation of assumptions).
- Cn is monotone and idempotent: $\mathrm{Cn}(\mathrm{Cn}(\Phi)) = \mathrm{Cn}(\Phi)$.

Proposition 1.28. *Let* Σ *be a propositional signature, let* $\varphi \in L(\Sigma)$ *be a propositional formula, and let* $\Phi \subseteq L(\Sigma)$ *be a set of propositional formulas. Then the following hold:*

1. $\varphi \in \mathrm{Cn}_\Sigma(\varphi)$.
2. $\Phi \subseteq \mathrm{Cn}_\Sigma(\Phi)$.
3. $\mathrm{Cn}_\Sigma(\varphi) = \mathrm{Cn}_\Sigma(\mathrm{Cn}_\Sigma(\varphi))$.
4. *For* $\Phi, \Phi' \subseteq L(\Sigma)$ *with* $\Phi \subseteq \Phi'$*, we have* $\mathrm{Cn}_\Sigma(\Phi) \subseteq \mathrm{Cn}_\Sigma(\Phi')$.

Proof. We use the semantic definition of deductive closure $\mathrm{Cn}_\Sigma(\cdot)$, that is, $\psi \in \mathrm{Cn}_\Sigma(\Theta)$ if and only if $\mathrm{Mod}_{\mathrm{PL}}(\Theta) \subseteq \mathrm{Mod}_{\mathrm{PL}}(\psi)$.

(1) $\varphi \in \mathrm{Cn}_\Sigma(\varphi)$.

By definition, we have $\mathrm{Mod}_{\mathrm{PL}}(\varphi) \subseteq \mathrm{Mod}_{\mathrm{PL}}(\varphi)$. Hence φ is a semantic consequence of itself, and therefore $\varphi \in \mathrm{Cn}_\Sigma(\varphi)$.

(2) $\Phi \subseteq \mathrm{Cn}_\Sigma(\Phi)$.

Let $\varphi \in \Phi$. Every interpretation $\omega \in \mathrm{Mod}_{\mathrm{PL}}(\Phi)$ satisfies, by definition, all formulas in Φ, and in particular φ. It follows that $\mathrm{Mod}_{\mathrm{PL}}(\Phi) \subseteq \mathrm{Mod}_{\mathrm{PL}}(\varphi)$, hence $\varphi \in \mathrm{Cn}_\Sigma(\Phi)$. Since φ was arbitrary, we conclude that $\Phi \subseteq \mathrm{Cn}_\Sigma(\Phi)$.

(3) $\mathrm{Cn}_\Sigma(\varphi) = \mathrm{Cn}_\Sigma(\mathrm{Cn}_\Sigma(\varphi))$.

First, by (2) with $\Phi = \mathrm{Cn}_\Sigma(\varphi)$, we immediately obtain $\mathrm{Cn}_\Sigma(\varphi) \subseteq \mathrm{Cn}_\Sigma(\mathrm{Cn}_\Sigma(\varphi))$.

For the reverse inclusion, let $\psi \in \mathrm{Cn}_\Sigma(\mathrm{Cn}_\Sigma(\varphi))$. Then, semantically, $\mathrm{Mod}_{\mathrm{PL}}(\mathrm{Cn}_\Sigma(\varphi)) \subseteq \mathrm{Mod}_{\mathrm{PL}}(\psi)$. Since, by definition of $\mathrm{Cn}_\Sigma(\varphi)$, every $\chi \in \mathrm{Cn}_\Sigma(\varphi)$ is a semantic consequence of φ, it follows that $\mathrm{Mod}_{\mathrm{PL}}(\varphi) \subseteq \mathrm{Mod}_{\mathrm{PL}}(\chi)$ for all $\chi \in \mathrm{Cn}_\Sigma(\varphi)$. Hence, $\mathrm{Mod}_{\mathrm{PL}}(\varphi) \subseteq \bigcap_{\chi \in \mathrm{Cn}_\Sigma(\varphi)} \mathrm{Mod}_{\mathrm{PL}}(\chi) = \mathrm{Mod}_{\mathrm{PL}}(\mathrm{Cn}_\Sigma(\varphi))$. Combining this with the assumption $\mathrm{Mod}_{\mathrm{PL}}(\mathrm{Cn}_\Sigma(\varphi)) \subseteq \mathrm{Mod}_{\mathrm{PL}}(\psi)$, we obtain $\mathrm{Mod}_{\mathrm{PL}}(\varphi) \subseteq \mathrm{Mod}_{\mathrm{PL}}(\psi)$, and thus $\psi \in \mathrm{Cn}_\Sigma(\varphi)$. This proves the reverse inclusion.

(4) Monotonicity: $\Phi \subseteq \Phi' \Rightarrow \mathrm{Cn}_\Sigma(\Phi) \subseteq \mathrm{Cn}_\Sigma(\Phi')$.

Since $\Phi \subseteq \Phi'$, every interpretation that satisfies all formulas in Φ' automatically also satisfies all formulas in Φ. Hence $\mathrm{Mod}_{\mathrm{PL}}(\Phi') \subseteq \mathrm{Mod}_{\mathrm{PL}}(\Phi)$. Now let $\psi \in \mathrm{Cn}_\Sigma(\Phi)$; this means that $\mathrm{Mod}_{\mathrm{PL}}(\Phi) \subseteq \mathrm{Mod}_{\mathrm{PL}}(\psi)$. By transitivity of set inclusion, it follows that $\mathrm{Mod}_{\mathrm{PL}}(\Phi') \subseteq \mathrm{Mod}_{\mathrm{PL}}(\psi)$, and therefore $\psi \in \mathrm{Cn}_\Sigma(\Phi')$. Since ψ was arbitrary, this yields the claimed inclusion. □

Brief Explanation (Didactic).

- Point (1) is the *reflexivity* of deductive closure: an assumption always follows from itself.
- Point (2) says that Cn *preserves* statements — everything assumed remains contained in the closure.
- Point (3) is *idempotence*: once a set has been closed, applying the closure again does not change it.
- Point (4) is the important property of *monotonicity*: adding more assumptions cannot invalidate earlier semantic consequences — it can only add new ones.

Small Example. Let $\Sigma = \{p, q\}$, $\Phi = \{p\}$, and $\Phi' = \{p, p \to q\}$. Then $\mathrm{Cn}(\Phi)$ contains, among other things, p, whereas $\mathrm{Cn}(\Phi')$ additionally contains q, so $\mathrm{Cn}(\Phi) \subseteq \mathrm{Cn}(\Phi')$.

Exercise for Students. Using the semantic definition, show that for arbitrary $\Phi \subseteq L(\Sigma)$, the set $\mathrm{Cn}_\Sigma(\Phi)$ is always closed under conjunction; that is, show that from $\psi_1, \psi_2 \in \mathrm{Cn}_\Sigma(\Phi)$ it follows that $\psi_1 \wedge \psi_2 \in \mathrm{Cn}_\Sigma(\Phi)$.

1.4.6 Theories and the Th/Mod Relationship

Definition 1.29. *For a set of interpretations $M \subseteq \Omega(\Sigma)$, we define the* theory *of M as*

$$\mathrm{Th}(M) \;=\; \{\varphi \in L(\Sigma) \mid M \subseteq \mathcal{M}(\varphi)\}.$$

Intuitively. $\mathrm{Th}(M)$ collects all formulas that are true in all situations from M — that is, the shared, necessary knowledge represented by the situations in M.

Theorem 1.30. *Let Σ be a propositional signature, let $\varphi \in L(\Sigma)$, and let $M \subseteq \Omega(\Sigma)$. Then the following hold:*

1. $\mathrm{Th}_\Sigma(\mathrm{Mod}_\Sigma(\varphi)) = \mathrm{Cn}_\Sigma(\varphi)$.

2. $\mathrm{Mod}_\Sigma(\mathrm{Th}_\Sigma(M)) \supseteq M$. *If M is* definable, *i.e. if $M = \mathrm{Mod}_\Sigma(\Psi)$ for a set Ψ of formulas, then in fact* $\mathrm{Mod}_\Sigma(\mathrm{Th}_\Sigma(M)) = M$.

3. $\mathrm{Cn}_\Sigma(\mathrm{Th}_\Sigma(M)) = \mathrm{Th}_\Sigma(M)$.

Proof. Let us briefly recall the definitions from the semantic perspective:

$$\begin{aligned}
\mathrm{Mod}_\Sigma(\Phi) &= \{\omega \in \Omega(\Sigma) \mid \forall \psi \in \Phi : \omega \models \psi\}, \\
\mathrm{Th}_\Sigma(M) &= \{\psi \in L(\Sigma) \mid M \subseteq \mathrm{Mod}_\Sigma(\psi)\}, \\
\mathrm{Cn}_\Sigma(\Phi) &= \{\psi \in L(\Sigma) \mid \mathrm{Mod}_\Sigma(\Phi) \subseteq \mathrm{Mod}_\Sigma(\psi)\}.
\end{aligned}$$

(1) $\mathrm{Th}_\Sigma(\mathrm{Mod}_\Sigma(\varphi)) = \mathrm{Cn}_\Sigma(\varphi)$.

"$\subseteq$". Let $\psi \in \mathrm{Th}_\Sigma(\mathrm{Mod}_\Sigma(\varphi))$. Then $\mathrm{Mod}_\Sigma(\varphi) \subseteq \mathrm{Mod}_\Sigma(\psi)$ by the definition of Th_Σ. This is exactly the semantic consequence relation $\varphi \models \psi$, and hence $\psi \in \mathrm{Cn}_\Sigma(\varphi)$.

"$\supseteq$". Let $\psi \in \mathrm{Cn}_\Sigma(\varphi)$. Then, by definition, $\mathrm{Mod}_\Sigma(\varphi) \subseteq \mathrm{Mod}_\Sigma(\psi)$. This means that ψ is true in all models of φ, and therefore ψ belongs to $\mathrm{Th}_\Sigma(\mathrm{Mod}_\Sigma(\varphi))$ by definition.

Thus, both inclusions have been established.

(2) $\mathrm{Mod}_\Sigma(\mathrm{Th}_\Sigma(M)) \supseteq M$ (and equality in the definable case).

Let $\omega \in M$. For every $\psi \in \mathrm{Th}_\Sigma(M)$, we have $M \subseteq \mathrm{Mod}_\Sigma(\psi)$ by the definition of Th_Σ. In particular, $\omega \in \mathrm{Mod}_\Sigma(\psi)$. Since this holds for all $\psi \in \mathrm{Th}_\Sigma(M)$, the valuation ω satisfies the entire set $\mathrm{Th}_\Sigma(M)$, and hence $\omega \in \mathrm{Mod}_\Sigma(\mathrm{Th}_\Sigma(M))$. Because ω was arbitrary, it follows that $M \subseteq \mathrm{Mod}_\Sigma(\mathrm{Th}_\Sigma(M))$.

Equality in the definable case. Assume that M is definable, i.e. that $M = \mathrm{Mod}_\Sigma(\Psi)$ for some set $\Psi \subseteq L(\Sigma)$. Then every $\psi \in \Psi$ belongs to $\mathrm{Th}_\Sigma(M)$, either because $\Psi \subseteq \mathrm{Th}_\Sigma(M)$ or directly by the definition. Therefore, $\mathrm{Mod}_\Sigma(\mathrm{Th}_\Sigma(M)) \subseteq \mathrm{Mod}_\Sigma(\Psi) = M$. Combined with the general inclusion $M \subseteq \mathrm{Mod}_\Sigma(\mathrm{Th}_\Sigma(M))$, we obtain the equality $\mathrm{Mod}_\Sigma(\mathrm{Th}_\Sigma(M)) = M$.

(3) $\mathrm{Cn}_\Sigma(\mathrm{Th}_\Sigma(M)) = \mathrm{Th}_\Sigma(M)$.

"$\supseteq$". In general, $\mathrm{Th}_\Sigma(M) \subseteq \mathrm{Cn}_\Sigma(\mathrm{Th}_\Sigma(M))$, since every set is contained in its deductive closure (reflexivity).

"$\subseteq$". Let $\psi \in \mathrm{Cn}_\Sigma(\mathrm{Th}_\Sigma(M))$. Then $\mathrm{Mod}_\Sigma(\mathrm{Th}_\Sigma(M)) \subseteq \mathrm{Mod}_\Sigma(\psi)$. From (2), we know that $M \subseteq \mathrm{Mod}_\Sigma(\mathrm{Th}_\Sigma(M))$. Combining these, we get $M \subseteq \mathrm{Mod}_\Sigma(\psi)$, hence $\psi \in \mathrm{Th}_\Sigma(M)$ by the definition of Th_Σ. This proves the reverse inclusion.

Taken together, the two inclusions yield the claimed equality. □

1.4.7 Independence from Superfluous Atoms

A common question is whether logical consequence depends on additional atomic variables that do not occur in the formula. The answer is no: if φ and ψ use only atoms from a subset $\Sigma_0 \subseteq \Sigma$, then the relation $\varphi \models \psi$ is independent of extensions of the signature that merely add new, unused atoms. Formally, this can be shown by a simple extension argument for interpretations: each interpretation over Σ_0 has exactly two extensions for every new atom, and these do not affect the truth of the original formulas.

Example 1.31 (The Same Formula over Different Signatures). *A propositional formula usually uses only some of the atoms available in a signature. Therefore, the same formula can be considered with respect to different signatures.*

Consider the formula

$$\varphi := P \to Q.$$

This formula uses only the atoms P and Q. It is therefore well-defined both with respect to the signature

$$\Sigma_1 = \{P, Q\}$$

and with respect to the larger signature

$$\Sigma_2 = \{P, Q, R\}$$

The difference appears in the possible interpretations. With respect to Σ_1, there are exactly four interpretations, *whereas with respect to Σ_2, there are already eight interpretations, since in addition a truth value for R must be specified.*

However, the truth of the formula φ depends only on the atoms P and Q. The truth value of R therefore has no influence on whether an interpretation satisfies φ.

In other words, additional atoms in the signature may be irrelevant for a formula. Such atoms are also called independent *or* superfluous *atoms for the formula in question.*

The Atoms Occurring in a Formula. A propositional formula usually uses only some of the propositional variables of a given signature. For many purposes, therefore, only the set of variables that actually occur is relevant.

Notation 1.32. *Let Σ be a propositional signature and let $\varphi \in L(\Sigma)$ be a formula. We write $\Sigma(\varphi)$ for the set of propositional variables occurring in φ. It is defined recursively as follows:*

$$\Sigma(\varphi) = \begin{cases} \{x\} & \text{if } \varphi = x \text{ with } x \in \Sigma, \\ \Sigma(\psi) & \text{if } \varphi = \neg\psi, \\ \Sigma(\psi) \cup \Sigma(\psi') & \text{if } \varphi = \psi \wedge \psi', \\ \Sigma(\psi) \cup \Sigma(\psi') & \text{if } \varphi = \psi \vee \psi'. \end{cases}$$

If $\Phi = \{\varphi_1, \ldots, \varphi_n\}$ is a set of formulas, we define

$$\Sigma(\Phi) = \Sigma(\varphi_1) \cup \cdots \cup \Sigma(\varphi_n).$$

This yields the minimal set of propositional variables needed to evaluate a formula meaningfully.

Example 1.33.

$$\Sigma((A \wedge (E \vee \neg C)) \wedge (\neg A \vee D)) = \{A, C, D, E\},$$

$$\Sigma(\{A,\ A \wedge B,\ C \vee \neg D\}) = \{A, B, C, D\}.$$

In particular, for every formula we always have

$$\varphi \in L(\Sigma(\varphi)),$$

i.e. every formula is already defined over the smallest signature that contains all of its variables.

Independence from the Choice of Signature. The truth of a formula depends only on the variables that actually occur in it. If additional variables are added to the signature, the logical consequence relations therefore remain unchanged.

Theorem 1.34. *Let Σ be a propositional signature and let $\varphi, \psi \in L(\Sigma)$ satisfy*

$$\varphi \models_\Sigma \psi.$$

Then for every signature Σ' with

$$\Sigma(\varphi) \cup \Sigma(\psi) \subseteq \Sigma'$$

we also have

$$\varphi \models_{\Sigma'} \psi.$$

Proof. We first show that adding new propositional variables does not change the consequence relation.

Step 1: Extension of the signature.

First, let

$$\Sigma' = \Sigma \cup \{A_{n+1}\}$$

with a new propositional variable A_{n+1}. Clearly, $\varphi, \psi \in L(\Sigma')$.

Let $\omega \in \Omega(\Sigma)$. We define two extensions of ω to interpretations over Σ':

$$\omega^{\text{true}}(x) = \omega(x) \ (x \in \Sigma), \qquad \omega^{\text{true}}(A_{n+1}) = \text{true},$$

$$\omega^{\text{false}}(x) = \omega(x) \ (x \in \Sigma), \qquad \omega^{\text{false}}(A_{n+1}) = \text{false}.$$

Conversely, from every interpretation $\omega' \in \Omega(\Sigma')$ we obtain a restriction to Σ:

$$\overline{\omega'}(x) = \omega'(x) \qquad (x \in \Sigma).$$

Since A_{n+1} occurs neither in φ nor in ψ, its truth value does not affect the evaluation of these formulas. Hence,

$$\mathrm{Mod}_{\Sigma'}(\varphi) = \{\omega^{\mathrm{true}}, \omega^{\mathrm{false}} \mid \omega \in \mathrm{Mod}_{\Sigma}(\varphi)\}. \tag{2.1}$$

Analogously,

$$\mathrm{Mod}_{\Sigma'}(\psi) = \{\omega^{\mathrm{true}}, \omega^{\mathrm{false}} \mid \omega \in \mathrm{Mod}_{\Sigma}(\psi)\}.$$

from the assumption

$$\mathrm{Mod}_{\Sigma}(\varphi) \subseteq \mathrm{Mod}_{\Sigma}(\psi)$$

it follows immediately that

$$\mathrm{Mod}_{\Sigma'}(\varphi) \subseteq \mathrm{Mod}_{\Sigma'}(\psi),$$

hence

$$\varphi \models_{\Sigma'} \psi.$$

By repeated application of this argument, we obtain

$$\varphi \models_{\Sigma''} \psi \quad \text{for all signatures } \Sigma'' \text{ with } \Sigma \subseteq \Sigma''.$$

Step 2: Reduction of the signature.
We now show that variables occurring in neither φ nor ψ can be removed from the signature.

To this end, let $x \in \Sigma$ with

$$x \notin \Sigma(\varphi) \cup \Sigma(\psi).$$

Then the truth value of x plays no role in the evaluation of φ and ψ. By the same argument as above, it follows from

$$\mathrm{Mod}_{\Sigma}(\varphi) \subseteq \mathrm{Mod}_{\Sigma}(\psi)$$

that also

$$\mathrm{Mod}_{\Sigma\setminus\{x\}}(\varphi) \subseteq \mathrm{Mod}_{\Sigma\setminus\{x\}}(\psi).$$

Thus, such a variable can be removed from the signature without changing the entailment relation.

Applying this argument repeatedly to all variables occurring in neither φ nor in ψ, we finally obtain

$$\mathrm{Mod}_{\Sigma(\varphi)\cup\Sigma(\psi)}(\varphi) \subseteq \mathrm{Mod}_{\Sigma(\varphi)\cup\Sigma(\psi)}(\psi),$$

hence

$$\varphi \models_{\Sigma(\varphi)\cup\Sigma(\psi)} \psi.$$

Combining this with Step 1 yields the claim. □

Example 1.35 (Dependence on the Signature). *Although logical consequence depends only on the variables occurring in the formulas, the notions* Mod, Cn, *and* Th *still depend on the particular signature under consideration.*

Consider the formula

$$\varphi := P.$$

First, let the signature be

$$\Sigma_1 = \{P\}.$$

Then there are exactly two interpretations. Only one of them satisfies φ, namely the interpretation with $P = true$. Hence,

$$\mathrm{Mod}_{\Sigma_1}(\varphi) = \{\omega \in \Omega(\Sigma_1) \mid \omega(P) = true\}.$$

Now consider the larger signature

$$\Sigma_2 = \{P, Q\}.$$

Then there are four interpretations. Two of them satisfy the formula φ, namely precisely those with $P = true$, independently of the truth value of Q. Thus,

$$\mathrm{Mod}_{\Sigma_2}(\varphi) = \{\omega \in \Omega(\Sigma_2) \mid \omega(P) = true\}.$$

The sets of models are therefore different, since they are subsets of different spaces of interpretations. Accordingly, the sets $Cn_\Sigma(\varphi)$ and $Th_\Sigma(M)$ also depend on the chosen signature.

1.4.8 A Short Checklist for Semantic Arguments

If you want to argue semantically—for example, to prove or refute $\Phi \models \psi$—check the following points in order:

1. Formulate $\mathcal{M}(\Phi)$ explicitly (if necessary, by means of conditions).
2. Show directly that $\mathcal{M}(\Phi) \subseteq \mathcal{M}(\psi)$, or provide a countermodel $\omega \in \mathcal{M}(\Phi) \setminus \mathcal{M}(\psi)$.
3. Use set-theoretic identities such as intersection, union, or complement in order to transform formulas semantically.

1.4.9 Exercise (Short)

Let $\Sigma = \{\text{Smoke}, \text{Heat}, \text{Alarm}\}$ and $\Phi = \{\text{Smoke} \to \text{Alarm}, \text{Heat} \to \text{Alarm}\}$. Show whether $\Phi \models \text{Smoke} \vee \text{Heat} \to \text{Alarm}$ holds. (*Hint:* check whether there is an ω that satisfies both assumptions but falsifies the conclusion formula.)

Note. In later chapters, we introduce syntactic proof systems (e.g. natural deduction and resolution). Two fundamental questions addressed there are: Are the rules *sound* with respect to semantic meaning? And are they *complete*, that is, do they make it possible to derive every semantic consequence syntactically as well?

Remark 1.36 (Concluding Remark). *Even though logical consequence $\varphi \models \psi$ depends only on the variables that actually occur in φ and ψ, the associated concepts of model classes* Mod_Σ*, deductive closure Cn_Σ, and theory Th_Σ remain signature-dependent. Changes to the signature can therefore affect the representation and size of the model classes without changing the logical consequence itself.*

1.5 Exercises and Solutions: Logical Consequence

Exercise A (Signature and Simple Classification). Let the signature be

$$\Sigma = \{A, B, C, D, E\}.$$

Consider the following formulas:

$$\begin{aligned}\varphi_1 &:= A \wedge (B \vee C),\\ \varphi_2 &:= \neg(B \wedge C),\\ \varphi_3 &:= (D \vee A) \wedge \neg D,\\ \varphi_4 &:= \neg(E \vee B).\end{aligned}$$

For each item, decide whether the following statement is true or false (a short justification is sufficient):

1. $\neg A \vee B \vee C$ is falsifiable.
2. $A \wedge \neg A$ is satisfiable.
3. $(D \vee A) \wedge \neg D \models_\Sigma A$.
4. $\Sigma(\varphi_4) = \{E, B\}$.
5. φ_3 is satisfiable.
6. How many interpretations (models) of φ_1 are there over Σ?

Solutions to Exercise A.

1. **True.** $\neg A \vee B \vee C$ is *falsifiable*, because there is a valuation that makes the formula false; for example, $A = \text{true}, B = \text{false}, C = \text{false}$ (with D, E arbitrary). Under this valuation, $\neg A \vee B \vee C = \text{false}$.
2. **False.** $A \wedge \neg A$ is *unsatisfiable* (a contradiction). No interpretation can assign both $A = \text{true}$ and $A = \text{false}$ at the same time.
3. **True.** Suppose $\omega \models (D \vee A) \wedge \neg D$. Then both $\neg D$ and $D \vee A$ hold, so A must be true. Hence every interpretation that satisfies $(D \vee A) \wedge \neg D$ also satisfies A. Therefore $(D \vee A) \wedge \neg D \models A$.

4. **True.** Clearly, only the atoms E and B occur in $\varphi_4 = \neg(E \vee B)$, so $\Sigma(\varphi_4) = \{E, B\}$.
5. **True.** $\varphi_3 = (D \vee A) \wedge \neg D$ is satisfiable; for example, the valuation $A = \text{true}, D = \text{false}$ (with the other atoms arbitrary) satisfies φ_3.
6. $\varphi_1 = A \wedge (B \vee C)$ requires $A = \text{true}$ and at least one of B, C to be true. The possibilities for (B, C) are $(1,0), (0,1), (1,1)$, i.e. 3 cases. The remaining atoms D, E are free (2 choices each), so there are $3 \cdot 2^2 = 12$ models in total.

Exercise B (Signatures, Models, Consequence). Let $\Sigma' = \{P, Q, R\}$. Given:

$$\psi_1 := P \vee Q, \qquad \psi_2 := \neg P \wedge R.$$

1. Give an example of an interpretation $\omega \in \Omega(\Sigma')$ such that $\omega \models \psi_1$ but $\omega \not\models \psi_2$.
2. Does $\psi_2 \models_{\Sigma'} R$ hold? Briefly justify your answer.
3. Determine $\Sigma(\psi_1)$ and $\Sigma(\psi_2)$.

Solutions to Exercise B.

1. Example: $\omega(P) = \text{true}$, $\omega(Q) = \text{false}$, $\omega(R) = \text{false}$. Then $\psi_1 = P \vee Q$ is true, while $\psi_2 = \neg P \wedge R$ is false (since $\neg P$ is false).
2. **Yes.** Every interpretation satisfying $\psi_2 = \neg P \wedge R$ has $R = \text{true}$. Therefore $\psi_2 \models R$.
3. $\Sigma(\psi_1) = \{P, Q\}$, $\Sigma(\psi_2) = \{P, R\}$.

Exercise C (Models and Theory). Let again $\Sigma = \{A, B, C, D, E\}$. Let $M \subseteq \Omega(\Sigma)$ be the set of all interpretations with $A = \text{true}$ and $B = \text{false}$ (the values of C, D, E are arbitrary).

1. Construct a formula or set of formulas Ψ such that $\mathrm{Mod}_\Sigma(\Psi) = M$.
2. Briefly describe the theory $\mathrm{Th}_\Sigma(M)$ in words.

Solutions to Exercise C.

1. A simple choice is $\Psi = \{A, \neg B\}$. Clearly, an interpretation satisfies Ψ exactly when $A = \text{true}$ and $B = \text{false}$, hence $\mathrm{Mod}_\Sigma(\Psi) = M$.
2. $\mathrm{Th}_\Sigma(M)$ consists of all formulas that are true in *every* interpretation in M. For example, A, $\neg B$, $A \vee C$, and $A \wedge \neg B$ belong to the theory; C does not necessarily belong to it, since C varies within M.

Exercise D (Short Proof / Explanatory Task). Explain in one or two sentences why the relation $\models$ is not materially affected by the choice of signature Σ, as long as all atoms occurring in the formulas under consideration are contained in the signature.

Solution to Exercise D. The evaluation of a formula depends only on the atoms that actually occur in it. If a signature contains additional atoms that do not appear in the formulas under consideration, interpretations can be extended or restricted accordingly without affecting the truth of those formulas. Hence, we have

$$\varphi \models_\Sigma \psi \iff \varphi \models_{\Sigma(\varphi)\cup\Sigma(\psi)} \psi.$$

Remark. These exercises cover a range of standard skills: classifying formulas (satisfiable / unsatisfiable / valid / falsifiable), constructing concrete models, determining $\Sigma(\varphi)$, working with sets of models, and developing a basic understanding of $\mathrm{Mod}, \mathrm{Th}, \mathrm{Cn}$.

1.6 Equivalence

Having introduced the notion of semantic consequence in the previous section, we now turn to a closely related relation between formulas: *equivalence.* Two formulas are equivalent if they have exactly the same models – that is, if they express the same content even though they may differ syntactically. Equivalences allow us to replace formulas by simpler ones and to define normal forms.

1.6.1 Basics

Definition 1.37. *Let Σ be a propositional signature. Two formulas $\varphi, \psi \in L(\Sigma)$ are called* equivalent, *written $\varphi \equiv_\Sigma \psi$, if and only if*

$$\varphi \models_\Sigma \psi \qquad \textit{and} \qquad \psi \models_\Sigma \varphi.$$

The following characterization in terms of model sets is often more convenient.

Proposition 1.38. *Let Σ be a propositional signature and let $\varphi, \psi \in L(\Sigma)$. Then*

$$\varphi \equiv_\Sigma \psi \quad \Longleftrightarrow \quad \mathrm{Mod}_\Sigma(\varphi) = \mathrm{Mod}_\Sigma(\psi).$$

Proof. If $\varphi \equiv_\Sigma \psi$, then $\varphi \models_\Sigma \psi$ and $\psi \models_\Sigma \varphi$, which immediately imply $\mathrm{Mod}_\Sigma(\varphi) \subseteq \mathrm{Mod}_\Sigma(\psi)$ and the reverse inclusion, hence equality. Conversely, from $\mathrm{Mod}_\Sigma(\varphi) = \mathrm{Mod}_\Sigma(\psi)$ it follows directly that both $\varphi \models_\Sigma \psi$ and $\psi \models_\Sigma \varphi$ hold. □

Corollary 1.39. *Let Σ be a propositional signature and let $\varphi, \psi \in L(\Sigma)$ satisfy $\varphi \equiv_\Sigma \psi$. Then, for every propositional signature Σ' with $\Sigma(\varphi) \cup \Sigma(\psi) \subseteq \Sigma'$, we have*

$$\varphi \equiv_{\Sigma'} \psi.$$

(In what follows, one often writes simply $\varphi \equiv \psi$ for $\varphi \equiv_{\Sigma(\varphi)\cup\Sigma(\psi)} \psi$.)

Proof. Set $\Sigma_0 := \Sigma(\varphi) \cup \Sigma(\psi)$, and let Σ' be a signature such that $\Sigma_0 \subseteq \Sigma'$. Consider the restriction map

$$\mathrm{res} : \Omega(\Sigma') \longrightarrow \Omega(\Sigma_0), \qquad \omega' \mapsto \omega'|_{\Sigma_0},$$

which assigns to each interpretation over Σ' its restriction to Σ_0. For every formula χ with $\Sigma(\chi) \subseteq \Sigma_0$, we have the obvious equivalence

$$\omega' \models_{\Sigma'} \chi \quad \Longleftrightarrow \quad \mathrm{res}(\omega') \models_{\Sigma_0} \chi,$$

since the truth of χ depends only on the valuations of the atoms occurring in χ.

Hence, for such formulas χ, the set of models is given as the preimage:

$$\mathrm{Mod}_{\Sigma'}(\chi) = \mathrm{res}^{-1}\left(\mathrm{Mod}_{\Sigma_0}(\chi)\right).$$

In particular, since $\Sigma(\varphi), \Sigma(\psi) \subseteq \Sigma_0$, we obtain

$$\begin{aligned}\mathrm{Mod}_{\Sigma'}(\varphi) &= \mathrm{res}^{-1}(\mathrm{Mod}_{\Sigma_0}(\varphi)),\\ \mathrm{Mod}_{\Sigma'}(\psi) &= \mathrm{res}^{-1}(\mathrm{Mod}_{\Sigma_0}(\psi)).\end{aligned}$$

From the assumption $\varphi \equiv_\Sigma \psi$, Proposition 1.32 yields $\mathrm{Mod}_{\Sigma_0}(\varphi) = \mathrm{Mod}_{\Sigma_0}(\psi)$. Applying res^{-1} to both sides, and noting that res^{-1} preserves equalities (and inclusions), we obtain

$$\mathrm{Mod}_{\Sigma'}(\varphi) = \mathrm{Mod}_{\Sigma'}(\psi).$$

By Proposition 1.32, this is exactly the statement that $\varphi \equiv_{\Sigma'} \psi$. □

Proposition 1.40. *Let Σ be a propositional signature. The relation $\equiv_\Sigma$ is an equivalence relation on $L(\Sigma)$, that is, it is*

1. *reflexive:* $\varphi \equiv_\Sigma \varphi$,
2. *symmetric:* $\varphi \equiv_\Sigma \psi \Rightarrow \psi \equiv_\Sigma \varphi$,
3. *transitive:* $\varphi \equiv_\Sigma \psi$ *and* $\psi \equiv_\Sigma \chi \Rightarrow \varphi \equiv_\Sigma \chi$.

Proof. These statements follow directly from the definition of $\varphi \equiv_\Sigma \psi$ as mutual semantic consequence;

alternatively, they also follow from the characterization in terms of sets of models, since equality of sets is an equivalence relation. □

An important and very useful result is that equivalent subformulas can be replaced arbitrarily without changing the semantics. We formulate this by means of a substitution operation.

Definition 1.41 (Substitution). *Let Σ be a signature, and let $\varphi, \psi, \psi' \in L(\Sigma)$. The notation $\varphi[\psi \mapsto \psi']$ denotes the formula obtained from φ by simultaneously replacing* all *occurrences of ψ in φ by ψ'. Formally, this is defined recursively over the structure of φ (atomic cases, negation, conjunction, disjunction).*

Lemma 1.42 (Substitution Lemma). *Let Σ be a signature, and let $\varphi, \psi, \psi' \in L(\Sigma)$ with $\psi \equiv_\Sigma \psi'$. Then*

$$\varphi \equiv_\Sigma \varphi[\psi \mapsto \psi'].$$

Proof. We prove the claim by structural induction on the formula φ.

Base case. Suppose that φ is an atom $p \in \Sigma$. Then there are two cases:

- If $\varphi = \psi$, then $\varphi[\psi \mapsto \psi'] = \psi'$, and $\varphi \equiv \varphi[\psi \mapsto \psi']$ follows from the assumption $\psi \equiv \psi'$.
- If $\varphi \neq \psi$, then $\varphi[\psi \mapsto \psi'] = \varphi$, so the claim is immediate.

Inductive step. Assume that the statement has already been proved for all proper subformulas of φ. We distinguish cases according to the outer form of φ:

- $\varphi = \neg\xi$. Then $\varphi[\psi \mapsto \psi'] = \neg(\xi[\psi \mapsto \psi'])$. By the induction hypothesis, $\xi \equiv \xi[\psi \mapsto \psi']$, hence $\mathrm{Mod}(\xi) = \mathrm{Mod}(\xi[\psi \mapsto \psi'])$. The model sets of $\neg\xi$ and $\neg(\xi[\psi \mapsto \psi'])$ are the corresponding complements of these sets, so they are equal as well. Thus $\neg\xi \equiv \neg(\xi[\psi \mapsto \psi'])$.
- $\varphi = \xi_1 \wedge \xi_2$. Then $\varphi[\psi \mapsto \psi'] = \xi_1[\psi \mapsto \psi'] \wedge \xi_2[\psi \mapsto \psi']$. By the induction hypothesis, $\xi_i \equiv \xi_i[\psi \mapsto \psi']$ for $i = 1, 2$. Since the model set of a conjunction is the intersection of the model sets of its conjuncts, the model sets are equal, and therefore the formulas are equivalent.
- $\varphi = \xi_1 \vee \xi_2$. This is analogous to the previous case, except that the model set of a disjunction is the union of the model sets of its disjuncts.

In all cases, substitution preserves the model set, and therefore $\varphi \equiv \varphi[\psi \mapsto \psi']$. □

Example 1.43. *Consider the signature Σ_{Alarm} = {Smoke, Heat, Motion, Alarm, Sprinkler}. To illustrate De Morgan's law, we consider the formulas*

$$\neg(\mathit{Alarm} \vee \mathit{Sprinkler}) \qquad \text{and} \qquad \neg\mathit{Alarm} \wedge \neg\mathit{Sprinkler}.$$

Semantic justification: For any interpretation $\omega \in \Omega(\Sigma_{Alarm})$, we have

$$\begin{aligned}
\omega \models \neg(\mathit{Alarm} \vee \mathit{Sprinkler}) &\iff \neg(\omega \models \mathit{Alarm} \\
&\qquad \text{or } \omega \models \mathit{Sprinkler}) \\
&\iff \neg(\omega \models \mathit{Alarm}) \\
&\qquad \text{and } \neg(\omega \models \mathit{Sprinkler}) \\
&\iff \omega \models \\
&\qquad (\neg \mathit{Alarm} \wedge \neg \mathit{Sprinkler})
\end{aligned}$$

It follows immediately that

$$\neg(\mathit{Alarm} \vee \mathit{Sprinkler}) \equiv \neg \mathit{Alarm} \wedge \neg \mathit{Sprinkler},$$

hence the De Morgan equivalence for the alarm signature.
■

In summary, equivalence is the natural semantic notion of identity for formulas and permits safe replacement rules (substitution of equivalent subformulas). In the next subsection, we discuss important concrete equivalences and the common normal forms (NNF, CNF, DNF) in more detail.

1.6.2 Specific Equivalences

Proposition 1.44. *Let Σ be a propositional signature and let $\varphi, \psi, \xi \in L(\Sigma)$. The following equivalences hold:*

1. $\varphi \wedge \varphi \equiv \varphi$ *($\wedge$-idempotence)*
2. $\varphi \vee \varphi \equiv \varphi$ *($\vee$-idempotence)*

3. $\neg\neg\varphi \equiv \varphi$ *(double negation)*

4. $\varphi \wedge \psi \equiv \psi \wedge \varphi$ *($\wedge$-commutativity)*

5. $\varphi \vee \psi \equiv \psi \vee \varphi$ *($\vee$-commutativity)*

6. $\varphi \wedge (\psi \wedge \xi) \equiv (\varphi \wedge \psi) \wedge \xi$ *($\wedge$-associativity)*

7. $\varphi \vee (\psi \vee \xi) \equiv (\varphi \vee \psi) \vee \xi$ *($\vee$-associativity)*

8. $\varphi \wedge (\psi \vee \xi) \equiv (\varphi \wedge \psi) \vee (\varphi \wedge \xi)$ *($\wedge$-distributivity)*

9. $\varphi \vee (\psi \wedge \xi) \equiv (\varphi \vee \psi) \wedge (\varphi \vee \xi)$ *($\vee$-distributivity)*

10. $\varphi \wedge (\varphi \vee \psi) \equiv \varphi$ *($\wedge$-absorption)*

11. $\varphi \vee (\varphi \wedge \psi) \equiv \varphi$ *($\vee$-absorption)*

12. $\neg(\varphi \wedge \psi) \equiv \neg\varphi \vee \neg\psi$ *(De Morgan's law 1)*

13. $\neg(\varphi \vee \psi) \equiv \neg\varphi \wedge \neg\psi$ *(De Morgan's law 2)*

14. $\varphi \rightarrow \psi \equiv \neg\psi \rightarrow \neg\varphi$ *(contraposition)*

Proof. We prove the statements semantically. Let $\omega \in \Omega(\Sigma)$ be an arbitrary interpretation throughout. Each equivalence follows because, for every ω, the left-hand side is true if and only if the right-hand side is true; hence the corresponding sets of models coincide.

1. $\varphi \wedge \varphi \equiv \varphi$.
   $\omega \models \varphi \wedge \varphi \iff (\omega \models \varphi \text{ and } \omega \models \varphi) \iff \omega \models \varphi.$

2. $\varphi \vee \varphi \equiv \varphi$.
   $\omega \models \varphi \vee \varphi \iff (\omega \models \varphi \text{ or } \omega \models \varphi) \iff \omega \models \varphi.$

3. $\neg\neg\varphi \equiv \varphi$.
   $\omega \models \neg\neg\varphi \iff \neg(\omega \models \neg\varphi) \iff \omega \models \varphi$.

4. $\varphi \wedge \psi \equiv \psi \wedge \varphi$.
   $\omega \models \varphi \wedge \psi \iff (\omega \models \varphi \text{ and } \omega \models \psi) \iff \omega \models \psi \wedge \varphi$.

5. $\varphi \vee \psi \equiv \psi \vee \varphi$. Analogous to (4), with $\vee$.

6. $\varphi \wedge (\psi \wedge \xi) \equiv (\varphi \wedge \psi) \wedge \xi$.
   $\omega \models \varphi \wedge (\psi \wedge \xi) \iff \omega \models \varphi$ and $(\omega \models \psi$ and $\omega \models \xi)$, which is equivalent to $(\omega \models \varphi$ and $\omega \models \psi)$ and $\omega \models \xi$, that is, $\omega \models (\varphi \wedge \psi) \wedge \xi$.

7. $\varphi \vee (\psi \vee \xi) \equiv (\varphi \vee \psi) \vee \xi$. Analogous to (6), with $\vee$.

8. $\varphi \wedge (\psi \vee \xi) \equiv (\varphi \wedge \psi) \vee (\varphi \wedge \xi)$.
   "$\subseteq$": If $\omega \models \varphi \wedge (\psi \vee \xi)$, then $\omega \models \varphi$ and $\omega \models \psi \vee \xi$. If $\omega \models \psi$, then $\omega \models \varphi \wedge \psi$; otherwise $\omega \models \xi$ and hence $\omega \models \varphi \wedge \xi$. In either case, $\omega \models (\varphi \wedge \psi) \vee (\varphi \wedge \xi)$.
   "$\supseteq$": If $\omega \models (\varphi \wedge \psi) \vee (\varphi \wedge \xi)$, then either $\omega \models \varphi \wedge \psi$ or $\omega \models \varphi \wedge \xi$; in both cases it follows that $\omega \models \varphi$ and $\omega \models \psi \vee \xi$, and therefore $\omega \models \varphi \wedge (\psi \vee \xi)$. Hence the two sets are equal.

9. $\varphi \vee (\psi \wedge \xi) \equiv (\varphi \vee \psi) \wedge (\varphi \vee \xi)$.
   "$\subseteq$": If $\omega \models \varphi \vee (\psi \wedge \xi)$, then either $\omega \models \varphi$ or $\omega \models \psi \wedge \xi$. In the first case, both $\omega \models \varphi \vee \psi$ and $\omega \models \varphi \vee \xi$ hold; in the second case, they do as well. Thus $\omega \models (\varphi \vee \psi) \wedge (\varphi \vee \xi)$.
   "$\supseteq$": If $\omega \models (\varphi \vee \psi) \wedge (\varphi \vee \xi)$, then $\omega \models \varphi \vee \psi$ and $\omega \models \varphi \vee \xi$. If $\omega \models \varphi$, then immediately $\omega \models$

$\varphi \vee (\psi \wedge \xi)$. If $\omega \not\models \varphi$, then the two disjunctions force $\omega \models \psi$ and $\omega \models \xi$, hence $\omega \models \psi \wedge \xi$, and therefore $\omega \models \varphi \vee (\psi \wedge \xi)$. Hence the two sets are equal.

10. $\varphi \wedge (\varphi \vee \psi) \equiv \varphi$.
If $\omega \models \varphi \wedge (\varphi \vee \psi)$, then $\omega \models \varphi$. Conversely, $\omega \models \varphi$ implies $\omega \models \varphi \vee \psi$, and therefore also $\omega \models \varphi \wedge (\varphi \vee \psi)$.

11. $\varphi \vee (\varphi \wedge \psi) \equiv \varphi$. Analogous to (10).

12. $\neg(\varphi \wedge \psi) \equiv \neg\varphi \vee \neg\psi$.
$\omega \models \neg(\varphi \wedge \psi) \iff \neg(\omega \models \varphi \wedge \psi) \iff \neg(\omega \models \varphi \text{ and } \omega \models \psi) \iff (\omega \not\models \varphi) \text{ or } (\omega \not\models \psi) \iff \omega \models \neg\varphi \vee \neg\psi$.

13. $\neg(\varphi \vee \psi) \equiv \neg\varphi \wedge \neg\psi$. Analogous to (12).

14. $\varphi \to \psi \equiv \neg\psi \to \neg\varphi$.
Use $\varphi \to \psi \equiv \neg\varphi \vee \psi$. Then

$$\begin{aligned} \varphi \to \psi &\equiv \neg\varphi \vee \psi \\ &\equiv \neg\varphi \vee \neg\neg\psi && \text{(double negation)} \\ &\equiv \neg\neg\psi \vee \neg\varphi && (\vee \text{ is commutative}) \\ &\equiv \neg\psi \to \neg\varphi. \end{aligned}$$

Thus the sets of models coincide.

This proves all the stated equivalences. □

Notation 1.45 (Parenthesization in Conjunctions/Disjunctions). *By Proposition 1.38 (4) and (6), the particular placement of parentheses in nested conjunctions is semantically irrelevant. Therefore, in what follows, we omit parentheses in pure conjunctions; for example,*

$$\begin{aligned} \varphi &= (\psi_1 \wedge \psi_2) \wedge (\psi_3 \wedge (\psi_4 \wedge \psi_5)) \\ &\textit{we write simply as} \\ &= \psi_1 \wedge \psi_2 \wedge \psi_3 \wedge \psi_4 \wedge \psi_5. \end{aligned}$$

Similarly, by Proposition 1.38 (5) and (7), we omit parentheses in nested disjunctions.

We recall the notions of a *contradiction* and a *tautology*: for a signature Σ and $\varphi \in L(\Sigma)$, we have

$$\begin{aligned} \varphi \text{ is a contradiction} &\iff \mathrm{Mod}_\Sigma(\varphi) = \varnothing, \\ \varphi \text{ is a tautology} &\iff \mathrm{Mod}_\Sigma(\varphi) = \Omega(\Sigma). \end{aligned}$$

Proposition 1.46. *Let Σ be a propositional signature and let $\varphi, \psi \in L(\Sigma)$.*

1. *If φ and ψ are contradictions, then $\varphi \equiv \psi$.*

2. *If φ and ψ are tautologies, then $\varphi \equiv \psi$.*

In words: any two (potentially syntactically different) contradictions are equivalent, and likewise any two tautologies.

Proof. We use the characterization of equivalence by equality of model sets (Prop. 1.32): $\varphi \equiv \psi$ if and only if $\mathrm{Mod}_\Sigma(\varphi) = \mathrm{Mod}_\Sigma(\psi)$.

(1) Let φ be a contradiction and let ψ be a contradiction. Then, by definition, $\mathrm{Mod}_\Sigma(\varphi) = \varnothing$ and $\mathrm{Mod}_\Sigma(\psi) = \varnothing$. Hence $\mathrm{Mod}_\Sigma(\varphi) = \mathrm{Mod}_\Sigma(\psi)$, and therefore $\varphi \equiv \psi$.

(2) Let φ be a tautology and let ψ be a tautology. Then, by definition, $\mathrm{Mod}_\Sigma(\varphi) = \Omega(\Sigma)$ and $\mathrm{Mod}_\Sigma(\psi) = \Omega(\Sigma)$. Hence $\mathrm{Mod}_\Sigma(\varphi) = \mathrm{Mod}_\Sigma(\psi)$, and therefore $\varphi \equiv \psi$.

This proves both statements. □

Since all contradictions have the same set of models (up to syntactic variation), we introduce the symbol $\bot$ for an arbitrary contradiction and write simply

$$\varphi \equiv \bot$$

if φ is a contradiction. Analogously, we use $\top$ for an arbitrary tautology and write

$$\varphi \equiv \top$$

if φ is a tautology.

Proposition 1.47. *Let Σ be a propositional signature and let $\varphi \in L(\Sigma)$ be a formula. Then the following hold:*

1. $\top \models_{\mathrm{PL}} \varphi \iff \varphi \equiv \top$.
2. $\varphi \models_{\mathrm{PL}} \bot \iff \varphi \equiv \bot$.
3. $\varphi \models_{\mathrm{PL}} \top$.
4. $\bot \models_{\mathrm{PL}} \varphi$.

Proof. Let Σ be a signature and let $\varphi \in L(\Sigma)$.

(1) $\top \models_{\text{PL}} \varphi \iff \varphi \equiv \top$.
By definition, $\top \models_{\text{PL}} \varphi$ holds if and only if

$$\text{Mod}_\Sigma(\top) \subseteq \text{Mod}_\Sigma(\varphi).$$

Since $\text{Mod}_\Sigma(\top) = \Omega(\Sigma)$, this is equivalent to $\text{Mod}_\Sigma(\varphi) = \Omega(\Sigma)$, that is, precisely $\varphi \equiv \top$.

(2) $\varphi \models_{\text{PL}} \bot \iff \varphi \equiv \bot$.
Analogously, $\varphi \models_{\text{PL}} \bot$ holds if and only if

$$\text{Mod}_\Sigma(\varphi) \subseteq \text{Mod}_\Sigma(\bot).$$

Since $\text{Mod}_\Sigma(\bot) = \varnothing$, this is equivalent to $\text{Mod}_\Sigma(\varphi) = \varnothing$, and hence $\varphi \equiv \bot$.

(3) $\varphi \models_{\text{PL}} \top$.
Since $\text{Mod}_\Sigma(\top) = \Omega(\Sigma)$ and for every formula $\text{Mod}_\Sigma(\varphi) \subseteq \Omega(\Sigma)$ holds, it follows immediately that $\text{Mod}_\Sigma(\varphi) \subseteq \text{Mod}_\Sigma(\top)$, also $\varphi \models_{\text{PL}} \top$.

(4) $\bot \models_{\text{PL}} \varphi$.
Since $\text{Mod}_\Sigma(\bot) = \varnothing$ and the empty set is a subset of every set, we have $\varnothing \subseteq \text{Mod}_\Sigma(\varphi)$ for all φ. Therefore, $\bot \models_{\text{PL}} \varphi$.

□

Remark 1.48. *Statement (4) is the well-known rule* ex falso quodlibet *("from falsehood, anything follows"), also called the* principle of explosion*: semantically, a contradiction entails every formula whatsoever.*

Notation 1.49. *If* $\top \models_{\mathrm{PL}} \varphi$ *holds for a formula* φ *(that is,* $\top \equiv \varphi$*), we also write this more briefly as*

$$\varnothing \models_{\mathrm{PL}} \varphi \qquad \textit{or simply} \qquad \models_{\mathrm{PL}} \varphi.$$

Here $\varnothing$ *is regarded as the empty set of formulas. For it we have*

$$\mathrm{Mod}_\Sigma(\varnothing) = \Omega(\Sigma) = \mathrm{Mod}_\Sigma(\top).$$

The following statement (Deduction Theorem) makes precise in what sense semantic inference at the meta-level coincides with its object-level formulation as an implication.

Theorem 1.50 (Deduction Theorem)**.** *Let* Σ *be a propositional signature and let* $\varphi, \psi \in L(\Sigma)$*. Then*

$$\varphi \models_{\mathrm{PL}} \psi \quad \Longleftrightarrow \quad \models_{\mathrm{PL}} \varphi \to \psi.$$

Proof. We prove the equivalence by a sequence of standard semantic transformations (where $\varphi \to \psi$ is understood as $\neg\varphi \vee \psi$).

$$
\begin{aligned}
& \varphi \models_{\mathrm{PL}} \psi \\
\iff\ & \mathrm{Mod}_\Sigma(\varphi) \subseteq \mathrm{Mod}_\Sigma(\psi) \qquad \text{(definition of } \models_{\mathrm{PL}} \text{)} \\
\iff\ & \forall \omega \in \Omega(\Sigma) : \ \omega \in \mathrm{Mod}_\Sigma(\varphi) \Rightarrow \omega \in \mathrm{Mod}_\Sigma(\psi) \\
\iff\ & \forall \omega \in \Omega(\Sigma) : \ \neg(\omega \in \mathrm{Mod}_\Sigma(\varphi)) \ \vee \ \omega \in \mathrm{Mod}_\Sigma(\psi) \\
\iff\ & \forall \omega \in \Omega(\Sigma) : \ \omega \in \mathrm{Mod}_\Sigma(\neg\varphi) \ \vee \ \omega \in \mathrm{Mod}_\Sigma(\psi) \\
\iff\ & \Omega(\Sigma) = \mathrm{Mod}_\Sigma(\neg\varphi) \cup \mathrm{Mod}_\Sigma(\psi) \\
\iff\ & \Omega(\Sigma) = \mathrm{Mod}_\Sigma(\neg\varphi \vee \psi) \\
\iff\ & \models_{\mathrm{PL}} \ \neg\varphi \vee \psi \\
\iff\ & \models_{\mathrm{PL}} \ \varphi \to \psi.
\end{aligned}
$$

In these steps, we used the fact that for every interpretation ω, the statement "$\omega \notin \mathrm{Mod}_\Sigma(\varphi)$ or $\omega \in \mathrm{Mod}_\Sigma(\psi)$" expresses exactly that ω is a model of $\neg\varphi \vee \psi$; the subsequent equality with $\Omega(\Sigma)$ says that $\neg\varphi \vee \psi$ is a tautology, which is equivalent to $\models_{\mathrm{PL}} \varphi \to \psi$. This proves the claim. □

Remark 1.51. *The deduction theorem explains why semantic entailment $\varphi \models \psi$ can often be expressed by the object-language tautology $\varphi \to \psi$.*

Semantically, saying that "φ entails ψ" means precisely that the implication $\varphi \to \psi$ is true in every possible interpretation. In other words, $\varphi \to \psi$ is a tautology.

This connection is frequently used both in theoretical proofs and in practical reductions (for example, when formulating proof systems).

1.6.3 Normal Forms

In the automatic processing of logical formulas, syntactic variation often hinders the efficient implementation of algorithms (e.g. SAT solvers, resolution, and symbolic manipulation). For this reason, it is common practice to transform formulas into a standardized *normal form*. The three most important normal forms in propositional logic are the *negation normal form* (NNF), the *conjunctive normal form* (CNF), and the *disjunctive normal form* (DNF). In this section, we give precise definitions, show why every formula can be translated into these forms equivalently, and provide short, constructive proofs or transformation procedures.

Definition 1.52. *Let Σ be a propositional signature. A formula $\varphi \in L(\Sigma)$ is in* negation normal form (NNF) *if and only if*

- *only the connectives $\neg, \wedge, \vee$ occur, and*
- *in every subformula of the form $\neg\psi$, the formula ψ is an atomic propositional variable.*

Theorem 1.53 (Existence of an NNF). *For every formula $\varphi \in L(\Sigma)$, there exists a formula ψ in NNF such that $\varphi \equiv \psi$.*

Constructive Proof. First replace every occurrence of $\rightarrow$ and $\leftrightarrow$ by

$$\varphi \rightarrow \psi \equiv \neg\varphi \vee \psi, \qquad \varphi \leftrightarrow \psi \equiv (\varphi \rightarrow \psi) \wedge (\psi \rightarrow \varphi).$$

After this step, the formula contains only the connectives $\neg, \wedge, \vee$.

Negation depth as a measure. For a formula χ, define the *negation depth* $h(\chi)$ as the maximal number of consecutive negation symbols occurring immediately in front of a non-atomic subformula. (If no such chain occurs, set $h(\chi) = 0$.)

Local replacement using De Morgan and substitution. If χ contains a subformula of the form $\neg(\xi \wedge \xi')$ or $\neg(\xi \vee \xi')$, we replace this subformula by the corresponding De Morgan form:

$$\neg(\xi \wedge \xi') \equiv \neg\xi \vee \neg\xi', \qquad \neg(\xi \vee \xi') \equiv \neg\xi \wedge \neg\xi'.$$

By the substitution lemma (Lemma 1.36), this replacement is admissible at any position in the formula, and thus yields an equivalent formula χ'.

Decrease of the measure. With each such replacement, the negation depth decreases strictly: the outermost chain of negations under consideration is distributed over the inner operands, so that $h(\chi') < h(\chi)$ holds. Hence no replacement can increase the negation depth.

Termination. Since $h(\chi)$ is a nonnegative integer-valued measure, repeated application of the replacement steps above leads after finitely many steps to a formula χ^* with $h(\chi^*) = 0$. Thus, in χ^* all negations occur only immediately in front of atomic propositions — that is, χ^* is a formula in negation normal form (NNF).

Correctness. Each replacement step preserves

equivalence because it is based on De Morgan's laws (Proposition 1.38 (12),(13)) and is justified by the substitution lemma. By transitivity of equivalence, we finally obtain $\varphi \equiv \chi^*$.

Therefore, the terminating algorithm described above constructively produces an equivalent formula in NNF.

□

Definition 1.54. *Let* Σ *be as above. A* literal *is either an atom* $p \in \Sigma$ *or its negation* $\neg p$*. A* clause *is a disjunctive formula* $l_1 \vee \cdots \vee l_m$ *consisting of literals (with* $m = 1$ *also allowed). A formula is in* conjunctive normal form (CNF) *if and only if it is a conjunction of clauses:*

$$\varphi = C_1 \wedge \cdots \wedge C_n, \qquad C_i \text{ clauses.}$$

Theorem 1.55 (Existence of a CNF)**.** *Every formula* $\varphi \in L(\Sigma)$ *is equivalent to a formula* ψ *in CNF.*

Proof (Existence of a CNF / constructive algorithm). We first transform the given formula φ into negation normal form (NNF); see Theorem 1.47. Then the resulting formula contains only the connectives $\wedge, \vee, \neg$, and every negation occurs directly in front of an atomic formula. We now show by structural induction on the structure of an NNF formula that every NNF formula has an equivalent formula in conjunctive normal form (CNF).

Base case. If φ is a literal (that is, an atomic formula p or $\neg p$), then φ is already a clause (a unit clause) and hence is in CNF.

Induction hypothesis. For all proper subformulas of smaller structural complexity, the claim holds: each such subformula has an equivalent CNF formula.

Induction step. Let φ be an NNF formula.

- $\varphi = \psi \wedge \xi$. By the induction hypothesis, there exist CNF formulas equivalent to $\psi \equiv \bigwedge_{i=1}^{n} C_i^{\psi}$ and $\xi \equiv \bigwedge_{j=1}^{m} C_j^{\xi}$ (that is, conjunctions of clauses). Then

$$\varphi \equiv \Big(\bigwedge_{i=1}^{n} C_i^{\psi}\Big) \wedge \Big(\bigwedge_{j=1}^{m} C_j^{\xi}\Big) = \bigwedge_{k} C_k,$$

which is again a conjunction of clauses and therefore is in CNF.

- $\varphi = \psi \vee \xi$. Again by the induction hypothesis, let $\psi \equiv \bigwedge_{i=1}^{n} C_i^{\psi}$ and $\xi \equiv \bigwedge_{j=1}^{m} C_j^{\xi}$, where all C_i^{ψ}, C_j^{ξ} are clauses (disjunctions of literals). We repeatedly apply the distributive laws to obtain a conjunction of clauses. For two clauses $C = \bigvee_{a=1}^{r} \ell_a$ and $D = \bigvee_{b=1}^{s} \ell'_b$, the formula $C \vee D$ is itself already a clause. The crucial case is the disjunction of two conjunctions:

$$\Big(\bigwedge_{i=1}^{n} C_i^{\psi}\Big) \vee \Big(\bigwedge_{j=1}^{m} C_j^{\xi}\Big) \equiv \bigwedge_{i=1}^{n} \bigwedge_{j=1}^{m} (C_i^{\psi} \vee C_j^{\xi}).$$

(This transformation follows by repeated application of the rule $(A \wedge B) \vee (C \wedge D) \equiv (A \vee C) \wedge (A \vee D) \wedge (B \vee C) \wedge (B \vee D)$, generalized to n, m.) Since for clauses C_i^{ψ} and C_j^{ξ} the disjunction $C_i^{\psi} \vee C_j^{\xi}$ is again a clause, the right-hand side is a conjunction of clauses, and hence a CNF formula.

Case of a Negation in Front of a CNF Formula: Suppose that ψ is already in CNF, that is,

$$\psi = \bigwedge_{i=1}^{n} C_i, \qquad \text{where} \qquad C_i = \bigvee_{j=1}^{m_i} \ell_{i,j}$$

for literals $\ell_{i,j}$. We now consider $\varphi = \neg\psi$ and transform φ step by step into an equivalent CNF formula.

$$\begin{aligned}
\varphi = \neg\Big(\bigwedge_{i=1}^{n} C_i\Big) &\overset{\text{De Morgan}}{\equiv} \bigvee_{i=1}^{n} \neg C_i \\
= \bigvee_{i=1}^{n} \neg\Big(\bigvee_{j=1}^{m_i} \ell_{i,j}\Big) &\overset{\text{De Morgan}}{\equiv} \bigvee_{i=1}^{n} \Big(\bigwedge_{j=1}^{m_i} \neg\ell_{i,j}\Big).
\end{aligned}$$

Thus, φ is initially given as a disjunction of conjunctions (that is, in DNF form). Write

$$\widetilde{C}_i := \bigwedge_{j=1}^{m_i} \neg\ell_{i,j} \qquad (i = 1, \ldots, n),$$

so that $\varphi \equiv \bigvee_{i=1}^{n} \widetilde{C}_i$, where each $\widetilde{C}_i$ is a term (that is, a conjunction of literals). By repeated application of the distributive laws (distributing $\vee$ over $\wedge$), this disjunction of conjunctions can be transformed into a conjunction of disjunctions. More precisely, we have:

$$\bigvee_{i=1}^{n} \widetilde{C}_i \equiv \bigwedge_{k_1=1}^{m_1} \bigwedge_{k_2=1}^{m_2} \cdots \bigwedge_{k_n=1}^{m_n} \bigvee_{i=1}^{n} \neg \ell_{i,k_i}.$$

The resulting clauses are of the form

$$\widehat{C}_{k_1,\ldots,k_n} = \bigvee_{i=1}^{n} \neg \ell_{i,k_i}, \qquad k_i \in \{1, \ldots, m_i\},$$

that is, each clause is obtained by selecting exactly one literal $\neg \ell_{i,k_i}$ from each original $\widetilde{C}_i$ and combining these n literals disjunctively. The full CNF is then the conjunction of all such clauses.

Remark. This transformation step is correct and yields a genuine CNF, but the number of clauses is in general $\prod_{i=1}^{n} m_i$, so the size of the formula may grow exponentially. For practical implementations, one therefore often uses alternative techniques (for example, the Tseitin transformation) to avoid this blow-up.

Termination and Correctness. The induction proceeds over the finite structure of the formula; hence every step is well defined, and after finitely many steps one obtains a CNF formula. Each transformation is based on equivalence-preserving rules (De Morgan's laws, associativity, commutativity, distributivity, and so on), so correctness, that is, equivalence to the original formula, is preserved.

Remark on Complexity. Note that repeated application of the distributive laws can, in the worst case, lead to an exponential increase in formula size. For practical and algorithmic purposes, one therefore often uses the Tseitin transformation: it produces a *satisfiability-equivalent* CNF with additional auxiliary variables in linear or polynomial time (typically in the context of SAT reductions). For theoretical existence proofs, however, the constructive conversion to CNF given above is sufficient. □

Definition 1.56. *A* term *is a conjunction of literals* $l_1 \wedge \cdots \wedge l_k$. *A formula is in* disjunctive normal form (DNF) *if and only if it is a disjunction of terms:*

$$\varphi = T_1 \vee \cdots \vee T_n, \qquad T_i \text{ terms.}$$

Theorem 1.57 (Existence of a DNF). *Let Σ be a propositional signature and $\varphi \in L(\Sigma)$. Then there exists a formula ψ in disjunctive normal form (DNF) such that $\varphi \equiv \psi$.*

Proof. We give a constructive derivation in two steps.

1. Transformation into NNF. First, we replace every occurrence of $\rightarrow$ and $\leftrightarrow$ by their definitions:

$$\varphi \rightarrow \psi \equiv \neg\varphi \vee \psi, \qquad \varphi \leftrightarrow \psi \equiv (\varphi \rightarrow \psi) \wedge (\psi \rightarrow \varphi),$$

and then use De Morgan's laws and double negation to push negations inward:

$$\neg(\alpha \wedge \beta) \equiv \neg\alpha \vee \neg\beta, \quad \neg(\alpha \vee \beta) \equiv \neg\alpha \wedge \neg\beta, \quad \neg\neg\alpha \equiv \alpha.$$

Each application of these transformations is equivalence-preserving and reduces the maximal nesting depth of negations in front of non-atomic subformulas. Hence the process terminates and yields an equivalent formula φ_{nnf} in negation normal form (NNF), that is, with negations occurring only directly in front of atoms.

2. Construction of the DNF (induction on the structure of φ_{nnf}). We show by structural induction that every formula in NNF is equivalent to a DNF.

Base case. If θ is a literal (an atom p or $\neg p$), then θ is already a term (a conjunction consisting of exactly one literal) and hence a trivial DNF.

Inductive step. Let θ now be an NNF formula, and assume that the claim holds for all proper subformulas of smaller structural complexity. There are three cases:

- $\theta = \alpha \vee \beta$. By the induction hypothesis, there exist DNFs $\alpha \equiv \bigvee_{i=1}^{m} T_i$ and $\beta \equiv \bigvee_{j=1}^{n} S_j$ with terms T_i, S_j. Then

  $$\theta \equiv \bigvee_{i=1}^{m} T_i \;\vee\; \bigvee_{j=1}^{n} S_j,$$

  which is a disjunction of terms and therefore in DNF.

- $\theta = \alpha \wedge \beta$. By the induction hypothesis, there exist DNFs $\alpha \equiv \bigvee_{i=1}^{m} T_i$ and $\beta \equiv \bigvee_{j=1}^{n} S_j$. Using distributivity, we obtain the equivalent formula

  $$(\bigvee_{i=1}^{m} T_i) \wedge (\bigvee_{j=1}^{n} S_j) \equiv \bigvee_{i=1}^{m} \bigvee_{j=1}^{n} (T_i \wedge S_j).$$

Since the conjunction of two terms is again a term, the right-hand side is a disjunction of terms, hence in DNF.

- The remaining case (a literal, already handled) is covered, since only literals, $\vee$, and $\wedge$ occur in NNF.

Thus, the induction yields an equivalent DNF for every NNF formula. Combined with Step 1, this proves the claim for arbitrary formulas φ.

□

Remark (Complexity). In the worst case, the constructive transformation may grow exponentially, since distributive expansion leads to a combinatorial blow-up. For practical applications (e.g. SAT reductions), one therefore often uses the *Tseitin transformation*, which introduces auxiliary variables in order to obtain a *satisfiability-equivalent* CNF with only linear growth in formula size.

Notation (Sets of Clauses / Sets of Terms). A CNF can be written compactly as a set of clauses, for example

$$A \wedge (\neg B \vee C) \wedge (\neg B \vee D) \quad \widehat{=} \quad \{\{A\}, \{\neg B, C\}, \{\neg B, D\}\}.$$

Analogously, a DNF can be written as a set of terms, where each term is represented as a set of literals.

Examples.

1. De Morgan's law (NNF example, alarm system):

$$\neg(\text{Alarm} \vee \text{Sprinkler}) \equiv \neg\text{Alarm} \wedge \neg\text{Sprinkler}.$$

2. Conversion into CNF (short example):

$$\varphi = \neg(A \wedge B) \vee \neg(C \vee D) \overset{\text{NNF}}{\equiv} (\neg A \vee \neg B) \vee (\neg C \wedge \neg D)$$

$$\equiv (\neg A \vee \neg B \vee \neg C) \wedge (\neg A \vee \neg B \vee \neg D),$$

hence a CNF: a conjunction of two clauses.

3. Conversion into DNF (short example):

$$(A \vee B) \wedge C \equiv (A \wedge C) \vee (B \wedge C).$$

hence a DNF: a disjunction of two terms.

Remarks.

- CNF and DNF are useful for different purposes: CNF, for instance, for SAT and resolution, DNF for explicit model enumeration and simple satisfiability checks.
- Direct application of the distributive laws may cause exponential growth; in practice, one therefore often uses:
  - the Tseitin transformation (introducing auxiliary variables) for compact CNF formulas, or

  - stepwise normalization and heuristics that help avoid such blow-ups.

- From a didactic point of view, the concept of NNF is helpful as an intermediate step: first push negations inward in a controlled way, then apply distributive transformations.

This completes the basic theory of normal forms — the concrete transformation rules (De Morgan's laws, double negation, distributivity, absorption, etc.) were already collected in Section 1.5.2 and are the tools by which the equivalences stated here can be realized constructively.

1.6.4 Exercises and Solutions — Equivalence

Consider the propositional signature

$$\Sigma = \{K, B, C, D, E, G, H, I\}$$

and the expressions

$$\varphi_4 = \neg((K \vee D) \wedge D), \qquad \psi_4 = K \vee D.$$

For each of the following statements, decide whether it is *true* or *false*. Give a brief justification.

1. $\neg\neg K \wedge G \wedge H \equiv_\Sigma (G \vee E) \wedge (\neg K \rightarrow D \vee G)$.

2. $I \wedge I \wedge D$ is a term.

3. $(E \vee K \wedge H) \wedge D \vee \neg D \wedge \neg E \wedge (\neg K \vee \neg H) \vee \neg E$ is in NNF.

4. $\varphi_4 \equiv_\Sigma \psi_4$ and ψ_4 is in CNF.

Solutions and brief explanations

1. **False.** $\neg\neg K \wedge G \wedge H$ (that is, K and G and H) *implies* $(G \vee E) \wedge (\neg K \rightarrow D \vee G)$: since G is true, $G \vee E$ is true; since $\neg K$ is false, the implication $\neg K \rightarrow D \vee G$ is automatically true. However, the converse does not hold (for example, $(G \vee E) \wedge (\neg K \rightarrow D \vee G)$ does not necessarily satisfy K or H). Hence we have only a one-way entailment, not an equivalence.

2. **True.** By definition, a *term* is a conjunction of literals. $I \wedge I \wedge D$ is a conjunction of atomic literals (idempotence may eliminate a duplicate), so it is a term.

3. **True.** In the given formula, negations occur only directly in front of atomic propositions ($\neg D$, $\neg E$, $\neg K$, $\neg H$); only the connectives $\wedge, \vee$ are used. Therefore, the formula is in negation normal form (NNF). (Note: if the parenthesization is ambiguous, the exact syntax should be clarified; here the usual precedence rules $\neg > \wedge > \vee$ apply.)

4. $\psi_4 = K \vee D$ is **in CNF**, because a single clause (a disjunction of literals) is a valid CNF formula (a conjunction with one clause). However, the statement $\varphi_4 \equiv \psi_4$ is **false**. In fact,

$$\varphi_4 = \neg((K \vee D) \wedge D) \equiv \neg D$$

(because $(K \vee D) \wedge D \equiv D$ by absorption), so $\varphi_4 \equiv \neg D$, whereas $\psi_4 \equiv K \vee D$. As a counterexample, the interpretation that assigns false to all atoms makes φ_4 equal to true (because $\neg D = \neg\text{false} = \text{true}$), but ψ_4 equal to false. Therefore, there is no equivalence.

Brief feedback (summary):

- (1) False — only one-way entailment ($\models$), not equality of the sets of models.
- (2) True — conjunction of literals.
- (3) True — negations appear only in front of atoms; only $\wedge, \vee$ are used.
- (4) Partly true: ψ_4 is in CNF, but $\varphi_4 \not\equiv \psi_4$.

Additional Short Review Exercises

1. Let $\Sigma = \{C, E, F, G, I\}$. Is it true that $\neg\neg(C \vee I) \in \text{Lit}(\Sigma)$?
   Answer: $\neg\neg(C \vee I)$ is *not* a literal.
   Reason: By definition, a literal is either an atomic proposition p or its single negation $\neg p$. The formula $\neg\neg(C \vee I)$ is the double negation of a *disjunction*, not of the form p or $\neg p$. Hence it is not a literal.

2. Let $\Sigma = \{J, K, B, C, F, G, I\}$. Is $\neg(C \wedge C \wedge I) \wedge G$ a clause?

Answer: $\neg(C \wedge C \wedge I) \wedge G$ is *not* a clause.
Reason: A clause is a disjunction of literals $(l_1 \vee \cdots \vee l_n)$. The given formula is a *conjunction*, namely $(\neg(C \wedge C \wedge I)) \wedge G$. By De Morgan's law, $\neg(C \wedge C \wedge I) \equiv \neg C \vee \neg I$, so the whole formula is equivalent to $(\neg C \vee \neg I) \wedge G$, that is, a conjunction of a disjunction and an atom. This is neither itself a pure disjunction of literals nor of the form of a single clause.

3. For the same $\Sigma = \{J, K, B, C, F, G, I\}$: is $J \vee B \wedge \neg F \wedge \neg I$ in DNF?
Answer: Yes, the formula is in DNF.
Reason: Under the usual operator precedence $(\neg > \wedge > \vee)$, the formula is read as

$$J \ \vee \ (B \wedge \neg F \wedge \neg I).$$

This is a disjunction of two terms: the first term is the literal J, and the second term is the conjunction $B \wedge \neg F \wedge \neg I$. Such a disjunction of terms is exactly the definition of DNF.

4. Let

$$\varphi_4 = K \wedge \neg(F \to K), \qquad \psi_4 = F \wedge \neg K \wedge K$$

over $\Sigma = \{J, K, B, C, F, G, I\}$. Do we have $\varphi_4 \equiv_\Sigma \psi_4$, and is ψ_4 in CNF?

Answer: Yes — $\varphi_4 \equiv \psi_4$, and ψ_4 is in CNF.
Reason: First rewrite:

$$\begin{aligned} F \to K &\equiv \neg F \lor K, \\ \neg(F \to K) &\equiv \neg(\neg F \lor K) \\ &\equiv F \land \neg K. \end{aligned}$$

Therefore,

$$\varphi_4 = K \land (F \land \neg K) \equiv F \land K \land \neg K.$$

Likewise,

$$\psi_4 = F \land \neg K \land K.$$

Both formulas are therefore equivalent to the conjunction $F \land (K \land \neg K)$, which is a contradiction (that is, $\bot$). Hence φ_4 and ψ_4 are equivalent. Moreover, ψ_4 is a conjunction of literals (three unit clauses), and thus a conjunction of clauses — so it is in CNF.

1.7 Algorithmic Perspective

In this section, we turn to concrete, algorithmically motivated questions in propositional logic. At the core is the issue of which logical problems can be decided algorithmically at all and how difficult such decisions are in terms of runtime and complexity. From a practical point of view, we are especially interested in how one can determine automatically whether one formula follows from another or whether a formula has a model at all.

The following subsections address (i) the most important decision problems, (ii) formal calculi for syntactic inference, and (iii) the resolution calculus as a central practical tool — and we begin here with (i).

1.7.1 Inference Problems

We formulate the relevant questions as *decision problems*. A decision problem yields, for a well-defined input (also called an *instance*), the answer Yes or No. If the answer for a given instance is Yes, we call the instance *positive*; otherwise, it is *negative*.

The three standard problems in propositional logic that we consider are the following:

Infer. Input: formulas $\varphi, \psi \in L(\Sigma)$.
Output: Yes if and only if $\varphi \models_{\mathrm{PL}} \psi$.

Taut. Input: formula $\varphi \in L(\Sigma)$.
Output: Yes if and only if $\varphi \equiv \top$ (i.e. φ is a tautology).

Sat. Input: formula $\varphi \in L(\Sigma)$.
Output: Yes if and only if $\varphi \not\equiv \bot$ (i.e. φ is satisfiable).

These three problems are closely related: an inference instance (φ, ψ) can be expressed by the tautology question for $\varphi \to \psi$ or, equivalently, by the satisfiability question for $\varphi \wedge \neg\psi$. This makes the treatment of the problems particularly transparent: an efficient algorithm for any one of the three can be transformed into an algorithm for the other two.

Example 1.58. *We consider several typical instances of the problems* Infer, Taut, *and* Sat.

- *For* $(A \wedge B, A)$, *we have:*

$$\begin{aligned} A \wedge B &\models_{\mathrm{PL}} A, \\ A \wedge B \to A & \text{ is a tautology,} \\ (A \wedge B) \wedge \neg A & \text{ is unsatisfiable.} \end{aligned}$$

  Thus, this is a positive instance of Infer *and* Taut, *but a negative instance of* Sat.

- *For* $(A \vee B, A)$, *we have:*

$$\begin{aligned} A \vee B &\not\models_{\mathrm{PL}} A, \\ A \vee B \to A & \text{ is not a tautology,} \\ (A \vee B) \wedge \neg A & \text{ is satisfiable.} \end{aligned}$$

  Thus, this is a negative instance of Infer *and* Taut, *but a positive instance of* Sat.

- *For* $((A \vee B) \wedge (A \vee \neg B), A)$, *we have:*

$$\begin{aligned} (A \vee B) \wedge (A \vee \neg B) &\models_{\mathrm{AL}} A, \\ (A \vee B) \wedge (A \vee \neg B) \to A & \text{ is a tautology.} \end{aligned}$$

  whereas

$$(A \vee B) \wedge (A \vee \neg B) \wedge \neg A$$

  is unsatisfiable. Hence, this is again a positive instance of Infer *and* Taut, *as well as a negative instance of* Sat.

Proposition 1.59. *Let Σ be a propositional signature and let $\varphi, \psi \in L(\Sigma)$. The following statements are equivalent:*

1. *(φ, ψ) is a positive instance of* Infer *(that is, $\varphi \models_{\mathrm{PL}} \psi$).*

2. *$\varphi \to \psi$ is a tautology (a positive instance of* Taut*).*

3. *$\varphi \wedge \neg\psi$ is unsatisfiable (a negative instance of* Sat*).*

Proof. We prove the three implications in succession.

(1) ⇒ (2). If $\varphi \models_{\mathrm{PL}} \psi$, then by the Deduction Theorem it follows that $\models_{\mathrm{PL}} \varphi \to \psi$, so $\varphi \to \psi$ is a tautology.

(2) ⇒ (3). If $\varphi \to \psi$ is tautological, then $\neg(\varphi \to \psi)$ is unsatisfiable. Since $\neg(\varphi \to \psi)$ is equivalent to $\varphi \wedge \neg\psi$, it follows that $\varphi \wedge \neg\psi$ is unsatisfiable.

(3) ⇒ (1). If $\varphi\wedge\neg\psi$ is unsatisfiable, then $\neg(\varphi\wedge\neg\psi)$ is tautological. But $\neg(\varphi\wedge\neg\psi) \equiv \varphi \to \psi$, hence $\models_{\mathrm{PL}} \varphi \to \psi$ and therefore $\varphi \models_{\mathrm{PL}} \psi$. □

Since the three problems can be reduced to one another, algorithmic research often focuses on Sat — the satisfiability problem.

A Naive Algorithm

An elementary but inefficient decision procedure for Sat searches through all possible interpretations:

NaiveSAT Algorithm.
Input: $\varphi \in L(\Sigma)$.
Output: 'Yes' iff φ is satisfiable.

1. For all $\omega \in \Omega(\Sigma)$ do:
2. If $\omega \models_{\mathrm{AL}} \varphi$, return 'Yes'.
3. End for.
4. Return 'No'.

Correctness is immediate: a model is found exactly when the formula is satisfiable. The practical drawback is equally clear: in the worst case, $2^{|\Sigma|}$ interpretations must be checked, which is exponential in the number of atomic propositions.

Lemma 1.60. *Let Σ be a signature, $\varphi \in L(\Sigma)$, and $\omega \in \Omega(\Sigma)$. The verification of $\omega \models_{\mathrm{AL}} \varphi$ can be decided in deterministic polynomial time (in the length of φ).*

Proof. The formula φ can be evaluated as a tree: atomic formulas are evaluated by ω, and the connectives $\wedge, \vee, \neg$ can be evaluated bottom-up in linear time (with respect to the size of the formula). □

Complexity of Sat

The naive algorithm shows only that Sat is easy to implement (but expensive to run). Formally, however, the problem is genuinely difficult: by Cook's theorem, Sat is NP-complete. In short, under the widely assumed hypothesis $P \neq NP$, there is no deterministic polynomial-time algorithm that solves Sat in all cases. From the relationship established above, it further follows that Infer and Taut (as a complementary problem) are also difficult: more precisely, Infer and Taut are coNP-complete.

Remark. Despite this negative complexity result, there are highly effective SAT solvers in practice (CDCL, DPLL variants, heuristic and incremental methods) that solve many realistic instances in a short time. Thus, theoretical intractability does not rule out useful algorithms for practical cases—it only describes the worst-case behavior.

1.7.2 Calculi — Syntactic Procedures for Inference

The semantic relation $\models_{\mathrm{AL}}$ states precisely when one formula follows from another ($\varphi \models_{\mathrm{AL}} \psi \iff \mathrm{Mod}_\Sigma(\varphi) \subseteq \mathrm{Mod}_\Sigma(\psi)$), but it is not particularly well suited for the practical computation of inferences. Calculi provide a syntactic alternative: they make it possible to generate new formulas from a formula (or a set of formulas) by applying transformation and derivation rules, without considering the set of all interpretations. In what follows, we introduce the formal notions, a small example calculus, and some remarks on soundness and completeness.

Definition 1.61. *Let Σ be a propositional signature. A* propositional calculus $\vdash$ *for Σ is a relation $\vdash \subseteq L(\Sigma) \times$*

$L(\Sigma)$.

Definition 1.62. *Let X be a set of (meta-logical) variables. An* inference rule *over X has the form*

$$\frac{\varphi_1, \ldots, \varphi_n}{\psi}\,[r] \qquad \textit{with } \varphi_i, \psi \in L(X).$$

The variables in X are placeholders: to apply the rule to concrete formulas, one uses a substitution $\sigma : X \to L(\Sigma)$.

Definition 1.63. *Let $r = \dfrac{\varphi_1, \ldots, \varphi_n}{\psi}$ be a rule over $X = \{x_1, \ldots, x_m\}$. For a set of formulas $\Phi \subseteq L(\Sigma)$, we say that r is applicable to Φ with substitution σ if and only if $\varphi_i[x_j \mapsto \sigma(x_j)] \in \Phi$ for all i. The application yields the extended set*

$$\Phi' = \Phi \cup \{\psi[x_j \mapsto \sigma(x_j) \,:\, 1 \leq j \leq m]\},$$

and we briefly write $\Phi \xrightarrow{r,\sigma} \Phi'$.

Notation. If σ is clear from the context, we omit it and simply write $\Phi \xrightarrow{r} \Phi'$.

We now consider a small, useful set of rules R that captures many elementary transformations:

$$\frac{\alpha \wedge \beta}{\alpha} \quad [\mathrm{CONJ}_1]$$

$$\frac{\alpha \wedge \beta}{\beta} \quad [\mathrm{CONJ}_2]$$

$$\frac{\alpha \vee \beta}{\beta \vee \alpha} \quad [\mathrm{COM}]$$

$$\frac{\alpha \vee (\beta \vee \gamma)}{(\alpha \vee \beta) \vee \gamma} \quad [\mathrm{ASSOC}]$$

$$\frac{\alpha \vee (\beta \wedge \gamma)}{(\alpha \vee \beta) \wedge (\alpha \vee \gamma)} \quad [\mathrm{DIST}]$$

$$\frac{\neg\neg\alpha}{\alpha} \quad [\mathrm{DNE}]$$

$$\frac{\neg(\alpha \vee \beta)}{\neg\alpha \wedge \neg\beta} \quad [\mathrm{DM1}]$$

$$\frac{\neg(\alpha \wedge \beta)}{\neg\alpha \vee \neg\beta} \quad [\mathrm{DM2}]$$

In what follows, we refer to this set of rules simply as R_1. The rules model well-known semantic equivalences (De Morgan's laws, distributivity, double negation, etc.) as well as the possibility of breaking conjunctions apart.

Example 1.64. *Let* $\Sigma = \{p, q, r\}$ *and* $\Phi = \{(p \wedge q), (\neg q \vee r) \wedge s\}$*. Using* $[\mathrm{CONJ}_1]$ *(substitution* $\alpha \mapsto p$*,* $\beta \mapsto q$*), we obtain*

$$\Phi \longrightarrow_{[\mathrm{CONJ}_1]} \Phi \cup \{p\}.$$

Applying $[\mathrm{CONJ}_1]$ *once more (to* $(\neg q \vee r) \wedge s$*) yields* $\Phi \cup \{p, \neg q \vee r\}$*.*

Definition 1.65. *Let* R *be a set of rules. The derivability relation associated with* R*, also called the* R*-calculus, is* *denoted by* $\vdash_R$ *and defined as follows:* $\varphi \vdash_R \psi$ *if and only if there exists a finite sequence of rule applications*

$$\{\varphi\} \longrightarrow_{r_1,\sigma_1} \Phi_1 \longrightarrow_{r_2,\sigma_2} \Phi_2 \longrightarrow \cdots \longrightarrow_{r_n,\sigma_n} \Phi_n$$

such that $\psi \in \Phi_n$.

Definition 1.66. *A calculus* $\vdash$ *is called*

1. sound *if* $\varphi \vdash \psi$ *always implies* $\varphi \models_{\mathrm{AL}} \psi$*;*

2. complete *if* $\varphi \models_{\mathrm{AL}} \psi$ *always implies* $\varphi \vdash \psi$.

Proposition 1.67. *The calculus* $\vdash_{R_1}$ *is sound: every formula* ψ *derivable from* φ *by means of* R_1 *satisfies* $\varphi \models_{\mathrm{AL}} \psi$.

Proof. Suppose that $\varphi \vdash_{R_1} \psi$ is obtained by a sequence of applications of rules from R_1, that is, there exist $\Phi_0 = \{\varphi\}, \Phi_1, \ldots, \Phi_n$ with $\Phi_{i-1} \overset{r_i,\sigma_i}{\longrightarrow} \Phi_i$ and $\psi \in \Phi_n$. We prove by induction on the number of applications i that every formula in Φ_i is a logical consequence of φ, i.e. that $\Phi_i \subseteq \mathrm{Cn}_\Sigma(\{\varphi\})$.

For $i = 0$, we have $\Phi_0 = \{\varphi\} \subseteq \mathrm{Cn}_\Sigma(\{\varphi\})$. For the induction step, consider the last rule application $\Phi_{i-1} \overset{r_i,\sigma_i}{\longrightarrow} \Phi_i$. The rule r_i is one of the rules listed in R_1; each of these rules is semantically valid, that is, its conclusion is semantically justified by its premises (for example, α follows semantically from $\alpha \wedge \beta$). By the induction hypothesis, all premises in Φ_{i-1} are already semantic consequences of φ, and hence the conclusion obtained by applying the rule is likewise a semantic consequence of φ. Therefore, $\Phi_i \subseteq \mathrm{Cn}_\Sigma(\{\varphi\})$. Finally, $\psi \in \Phi_n$ implies $\psi \in \mathrm{Cn}_\Sigma(\{\varphi\})$, and thus $\varphi \models_{\mathrm{AL}} \psi$. □

Incompleteness of R_1. The calculus R_1 is in general *not* complete. A classical example is:

$$(p \vee \neg q) \wedge (p \vee q) \models_{\mathrm{AL}} p,$$

because every interpretation that satisfies $(p \vee \neg q)$ and $(p \vee q)$ must assign truth to p. Nevertheless, p cannot necessarily be generated from $(p \vee \neg q) \wedge (p \vee q)$ using only the rules of R_1, because R_1 contains no direct rule for the combinatorial case analysis of disjunctions (a resolution-like rule). This motivates stronger calculi that capture such combinations, such as *resolution*, which we discuss in the following section.

Summary and Remarks.

- Calculi are tools that approximate or exactly characterize semantic relations ($\models$) by means of purely syntactic manipulations.
- Soundness is comparatively easy to achieve (the rules must be semantically justified).
- Completeness is more difficult — additional rules (e.g. resolution or case distinction rules) are usually required in order to capture all semantic consequences syntactically.
- For practical methods (automata / SAT solvers, theorem provers), one often combines normal-form transformations with efficient, problem-oriented rule application.

1.7.3 Resolution Calculus

The calculi discussed so far have shown that it is often useful to first transform formulas into a simple syntactic form (e.g. into CNF / clause-set form) and then work with only a small number of carefully chosen rules. The resolution calculus is a particularly characteristic example of such a method: it is both highly efficient and powerful. It requires only *one* central rule, namely the resolution rule, and in propositional logic it is sound and complete as a refutation method for satisfiability.

Definition 1.68 (Clause-Set Form). *A formula φ is said to be in clause-set form if it is represented as a set $\varphi = \{C_1, \dots, C_n\}$ of clauses, where each clause is a disjunction of literals. We identify a clause C with the set of its literals.*

Definition 1.69 (Resolvent / Resolution Rule). *Let ℓ be a literal (for example p or $\neg p$). Let C_1, C_2 be two clauses such that $\ell \in C_1$ and $\overline{\ell} \in C_2$, where $\overline{\ell}$ denotes the complementary form of ℓ, that is, $\overline{p} = \neg p$ and $\overline{\neg p} = p$. Then*

$$\mathrm{Res}_\ell(C_1, C_2) := (C_1 \setminus \{\ell\}) \cup (C_2 \setminus \{\overline{\ell}\})$$

is called the resolvent *of C_1 and C_2 with respect to ℓ. The corresponding derivation rule (on the metalevel) is:*

$$\frac{C_1\,,\ C_2}{\mathrm{Res}_\ell(C_1, C_2)}\ [\mathrm{Res}].$$

A clause containing exactly one literal is called a unit clause*; deriving such a clause is called unit resolution.*

Examples.

$$\begin{aligned}
&\mathrm{Res}_p(\{p,q\},\{\neg p,r\}) = \{q,r\},\\
&\mathrm{Res}_q(\{p,q\},\{\neg q\}) = \{p\},\\
&\mathrm{Res}_p(\{p\},\{\neg p\}) = \varnothing \quad \text{(empty clause)}.
\end{aligned}$$

Semantically, the empty clause $\varnothing$ represents a *contradiction* (no interpretation can satisfy it). The aim of a resolution procedure is therefore to derive, from a clause set Φ, as many new clauses as possible by repeatedly applying the resolution rule; if the empty clause is eventually derived, then Φ is unsatisfiable.

Proposition 1.70 (Soundness of the Resolution Rule). *Let C_1, C_2 be clauses and let ℓ be a literal such that $\ell \in C_1$ and $\overline{\ell} \in C_2$. Then, semantically,*

$$\{C_1, C_2\} \models_{\mathrm{AL}} \mathrm{Res}_\ell(C_1, C_2).$$

In particular, resolution cannot turn a satisfiable clause set into an unsatisfiable one.

Proof. Let ω be an interpretation such that $\omega \models C_1$ and $\omega \models C_2$. Since $\omega \models C_1$, there exists a literal $L \in C_1$ with $\omega \models L$. If $L \neq \ell$, then $L \in \mathrm{Res}_\ell(C_1, C_2)$, and hence $\omega \models \mathrm{Res}_\ell(C_1, C_2)$. If $L = \ell$, then $\omega \not\models \overline{\ell}$; because $\omega \models C_2$, some other literal $L' \in C_2 \setminus \{\overline{\ell}\}$ must therefore be true. Hence there exists $L' \in C_2 \setminus \{\overline{\ell}\}$ with $\omega \models L'$. Then $L' \in \mathrm{Res}_\ell(C_1, C_2)$, and again $\omega \models \mathrm{Res}_\ell(C_1, C_2)$. Therefore, every interpretation that satisfies both C_1 and C_2 also satisfies the resolvent, which proves the claim. □

Corollary 1.71. *If Φ is a clause set and $C_1, C_2 \in \Phi$ admit a resolution step, then Φ is equivalent to $\Phi \cup \{\mathrm{Res}_\ell(C_1, C_2)\}$ (a semantic extension without loss of models).*

Proof. Let $R := \mathrm{Res}_\ell(C_1, C_2)$. Then we have

$$\mathrm{Mod}_\Sigma(\Phi \cup \{R\}) = \mathrm{Mod}_\Sigma(\Phi) \cap \mathrm{Mod}_\Sigma(\{R\}),$$

since the model set of a union of formulas is the intersection of the corresponding model sets. Because $C_1, C_2 \in \Phi$, it follows immediately from the definition of $\mathrm{Mod}_\Sigma(\Phi)$ that

$$\mathrm{Mod}_\Sigma(\Phi) \subseteq \mathrm{Mod}_\Sigma(\{C_1, C_2\}).$$

By the soundness of the resolution rule (see the proposition above), it follows that

$$\mathrm{Mod}_\Sigma(\{C_1, C_2\}) \subseteq \mathrm{Mod}_\Sigma(\{R\}).$$

Combining these inclusions, we obtain

$$\mathrm{Mod}_\Sigma(\Phi) \subseteq \mathrm{Mod}_\Sigma(\{R\}),$$

and therefore

$$\mathrm{Mod}_\Sigma(\Phi) \cap \mathrm{Mod}_\Sigma(\{R\}) = \mathrm{Mod}_\Sigma(\Phi).$$

Hence,

$$\mathrm{Mod}_\Sigma(\Phi \cup \{R\}) = \mathrm{Mod}_\Sigma(\Phi),$$

so Φ and $\Phi \cup \{R\}$ have the same models, which proves the claim. □

Abbreviated Notation If it is clear from the context which clauses and which literal are used in a resolution step, we write more briefly

$$\Phi \xrightarrow{\text{RES}} \Phi'$$

for a single application of the resolution rule, and

$$\Phi \xrightarrow{\text{RES}} \cdots \xrightarrow{\text{RES}} \Phi'$$

for a sequence of resolutions that leads from the clause set Φ to Φ'. If no further resolution can be applied, we call the resulting clause set *fully resolved.* In brief notes, we also simply write $\Phi \rightsquigarrow \Phi'$.

Every formula ϕ can be resolved uniquely, completely, and in a terminating manner.

Proposition 1.72. *Let Σ be a propositional signature, and let φ be a clause set over Σ. Suppose there are two resolution sequences*

$$\varphi \xrightarrow{\text{RES}} \varphi_1 \xrightarrow{\text{RES}} \cdots \xrightarrow{\text{RES}} \varphi_n,$$
$$\varphi \xrightarrow{\text{RES}} \varphi'_1 \xrightarrow{\text{RES}} \cdots \xrightarrow{\text{RES}} \varphi'_m.$$

such that both φ_n and φ'_m are fully resolved. Then $\varphi_n = \varphi'_m$.

Proof. We first show the inclusion $\varphi_n \subseteq \varphi'_m$. The converse direction then follows by exactly the same argument with the roles exchanged, yielding the desired equality.

First note that each application of the resolution rule *adds* a new clause and never removes an existing one. Hence we always have $\varphi \subseteq \varphi_i$ and $\varphi \subseteq \varphi'_j$ for all intermediate steps φ_i, φ'_j that arise.

We prove $\varphi_n \subseteq \varphi'_m$ by induction on n.

Base case. Let $n = 1$. Let $C \in \varphi_1$. Then there are two cases:

- $C \in \varphi$. Since $\varphi \subseteq \varphi'_m$, it follows that $C \in \varphi'_m$.
- $C \notin \varphi$. Then C was produced in the single resolution step as the resolvent $C = \mathrm{Res}_\ell(C', C'')$ with $C', C'' \in \varphi$. Since $\varphi \subseteq \varphi'_m$, the clauses C', C'' also belong to φ'_m. Because φ'_m is fully resolved, it contains the resolvents of every applicable pair of clauses; therefore $\mathrm{Res}_\ell(C', C'') = C$ belongs to φ'_m.

This proves the base case.

Induction step. Assume that the statement has been proved for all sequences of length $< n$, and consider a sequence of length n with final stage φ_n. Let $C \in \varphi_n$. If $C \in \varphi_{n-1}$, then by the induction hypothesis it follows that $C \in \varphi'_m$. If $C \notin \varphi_{n-1}$, then C was generated in the last application as the resolvent $C = \mathrm{Res}_\ell(D_1, D_2)$ with $D_1, D_2 \in \varphi_{n-1}$. By the induction hypothesis, D_1 and D_2 belong to φ'_m, and since φ'_m is complete (closed under resolution), it follows again that $C = \mathrm{Res}_\ell(D_1, D_2) \in \varphi'_m$. Thus every case is covered, and $C \in \varphi'_m$.

Hence $\varphi_n \subseteq \varphi'_m$. By exchanging the roles of φ_n and φ'_m, we obtain symmetrically that $\varphi'_m \subseteq \varphi_n$. Therefore, $\varphi_n = \varphi'_m$. □

Proposition 1.73 (Finiteness / Termination). *Let Σ be a finite signature and Φ a set of clauses over Σ. Every complete resolution process reaches a point after finitely many steps at which no new clauses are generated; in other words, the process always terminates.*

Proof. A clause is determined by specifying, for each atomic formula $p \in \Sigma$, whether p occurs positively, negatively, both positively and negatively, or not at all. For each atom there are only finitely many possibilities (exactly four), so there are at most $4^{|\Sigma|}$ different clauses over Σ in total. Each application of the resolution rule either adds no new clause (because the resolvent is already present) or adds a clause that has not appeared before. Since the number of distinct possible clauses is finite, this process cannot produce a new clause infinitely many times. After finitely many steps, all possible resolvents have been generated and the process is complete — a fully resolved clause set has been reached. □

Theorem 1.74 (Characterization of Satisfiability via Resolution). *Let Σ be a propositional signature, and let $\varphi \in L(\Sigma)$ be a formula in clause-set form. Assume moreover that $\varphi \rightsquigarrow \varphi'$ is a fully resolved sequence (that is, φ' is obtained from φ by finitely many applications of the resolution rule and cannot be further resolved afterwards). Then the following holds:*

$$\varphi \text{ is satisfiable} \quad \Longleftrightarrow \quad \varnothing \notin \varphi'.$$

In particular, φ is unsatisfiable if and only if the empty clause is produced in some resolution step.

Proof. First, note that every resolvent added by resolution leaves the set of models unchanged (by the proposition on the correctness of the resolution rule). Hence $\varphi \equiv \varphi'$, and therefore $\mathrm{Mod}_\Sigma(\varphi) = \mathrm{Mod}_\Sigma(\varphi')$.

"$\Leftarrow$": If $\varnothing \in \varphi'$, then φ' is clearly unsatisfiable, since a clause set containing the empty clause has no model. Because $\varphi \equiv \varphi'$, it follows that φ is also unsatisfiable.

"$\Rightarrow$": Now assume that φ is unsatisfiable. We show that this necessarily implies $\varnothing \in \varphi'$. The proof proceeds by complete induction on $n := |\Sigma|$.

Base case ($n = 0$). If $|\Sigma| = 0$, there is only the atomic case: either $\varphi = \varnothing$ (in which case φ is satisfiable) or $\varphi = \{\varnothing\}$ (in which case φ is already unsatisfiable and $\varnothing \in \varphi$). Since we assume that φ is unsatisfiable, it follows immediately that $\varnothing \in \varphi'$.

Induction step. Let $n \geq 0$ and assume that the statement holds for all signatures of size n. Let Σ be such that $|\Sigma| = n + 1$, and let φ be an unsatisfiable clause set over Σ. Choose an arbitrary atom $x \in \Sigma$ and construct the two clause sets over $\Sigma \setminus \{x\}$:

$$\varphi_x := \{\, C \setminus \{x\} \mid C \in \varphi,\ \neg x \notin C \,\},$$
$$\varphi_{\neg x} := \{\, C \setminus \{\neg x\} \mid C \in \varphi,\ x \notin C \,\}.$$

Intuitively, φ_x is the formula obtained by assuming that x is true (clauses satisfied by $\neg x$ disappear, and occurrences of x are removed from the others). Analogously, $\varphi_{\neg x}$ is the formula obtained under the assumption that x is false.

Since φ is unsatisfiable, neither φ_x nor $\varphi_{\neg x}$ can be satisfiable (for instance, if φ_x were satisfiable, then a model of φ_x could be extended to a model of φ by setting $\omega(x) =$ false). Thus φ_x and $\varphi_{\neg x}$ are also unsatisfiable and are defined over the smaller signature $\Sigma \setminus \{x\}$. By the induction hypothesis, it therefore follows that for fully resolved results $\varphi_x \rightsquigarrow \psi_x$ and $\varphi_{\neg x} \rightsquigarrow \psi_{\neg x}$, we must have $\varnothing \in \psi_x$ and $\varnothing \in \psi_{\neg x}$.

We now show that these facts already imply that the empty clause must be produced in a complete resolution of φ. The empty clause is obtained from resolvents; by the induction hypothesis, there are resolutions in the fully resolved sequences of φ_x and $\varphi_{\neg x}$, respectively, that produce the empty clause. Every clause used in the resolution of φ_x comes either directly from φ or from a clause of the form $C \setminus \{x\}$ with $C \in \varphi$ (and analogously for $\varphi_{\neg x}$). It follows that, using the same initial clauses in φ and exploiting the possibilities of resolution over x and $\neg x$, one can first derive the unit clause $\{x\}$ and, respectively, $\{\neg x\}$ in the resolution of φ. Indeed, if in φ_x a sequence of resolutions from $C_1 \setminus \{x\}, \ldots, C_m \setminus \{x\}$ yields the empty clause, then the corresponding initial clauses in φ occur either unchanged or with the additional literal x, and repeated applications of the resolution rule then eventually yield the unit clause $\{x\}$ in φ. Analogously, the clause $\{\neg x\}$ is derived from the resolution that produces the empty clause in $\varphi_{\neg x}$.

Therefore, if in a (complete) resolution of φ both unit clauses $\{x\}$ and $\{\neg x\}$ are derivable, then resolving these two clauses yields the empty clause. Since φ' is fully resolved, it contains all clauses derivable from φ; hence $\varnothing \in \varphi'$ must follow.

This proves that the unsatisfiability of φ implies $\varnothing \in \varphi'$. Combined with the first implication, we obtain the equivalent characterization

$$\varphi \text{ is satisfiable} \iff \varnothing \notin \varphi'.$$

□

Example 1.75 (Early Termination in Resolution — A Short Alternative Example). *Consider the signature* $\Sigma = \{P, Q, R\}$ *and the clause set*

$$\Phi = \{\{P, Q\},\ \{\neg P\},\ \{\neg Q\}\}.$$

We now apply resolution step by step, highlighting the resolvent generated at each stage:

$$\begin{aligned}
&\Phi = \{\{P, Q\},\ \{\neg P\},\ \{\neg Q\}\} \\
&\quad \xrightarrow{\text{RES}} \{\{P, Q\},\ \{\neg P\},\ \{\neg Q\},\ \underbrace{\{Q\}}_{\mathrm{Res}_{\neg P}(\{P,Q\},\{\neg P\})}\} \\
&\quad \xrightarrow{\text{RES}} \{\{P, Q\},\ \{\neg P\},\ \{\neg Q\},\ \{Q\},\ \underbrace{\varnothing}_{\mathrm{Res}_{\neg Q}(\{Q\},\{\neg Q\})}\}.
\end{aligned}$$

The two resolution steps are, in detail:

$$\mathrm{Res}_{\neg P}(\{P, Q\}, \{\neg P\}) = \{Q\}, \qquad \mathrm{Res}_{\neg Q}(\{Q\}, \{\neg Q\}) = \varnothing.$$

Thus, after only two steps, we have already produced a subset Φ'' *of resolvents (namely, the right-hand side of the last line) that contains the empty clause, that is,* $\varnothing \in \Phi''$. *Since any further complete resolution process can only add more resolvents, we always have* $\Phi'' \subseteq \Phi'$ *for any fully resolved result* Φ' *obtained from* Φ. *Consequently,* $\varnothing \in \Phi'$ *as well, and by the resolution theorem,* Φ *is unsatisfiable.*

Remark. *This example illustrates an important practical principle:* it is not always necessary to compute all possible resolvents. *As soon as the empty clause appears anywhere in the resolution sequence, the process can be stopped — the original formula is then definitively unsatisfiable. In practice, this can save a great deal of computational effort.*

Definition 1.76 (Resolution Derivability). *Let* Σ *be a propositional signature, and let* $\varphi, \psi \in L(\Sigma)$. *Let* ξ *be a formula in clause-set form such that* $\xi \equiv \varphi \wedge \neg\psi$, *and let* ξ' *be a fully resolved result with* $\xi \leadsto \xi'$ *(cf. the theorem on the characterization of satisfiability via resolution). We write*

$$\varphi \vdash_{\mathrm{Res}} \psi \qquad \textit{iff} \qquad \varnothing \in \xi'.$$

That is, ψ *is derivable from* φ *by resolution if and only if the empty clause is generated during the resolution of* $\varphi \wedge \neg\psi$.

Corollary 1.77 (Soundness and Completeness of the Resolution Calculus). *The relation* $\vdash_{\mathrm{Res}}$ *is sound and* *complete for propositional logic; that is, for all* $\varphi, \psi \in L(\Sigma)$, *we have*

$$\varphi \vdash_{\text{Res}} \psi \iff \varphi \models_{\text{AL}} \psi.$$

Proof. Let $\varphi, \psi \in L(\Sigma)$ be arbitrary, and let ξ be, as above, a clause set with $\xi \equiv \varphi \wedge \neg\psi$, and let $\xi \rightsquigarrow \xi'$ be a fully resolved result.

Soundness. Assume that $\varphi \vdash_{\text{Res}} \psi$. Then $\varnothing \in \xi'$, so ξ (and hence $\varphi \wedge \neg\psi$) is unsatisfiable. Therefore $\varphi \wedge \neg\psi \equiv \bot$, and thus $\varphi \wedge \neg\psi$ has no models. By the characterization of semantic consequence (deduction/proposition on Infer/Taut/Sat), it follows that $\varphi \models_{\text{PL}} \psi$.

Completeness. Assume that $\varphi \models_{\text{PL}} \psi$. Then $\varphi \wedge \neg\psi$ is unsatisfiable. By the resolution theorem (the theorem on the characterization of satisfiability via resolution), a complete resolution process over an equivalent clause set must generate the empty clause. That is, for the clause set ξ equivalent to $\varphi \wedge \neg\psi$, we have $\xi \rightsquigarrow \xi'$ with $\varnothing \in \xi'$. Hence $\varphi \vdash_{\text{Res}} \psi$.

Thus $\varphi \vdash_{\text{Res}} \psi$.

The claim follows from both directions. □

Remark. Formally, the practical use of $\vdash_{\text{Res}}$ depends on transforming $\varphi \wedge \neg\psi$ into an (equivalent) set of clauses (existence of a CNF / clause-set form). In practice, one typically uses a Tseitin translation or a similar method in order to keep the transformation costs manageable.

Example 1.78.

$$\Sigma = \{A, B, C\}, \qquad \varphi = (A \vee B) \wedge (\neg A \vee C), \qquad \psi = B \vee C.$$

Semantic justification. *Let ω be an arbitrary interpretation such that $\omega \models \varphi$.*

- *If $\omega(A) = true$, then $\neg A \vee C$ implies that $\omega(C) = true$, and hence $\omega \models B \vee C$.*
- *If $\omega(A) = false$, then $A \vee B$ implies that $\omega(B) = true$, and again $\omega \models B \vee C$.*

In both cases, we obtain $\omega \models B \vee C$. Therefore $\varphi \models_{\mathrm{PL}} \psi$.

Resolution (syntactic proof). *Consider $\xi = \varphi \wedge \neg\psi$. Since $\neg\psi = \neg(B \vee C) = \neg B \wedge \neg C$, the formula ξ is already in clause-set form:*

$$\xi = \{\,\{A, B\},\ \{\neg A, C\},\ \{\neg B\},\ \{\neg C\}\,\}.$$

We now apply resolution step by step:

$$\begin{aligned} \{A, B\},\ \{\neg B\} &\implies \mathrm{Res}_B(\{A, B\}, \{\neg B\}) = \{A\}, \\ \{A\},\ \{\neg A, C\} &\implies \mathrm{Res}_A(\{A\}, \{\neg A, C\}) = \{C\}, \\ \{C\},\ \{\neg C\} &\implies \mathrm{Res}_C(\{C\}, \{\neg C\}) = \varnothing. \end{aligned}$$

Since the empty clause $\varnothing$ can be derived, ξ is unsatisfiable. Because $\xi \equiv \varphi \wedge \neg\psi$, it follows that

$$\varphi \vdash_{\mathrm{Res}} \psi.$$

Conclusion. *In this example, the semantic and syntactic perspectives coincide:*

$$\varphi \models_{\mathrm{PL}} \psi \iff \varphi \vdash_{\mathrm{Res}} \psi.$$

1.7.4 Exercises and Solutions

Exercise 1.79. *Consider the propositional signature*

$$\Sigma = \{J, K, A, C, D, G, I\}$$

and the clause set

$$\varphi_3 = \left\{ \begin{array}{l} \{J, C, \neg I\}, \{G, \neg J\}, \{\neg A, G, \neg I\}, \\ \{J, \neg A, \neg I\}, \{A, \neg D, G\}, \\ \{C, G, \neg I\}, \{J, \neg D, G, \neg I\} \end{array} \right\}.$$

Determine for each of the following statements whether it is true or false.

1. $\{K, \neg G\}$ *and* $\{K, I\}$ *are resolvable.*

2. $\{\neg A\}$ *and* $\{A, \neg D\}$ *are resolvable.*

3. φ_3 *is fully resolved.*

4. $\{\{J, \neg C\}, \{\neg G\}, \{A, \neg C, G\}, \{K, G\}, \{\neg K, G\}\} \not\equiv \bot$.

Solution 1.80. 1. ***False.*** *The two clauses do contain* K*, but they do not contain a pair of complementary literals. For a resolution step, one clause would have to contain* L *and the other* $\neg L$*. That is not the case here.*

2. ***True.*** *The clauses are resolvable with respect to the literal A. We have*

$$\text{Res}_{\neg A}(\{\neg A\}, \{A, \neg D\}) = \{\neg D\}.$$

Hence the resolution rule is applicable.

3. ***False.*** *The clause set is not yet fully resolved. In fact, further resolution steps are still possible, for example*

$$\text{Res}_{J}(\{J, \neg D, G, \neg I\}, \{G, \neg J\}) = \{\neg D, G, \neg I\}.$$

Thus, at least one additional resolvent can still be derived.

4. ***False.*** *The formula is unsatisfiable, and therefore in particular*

$$\{\{J, \neg C\}, \{\neg G\}, \{A, \neg C, G\}, \\ \{K, G\}, \{\neg K, G\}\} \equiv \bot$$

One can derive the empty clause, for instance, as follows:

$$\text{Res}_{\neg K}(\{\neg K, G\}, \{K, G\}) = \{G\}, \\ \text{Res}_{\neg G}(\{\neg G\}, \{A, \neg C, G\}) = \{A, \neg C\}$$

and further

$$\text{Res}_{\neg G}(\{\neg G\}, \{K, G\}) = \{K\}, \\ \text{Res}_{\neg G}(\{\neg G\}, \{\neg K, G\}) = \{\neg K\}$$

finally,

$$\mathrm{Res}_{\neg K}(\{\neg K\}, \{K\}) = \varnothing.$$

Therefore, the clause set is unsatisfiable and hence equivalent to $\bot$.

Brief feedback.

- *(1) No complementary pair of literals is present.*
- *(2) Resolution on A yields $\{\neg D\}$.*
- *(3) Not yet fully resolved; further resolvents are possible.*
- *(4) Unsatisfiable, since the empty clause can be derived.*

1.8 SAT Solvers

We now consider the algorithmic satisfiability problem for propositional formulas. At the center of this discussion are so-called SAT solvers, that is, procedures that decide whether a formula is satisfiable. The classical DPLL algorithm provides the foundation for many modern implementations. Before introducing it, however, we need one further important tool: satisfiability equivalence. It allows formulas to be transformed or simplified in such a way that their satisfiability remains unchanged. This is especially useful when formulas are to be converted into conjunctive normal form.

In the following subsection, we first introduce the notion of satisfiability equivalence and show, by means of simple examples, how formulas can be modified in a satisfiability-equivalent way. We then turn to Tseitin encoding, which makes it possible to transform formulas efficiently into CNF. After that, we present the DPLL algorithm as a fundamental search procedure for SAT.

1.8.1 Satisfiability Equivalence

Two formulas are called satisfiability-equivalent if they cannot be distinguished with respect to satisfiability: either both are satisfiable or both are unsatisfiable. Unlike logical equivalence, this does not mean that the two formulas have the same truth value under every interpretation; rather, it concerns only whether a model exists at all.

Definition 1.81. *Let Σ be a propositional signature and let $\varphi, \psi \in L(\Sigma)$. Then φ and ψ are called* satisfiability-equivalent, *written*

$$\varphi \cong_\Sigma \psi,$$

if

$$\mathrm{Mod}_\Sigma^{PL}(\varphi) \neq \emptyset \;\Leftrightarrow\; \mathrm{Mod}_\Sigma^{PL}(\psi) \neq \emptyset.$$

For brevity, we also write $\varphi \cong \psi$ whenever the underlying signature is clear from the context.

Every pair of logically equivalent formulas is automatically satisfiability-equivalent. The converse, however, does not hold in general: two formulas may have the same satisfiability status while differing in semantic meaning.

Example 1.82. *A very simple form of satisfiability equivalence arises from renaming variables. Let Σ be a propositional signature, let $\varphi \in L(\Sigma)$, and let $p, q \in \Sigma$ be chosen such that p occurs in φ whereas q does not. Then*

$$\varphi \cong \varphi[p \mapsto q].$$

For example, let

$$\varphi = (P \wedge (\neg R \vee S)) \wedge (\neg P \vee T)$$

and, by replacing P with Q,

$$\varphi[P \mapsto Q] = (Q \wedge (\neg R \vee S)) \wedge (\neg Q \vee T).$$

From every model of φ, we immediately obtain a model of $\varphi[P \mapsto Q]$ by simply assigning to Q the same truth value that P has. Conversely, the construction works in exactly the same way. Hence the two formulas are satisfiability-equivalent, although in general they are not logically equivalent.

Pure Literal Elimination. For formulas in clause-set form, one particularly useful simplification is *pure literal elimination.* A literal is called *pure* if its complementary literal does not occur anywhere in the given formula. In that case, all clauses containing this literal can be removed without changing satisfiability.

Proposition 1.83. *Let Σ be a propositional signature, let $\varphi \in L(\Sigma)$ be a formula in clause-set form, and let $l \in \mathrm{Lit}(\Sigma)$. If $l \notin C$ for all $C \in \varphi$, then*

$$\varphi \cong \{C \in \varphi \mid \bar{l} \notin C\}.$$

Proof. Let

$$\varphi' := \{C \in \varphi \mid \bar{l} \notin C\}.$$

Without loss of generality, let $l = x$ for some $x \in \Sigma$, so that $\bar{l} = \neg x$. The case $l = \neg x$ is analogous.

We must show:

$$\mathrm{Mod}_{\Sigma}^{PL}(\varphi) \neq \emptyset \;\Leftrightarrow\; \mathrm{Mod}_{\Sigma}^{PL}(\varphi') \neq \emptyset.$$

- If $\omega \in \mathrm{Mod}_{\Sigma}^{PL}(\varphi)$, then, since $\varphi' \subseteq \varphi$, we also have $\omega \in \mathrm{Mod}_{\Sigma}^{PL}(\varphi')$.

- Conversely, let $\omega \in \mathrm{Mod}_{\Sigma}^{PL}(\varphi')$. We define an interpretation $\omega' \in \Omega(\Sigma)$ by

  $$\omega'(y) = \omega(y) \quad \forall y \in \Sigma \setminus \{x\}, \qquad \omega'(x) = \text{true}.$$

  We show that $\omega' \models^{PL} \varphi$.

  To this end, let $C \in \varphi$. If $\bar{l} = \neg x \in C$, then $\omega' \models^{PL} C$, because $\omega'(x) = \text{true}$. If $\bar{l} \notin C$, then $C \in \varphi'$. Since $l = x$ does not occur in any clause of φ, neither x nor $\neg x$ occurs in C. Hence ω and ω' agree on all variables occurring in C. Since $\omega \models^{PL} C$, it follows that $\omega' \models^{PL} C$ as well.

Thus $\omega' \in \mathrm{Mod}_\Sigma^{PL}(\varphi)$, and therefore

$$\mathrm{Mod}_\Sigma^{PL}(\varphi) \neq \emptyset.$$

Hence $\varphi \cong \varphi'$. □

In other words, if a variable occurs in a clause set with only one polarity, then the corresponding clauses can safely be removed. This observation is systematically exploited later in SAT procedures.

Example 1.84. *Consider the clause set*

$$\psi = \{\{P, R\}, \{\neg P, S\}, \{S, T\}, \{\neg T, U\}, \{S, U\}\}.$$

Here, S is a pure literal, since $\neg S$ does not occur in ψ. Therefore, we may delete the clauses containing S and obtain

$$\psi' = \{\{P, R\}, \{\neg T, U\}\}.$$

We have $\psi \cong \psi'$. In fact, ψ' is satisfiable, for instance under an assignment with

$$\omega(P) = \omega(R) = \omega(U) = \text{true}, \qquad \omega(T) = \text{false},$$

and the variable S can then be assigned arbitrarily, for example as true, so that ψ is satisfied as well.

Satisfiability equivalence is therefore a fundamental tool for the preprocessing of SAT instances. It allows formulas to be simplified without changing the answer to the question of satisfiability. In the next subsection, we show how this idea can be used systematically with Tseitin encoding for the transformation into CNF.

1.8.2 Tseitin Encoding

When working with resolution procedures, conjunctive normal form is particularly useful. The obvious idea of converting an arbitrary formula into CNF simply by repeatedly applying the equivalences from Section 1.5.3 has, however, a drawback: the resulting formula can be much longer than the original one. This is undesirable for algorithmic procedures, since the syntactic blow-up can be exponential in the worst case.

To make this problem more precise, we first introduce a simple measure of size.

Definition 1.85 (Length of a Formula). *Let Σ be a propositional signature and let $\varphi \in L(\Sigma)$. The length $|\varphi|$ is defined inductively by*

$$|\varphi| := \begin{cases} 1, & \text{if } \varphi = x \in \Sigma, \\ 1 + |\psi|, & \text{if } \varphi = \neg\psi, \\ 1 + |\psi| + |\psi'|, & \text{if } \varphi = \psi \wedge \psi', \\ 1 + |\psi| + |\psi'|, & \text{if } \varphi = \psi \vee \psi'. \end{cases}$$

Here we count only the connectives $\neg$, $\wedge$, and $\vee$. Formulas containing $\rightarrow$ or $\leftrightarrow$ are first reduced to these connectives by their defining equivalences.

Example 1.86. *We consider a slightly smaller signature*

$$\Sigma_{\text{House}} = \{\textit{Door}, \textit{Window}, \textit{Light}\}.$$

Then, for example,

$$|\neg(Door \vee Window)| = 4, \qquad |(Door \wedge \neg Window) \vee Light| = 6.$$

For the formula

$$(Door \vee Window) \wedge \neg Light$$

we obtain, by the definition of length,

$$|Door \vee Window| = 3, \qquad |\neg Light| = 2,$$

and hence in total

$$|(Door \vee Window) \wedge \neg Light| = 1 + 3 + 2 = 6.$$

The naive conversion into CNF can grow dramatically. This becomes especially clear for the family

$$\varphi_n = (A_1 \wedge B_1) \vee (A_2 \wedge B_2) \vee \cdots \vee (A_n \wedge B_n).$$

The length of φ_n grows only linearly in n, but a naive application of the distributive laws yields a CNF with exponentially many clauses: for every choice of exactly one atom from each pair (A_i, B_i), a separate clause is produced. The result is therefore much longer than the original formula.

For practical procedures, a different idea is therefore more useful: instead of replacing every subformula directly by an equivalent CNF, we introduce new auxiliary variables for subformulas. These new variables "store" the truth value of the corresponding subformula. In this way, we obtain a formula in CNF that is not necessarily equivalent to the original formula, but is *equisatisfiable* with it: that is, each formula is satisfiable if and only if the other one is satisfiable.

This construction is called *Tseitin encoding.*

Example 1.87. *Consider the formula*

$$\varphi = (\mathit{Alarm} \wedge (\mathit{Sensor} \vee \neg \mathit{Power})) \vee (\neg \mathit{Alarm} \wedge \mathit{Siren}).$$

A naive conversion into CNF would cause the formula to blow up considerably. Instead, we introduce new auxiliary variables for its subformulas. For example, we may mark the following subformulas:

$$\begin{aligned}
\varphi_1 &= \varphi, \\
\varphi_2 &= \mathit{Alarm} \wedge (\mathit{Sensor} \vee \neg \mathit{Power}), \\
\varphi_3 &= \neg \mathit{Alarm} \wedge \mathit{Siren}.
\end{aligned}$$

as well as

$$\begin{aligned}
\varphi_4 &= \mathit{Alarm}, \\
\varphi_5 &= \mathit{Sensor} \vee \neg \mathit{Power}, \\
\varphi_6 &= \neg \mathit{Alarm}, \\
\varphi_7 &= \mathit{Siren}.
\end{aligned}$$

We then introduce new propositional variables

$$Q_1, \ldots, Q_7$$

and encode the structure of the formula by the equivalences

$$Q_1 \leftrightarrow (Q_2 \vee Q_3), \qquad Q_2 \leftrightarrow (Q_4 \wedge Q_5), \qquad Q_3 \leftrightarrow (Q_6 \wedge Q_7),$$

$$Q_4 \leftrightarrow Alarm,$$
$$Q_5 \leftrightarrow (Sensor \vee \neg Power),$$
$$Q_6 \leftrightarrow \neg Alarm,$$
$$Q_7 \leftrightarrow Siren.$$

Finally, we require the root variable Q_1 *to be true. This yields a CNF that is not necessarily equivalent to* φ*, but is satisfiable exactly when* φ *is satisfiable.*

Definition 1.88 (Tseitin Encoding). *Let* Σ *be a propositional signature and* $\varphi \in L(\Sigma)$*. We define the Tseitin encoding* $T(\varphi)$ *in such a way that it introduces a new variable for each relevant subformula. The root variable of the encoding is required to be true.*

Conceptually, this is done recursively: for each subformula ψ*, we introduce a fresh variable* x_ψ *and add clauses describing the relationship between* x_ψ *and the immediate subformulas of* ψ*. For the base cases and the* *three connectives, this yields the following building blocks:*

$$\begin{aligned}
T_X(x,y) &= (\neg x \vee y) \wedge (x \vee \neg y) \\
T_X(x,\neg\psi) &= (\neg x \vee \neg x') \wedge (x \vee x') \\
&\quad \wedge\ T_{X\cup\{x'\}}(x',\psi), \\
T_X(x,\psi\wedge\psi') &= (\neg x \vee x') \wedge (\neg x \vee x'') \\
&\quad \wedge\ (\neg x' \vee \neg x'' \vee x) \\
&\quad \wedge\ T_{X\cup\{x',x''\}}(x',\psi) \\
&\quad \wedge\ T_{X\cup\{x',x''\}\cup\Sigma(\xi)}(x'',\psi'), \\
T_X(x,\psi\vee\psi') &= (\neg x' \vee x) \wedge (\neg x'' \vee x) \\
&\quad \wedge\ (\neg x \vee x' \vee x'') \\
&\quad \wedge\ T_{X\cup\{x',x''\}}(x',\psi) \\
&\quad \wedge\ T_{X\cup\{x',x''\}\cup\Sigma(\xi)}(x'',\psi').
\end{aligned}$$

Here, x' and x'' are fresh variables. The full encoding is then given by

$$T(\varphi) = x_\varphi \wedge T_{\Sigma(\varphi)\cup\{x_\varphi\}}(x_\varphi,\varphi).$$

where

$$x \notin \Sigma(\varphi), \qquad x', x'' \notin X, \qquad \varphi = y \in \Sigma.$$

Theorem 1.89 (Tseitin Encoding)**.** *Let Σ be a propositional signature and let $\varphi \in L(\Sigma)$. Then $T(\varphi)$ is in conjunctive normal form, and moreover*

$$\varphi \sim T(\varphi).$$

Here, $\varphi \sim T(\varphi)$ means that φ and $T(\varphi)$ are equisatisfiable.

Proof. We first show that $T(\varphi)$ is in CNF. This follows directly from the recursive definition of the Tseitin encoding:

- For atomic formulas,

$$T_X(x, y) = (\neg x \vee y) \wedge (x \vee \neg y)$$

  is already a conjunction of clauses.

- The formulas $T_X(x, \neg\psi)$, $T_X(x, \psi \wedge \psi')$, and $T_X(x, \psi \vee \psi')$ are each constructed as conjunctions of clauses together with further expressions of the form $T(\cdot, \cdot)$.

- Since the encoding is defined recursively over subformulas, an induction on the structure of φ shows that all occurring components are already in CNF.

- Finally,

$$T(\varphi) = x_\varphi \wedge T_{\Sigma(\varphi)\cup\{x_\varphi\}}(x_\varphi, \varphi)$$

  is also a conjunction of clauses, and hence is in CNF as well.

We now prove equisatisfiability. Set

$$\Sigma' := \Sigma(T(\varphi)).$$

We first show that

$$\mathrm{Mod}_{\Sigma'}(\varphi) \neq \varnothing \quad \Longrightarrow \quad \mathrm{Mod}_{\Sigma'}(T(\varphi)) \neq \varnothing.$$

To this end, let $\omega \in \Omega(\Sigma')$ with $\omega \models \varphi$. We define an extension ω' of ω to the auxiliary variables that occur in $T(\varphi)$ by

$$\omega'(A_i) = \omega(A_i) \qquad (i = 1, \ldots, n),$$

and, for each new variable x_ψ introduced in $T(\varphi)$ for a subformula ψ,

$$\omega'(x_\psi) = [\![\psi]\!]_\omega.$$

We prove by structural induction on the subformulas ψ of φ that

$$\omega'(x_\psi) = [\![\psi]\!]_\omega$$

for all subformulas ψ. For atomic formulas, this is immediate from the definition of ω'. For negation, conjunction, and disjunction, the claim follows from the recursive form of the Tseitin encoding: the corresponding clauses enforce exactly the truth table of the respective connective. Hence ω' satisfies all equivalences in $T(\varphi)$ between auxiliary variables and subformulas, so $\omega' \models T(\varphi)$. Therefore, $T(\varphi)$ is satisfiable.

For the converse implication,

$$\mathrm{Mod}_{\Sigma'}(T(\varphi)) \neq \varnothing \quad \Longrightarrow \quad \mathrm{Mod}_{\Sigma'}(\varphi) \neq \varnothing$$

let $\omega \in \Omega(\Sigma')$ satisfy $\omega \models T(\varphi)$. Again, we show by structural induction on the subformulas ψ of φ that

$$\omega \models x_\psi \quad \Longleftrightarrow \quad \omega \models \psi.$$

The base case for atomic formulas follows directly from the clauses

$$(\neg x_p \vee p) \quad \text{and} \quad (x_p \vee \neg p),$$

which together express exactly $x_p \leftrightarrow p$. In the inductive step, we consider the three possible forms of a subformula:

- If $\psi = \neg\xi$, then $T(\varphi)$ enforces the equivalence $x_\psi \leftrightarrow \neg x_\xi$.
- If $\psi = \xi \wedge \xi'$, then $T(\varphi)$ enforces the equivalence $x_\psi \leftrightarrow (x_\xi \wedge x_{\xi'})$.
- If $\psi = \xi \vee \xi'$, then $T(\varphi)$ enforces the equivalence $x_\psi \leftrightarrow (x_\xi \vee x_{\xi'})$.

In all cases, the induction hypothesis implies that

$$\omega \models x_\psi \quad \Longleftrightarrow \quad \omega \models \psi.$$

In particular, this holds for the root subformula $\psi = \varphi$. Since $T(\varphi)$ contains the unit clause x_φ and $\omega \models T(\varphi)$, it follows that $\omega \models x_\varphi$. Hence $\omega \models \varphi$ as well, and therefore φ is satisfiable.

Combining both directions, we obtain

$$\varphi \sim T(\varphi).$$

□

Proposition 1.90. *Let Σ be a propositional signature and let $\varphi \in L(\Sigma)$. Then the length of the Tseitin encoding grows at most linearly:*

$$|T(\varphi)| \in O(|\varphi|).$$

Remark 1.91. *The Tseitin encoding is so important for algorithmic applications because it avoids the exponential blow-up of the naive CNF transformation. Thus, for the question of satisfiability, it is sufficient to replace the original formula by an equisatisfiable CNF using fresh auxiliary variables.*

Remark 1.92. *From an algorithmic point of view, we may therefore often tacitly assume in what follows that formulas are given in CNF or already in clause-set form. This is exactly the perspective that is particularly useful for the resolution calculus.*

1.8.3 The DPLL Algorithm

The DPLL algorithm is a recursive search procedure for the satisfiability problem in clause-set form. The basic idea is simple: one chooses a propositional variable that has not yet been assigned and considers the two cases in which it is true or false, respectively. In this way, a search tree is generated. To prevent the search from branching unnecessarily, two natural simplifications are carried out at each step: unit clauses are propagated and pure literals are eliminated.

Proposition 1.93 (Splitting on a Literal). *Let Φ be a clause set over a signature Σ, and let $x \in \Sigma$. Then the following holds:*

$$\begin{aligned}\mathrm{Mod}_\Sigma(\Phi) \neq \varnothing \iff & \mathrm{Mod}_\Sigma(\Phi \cup \{\{x\}\}) \neq \varnothing \\ \textit{or} \quad & \mathrm{Mod}_\Sigma(\Phi \cup \{\{\neg x\}\}) \neq \varnothing.\end{aligned}$$

Proof. Every interpretation makes either x true or $\neg x$ true. Thus, if Φ is satisfiable, then a model of Φ also satisfies either $\{x\}$ or $\{\neg x\}$. Conversely, it is clear that every satisfying assignment of $\Phi \cup \{\{x\}\}$ or $\Phi \cup \{\{\neg x\}\}$ is also a model of Φ. □

Unit Propagation. If a clause set contains a unit clause $\{l\}$, then all clauses containing l may be deleted, and in all remaining clauses the complementary literal $\neg l$ may be removed. This simplification does not change satisfiability.

By the soundness of the resolution rule (Proposition 1.64), we obtain:

Corollary 1.94. *Let Φ be a clause set. Then Φ is satisfiability-equivalent to the formula set obtained from Φ by exhaustive unit propagation.*

Pure Literal Elimination. If a literal l occurs in a clause set but its complement $\neg l$ does not, then all clauses containing l can be removed. This simplification also preserves satisfiability.

By Proposition 1.85 on splitting with respect to a literal, we obtain:

Corollary 1.95. *Let Φ be a clause set. Then Φ is satisfiability-equivalent to the formula set obtained from Φ by exhaustive pure literal elimination.*

In each recursive step of the DPLL procedure, these two simplifications are applied first. After that, one checks the following termination conditions:

- If no clause remains, that is, if $\Phi = \varnothing$, then the formula is satisfiable.
- If Φ contains the empty clause $\varnothing$, then the formula is unsatisfiable.

If neither condition applies, a literal l is selected and the search continues in the two branches $\Phi \cup \{\{l\}\}$ and $\Phi \cup \{\{\neg l\}\}$.

Theorem 1.96 (Correctness of DPLL). *Let Σ be a propositional signature and let Φ be a clause set over Σ. Then:*

$$\Phi \textit{ is satisfiable} \iff \mathsf{DPLL}(\Phi) = \mathsf{\textit{Yes}}.$$

Proof. We prove both directions. The key idea is that the two simplification steps

$$\mathsf{UnitProp} \qquad \text{and} \qquad \mathsf{LitElim}$$

preserve satisfiability, and that the subsequent case distinction on a literal splits satisfiability completely into two subcases.

We carry out the proof by induction on the number of propositional variables that still occur in the current clause set.

Preparation. For a clause set Φ, define

$$\Phi^* := \mathsf{LitElim}(\mathsf{UnitProp}(\Phi)).$$

By the preceding corollaries, we have

$$\Phi \sim \Phi^*.$$

In particular, Φ is satisfiable if and only if Φ^* is satisfiable.

Base case. If Φ contains no propositional variables any longer, then after applying UnitProp and LitElim only one of the following two situations can occur:

$$\Phi^* = \varnothing \qquad \text{or} \qquad \varnothing \in \Phi^*.$$

In the first case, Φ^* is satisfiable and DPLL returns Yes. In the second case, a contradiction is present and DPLL returns No. This establishes the claim in the base case.

Induction step. Assume that the statement has been proved for all clause sets containing fewer than n distinct propositional variables, and let Φ now be a clause set with n variables. Again set

$$\Phi^* := \mathsf{LitElim}(\mathsf{UnitProp}(\Phi)).$$

If $\Phi^* = \varnothing$, then Φ is satisfiable and DPLL immediately returns Yes. If $\varnothing \in \Phi^*$, then Φ is unsatisfiable and DPLL returns No. So it remains to consider the case where neither $\Phi^* = \varnothing$ nor $\varnothing \in \Phi^*$ holds.

Then DPLL chooses a literal l whose complement also occurs in Φ^*, and considers the two extensions

$$\Phi^* \cup \{\{l\}\} \qquad \text{and} \qquad \Phi^* \cup \{\{\neg l\}\}.$$

By Proposition 1.85, Φ^* is satisfiable if and only if at least one of these two clause sets is satisfiable. After the subsequent application of UnitProp and LitElim, the variable of l or, respectively, $\neg l$ is fixed in each case; hence the resulting clause sets each contain strictly fewer distinct variables than Φ^*. Therefore, we may apply the induction hypothesis to both recursive calls.

Thus, if one of the two recursive calls is satisfiable, DPLL returns Yes there and hence also for Φ. Conversely, if DPLL returns Yes for Φ, then one of the two recursive calls must return Yes; by the induction hypothesis, at least one of the two branches is therefore satisfiable, and hence so are Φ^* and Φ.

This proves:

$$\Phi \text{ is satisfiable} \quad \Longleftrightarrow \quad \mathsf{DPLL}(\Phi) = \mathsf{Yes}.$$

□

Heuristics. The choice of the branching literal does not affect correctness, but it does influence the running time. In practice, one therefore uses heuristics, for example choosing the most frequently occurring literal or one that affects as many clauses as possible.

Example 1.97 (A satisfiable example). *Consider the clause set*

$$\Phi = \{\{P, Q\}, \{\neg P, R\}, \{\neg Q, R\}, \{\neg R, S\}\}.$$

The literal S occurs only positively. Pure literal elimination therefore removes the clause $\{\neg R, S\}$. What remains is

$$\Phi_1 = \{\{P, Q\}, \{\neg P, R\}, \{\neg Q, R\}\}.$$

If we now choose R, then in the branch $R = true$ the formula is immediately simplified to

$$\{\{P, Q\}\}.$$

A further choice of P then yields the empty clause set, and hence a satisfying assignment. For example,

$$\omega(P) = true, \quad \omega(R) = true, \quad \omega(S) = true$$

and $\omega(Q) = arbitrary$ is a model of Φ.

Example 1.98 (An unsatisfiable example). *Consider the clause set*

$$\Psi = \{\{P, Q\}, \{P, \neg Q\}, \{\neg P, Q\}, \{\neg P, \neg Q\}\}.$$

There are neither unit clauses nor pure literals. If we choose P, both branches lead to unsatisfiability:

- *In the branch $P = true$, the clauses $\{Q\}$ and $\{\neg Q\}$ remain. This yields the empty clause.*
- *In the branch $P = false$, one likewise obtains $\{Q\}$ and $\{\neg Q\}$, and thus again the empty clause.*

Hence Ψ is unsatisfiable and DPLL outputs No.

Search trees. For illustration, the recursive calls are often written as a search tree: each node carries the clause set currently under consideration, and each edge is labeled with the assumption made. A leaf labeled with $\top$ stands for satisfiable, while a leaf labeled with $\bot$ stands for unsatisfiable. In this way, the entire course of the procedure can be displayed clearly.

Remark 1.99. *The DPLL algorithm forms the basis of many modern SAT solvers. In practice, additional techniques such as clause learning, more sophisticated heuristics, and non-chronological backtracking are used. The core idea, however, remains the same: systematic search combined with consistent simplification.*

1.8.4 Exercises and Solutions

Exercise 1.100. *Consider the propositional signature*

$$\Sigma = \{K, F, G, H, I\}$$

and the following expressions:

$$\begin{aligned} \varphi_2 &= \neg(F \wedge \neg H), \\ \psi_2 &= \{\, Q_1 \leftrightarrow \neg Q_2,\ Q_4 \leftrightarrow \neg Q_5,\ Q_3 \leftrightarrow F, \\ &\qquad Q_2 \leftrightarrow Q_3 \wedge Q_4,\ Q_5 \leftrightarrow H \,\}. \end{aligned}$$

$$\begin{aligned} \varphi_3 &= \{\{\neg F\}, \{\neg H\}, \{K, F\}\}, \\ \varphi_3' &= \{\{\neg F\}, \{\neg H\}, \{K, F\}\}. \end{aligned}$$

$$\varphi_4 = \{\{\neg G, \neg H\}, \{G, H\}, \{\neg H\}, \{\neg K, \neg G\}\},$$
$$\varphi_4' = \{\{\neg G, \neg H\}, \{G, H\}, \{\neg H\}\}.$$

Determine for each of the following statements whether it is true or false.

1. *It holds that* $|\neg\neg(\neg F \wedge H)| = 5$.

2. ψ_2 *is the Tseitin encoding of* φ_2.

3. $\mathsf{UnitProp}(\varphi_3) = \varphi_3'$.

4. $\mathsf{LitElim}(\varphi_4) = \varphi_4'$.

Solution 1.101. 1. ***False.*** *We have*

$$|\neg F| = 2,$$
$$|\neg F \wedge H| = 1 + 2 + 1 = 4,$$
$$|\neg\neg(\neg F \wedge H)| = 1 + 1 + 4 = 6.$$

Hence the stated length is not 5, but 6.

2. ***False.*** *In a Tseitin encoding, the root variable representing the entire original formula must additionally be required to be true. Here, the unit clause, or equivalently the requirement for* Q_1, *is missing. Therefore,* ψ_2 *is not a complete Tseitin encoding of* φ_2.

3. ***False.*** *From the unit clause* $\{\neg F\}$, *full unit propagation implies that the literal* F *is removed from the clause* $\{K, F\}$. *This yields the clause* $\{K\}$. *Thus,*

$$\mathsf{UnitProp}(\varphi_3) = \{\{\neg F\}, \{\neg H\}, \{K\}\},$$

and therefore not φ_3'.

4. ***True.*** *The literal* $\neg K$ *occurs only negatively in* φ_4 *and is therefore a pure literal. In pure literal elimination, all clauses containing* $\neg K$ *are removed. Here, this is exactly the clause* $\{\neg K, \neg G\}$. *Hence,*

$$\mathsf{LitElim}(\varphi_4) = \{\{\neg G, \neg H\}, \{G, H\}, \{\neg H\}\} = \varphi_4'.$$

Brief feedback.

- *(1) Length is counted syntactically, not via equivalent transformations.*
- *(2) Tseitin encoding requires the root variable as an additional condition.*
- *(3) Unit propagation additionally produces the clause* $\{K\}$.
- *(4) Precisely the pure literal* $\neg K$ *is eliminated.*

Chapter 2

Predicate Logic

2.1 Introduction and Overview

Predicate logic extends propositional logic by making it possible not only to treat entire statements as indivisible units, but also to make the internal structure of statements explicit. This allows properties of objects and relations between objects to be expressed precisely. Instead of merely stating whether a complete proposition is true or false, one can now also formulate *which* object has which property and *how* different objects are related to one another.

One simple example is the statement:

> "Every machine in the production line that is switched on has at least one active sensor."

Here, the structure of the statement becomes clearly

visible: it concerns objects of the type machine, a property such as "being switched on," and a relation between a machine and a sensor. Such statements can be described only awkwardly, or not adequately at all, by means of propositional logic. Predicate logic remedies this by introducing variables, functions, relations, and quantifiers.

The elementary building blocks of this logic are, first of all, *terms*, which are used to denote objects in the world under consideration or to compose them from other objects. On this basis, one formulates *atomic statements*, for example that an object has a certain property or that several objects stand in a relation to one another. With the quantifiers $\forall$ and $\exists$, one can finally express that a statement holds for all objects or for at least one object. It is precisely this quantification that makes predicate logic significantly more expressive than propositional logic.

With this greater expressive power, however, the theoretical complexity also increases. Many questions that can be handled quite well in propositional logic become fundamentally more difficult in predicate logic, or even undecidable. For this reason, logic often considers not only the formal language itself, but also suitable fragments and restrictions. This is precisely where the special significance of predicate logic becomes apparent: on the one hand, it is expressive enough to model a great many mathematical and computational situations; on the other hand, it is demanding enough to reveal central limitations of automated methods.

As in propositional logic, we again distinguish clearly between syntax and semantics. Syntax determines which formulas may be formed at all. Semantics describes how these formulas are to be interpreted in a given structure. Such structures assign meanings to the symbols of the language and thereby determine when a formula is satisfied. On this basis, one can then investigate models, consequence, equivalence, and satisfiability. Many of these notions will reappear throughout the chapter in an analogous, but considerably richer, form.

This chapter is structured as follows: In Section 2.1, we introduce the syntax and semantics of first-order predicate logic. There we clarify how terms, formulas, and structures are built and how truth in a structure is to be understood. In Section 2.2, we examine the logical equivalence of formulas and consider in what sense predicate logic is independent of the chosen signature. In Section 2.3, we turn to normal forms and show how formulas can be brought into suitable standard forms. We then treat the resolution principle for predicate logic in Section 2.4, thereby introducing an important tool for automated reasoning. In Section 2.5, we study the description logic $\mathcal{ALC}$, which is a particularly important and practically relevant logical formalism. Finally, in Section 2.6, we look at second-order predicate logic and compare its expressive power with that of first-order logic.

The aim of this chapter is to develop the fundamental concepts and methods in such a way that predicate logic appears not merely as an abstract theory, but as a precise tool for modeling, analyzing, and automatically processing complex states of affairs.

2.2 First-Order Predicate Logic: Syntax and Semantics

First-order predicate logic extends propositional logic with a much finer-grained linguistic structure. Whereas propositional logic treats entire statements as indivisible units, predicate logic allows us to speak about *objects*, *properties*, and *relations between objects*. This makes it possible to describe situations from mathematics, computer science, and many other fields far more precisely.

One simple example is the sentence:

> "Every active server has at least one open port."

Here, the issue is not merely whether the statement as a whole is true or false, but which objects in the domain are being referred to and which properties or relations hold among them. Statements of this kind can be captured only very crudely in propositional logic, whereas predicate logic allows them to be formulated directly and systematically.

The language of predicate logic contains several kinds of symbols. Some symbols denote properties or relations, others stand for functions or fixed objects. In addition, there are variables that allow us to speak about objects in general, as well as quantifiers such as "for all" and "there exists." Terms serve to name objects within the language or to build new object-denoting expressions from given ones. Formulas are then constructed from these terms by means of relational expressions and logical connectives.

As in propositional logic, we make a clear distinction here between *syntax* and *semantics*. Syntax describes how well-formed terms and formulas are built. Semantics explains how these formulas are to be interpreted in a concrete structure. Such a structure specifies the domain from which the objects are taken and how the symbols used in the language are to be understood. Only through this interpretation is it determined when a formula is true.

In the remainder of this section, we proceed step by step. In Subsection 2.1.1, we first introduce signatures, that is, the vocabulary of the language. We then explain in Subsection 2.1.2 how terms are formed and how they are to be read. Subsection 2.1.3 deals with the formation of formulas, in particular with the role of quantifiers. Finally, in Subsection 2.1.4, we develop the semantics: we explain what a structure is, how terms are evaluated, and when a formula is satisfied in a structure.

The aim of this section is to provide the foundation for understanding predicate logic not merely as a formal symbolic system, but as a precise instrument for modeling statements about objects and their relationships in a rigorous way. These foundations will be used repeatedly in the sections that follow, both in semantic investigations and in later algorithmic and logical developments.

2.2.1 Signatures

In propositional logic, the basic symbols of the language were still very simple: each atomic statement stood on its own. In predicate logic, the situation is richer. Here we distinguish between symbols that describe relations or properties and symbols that combine objects to form new objects. This linguistic stock of symbols is called a *signature*, or equivalently, a *vocabulary*.

A signature therefore specifies which non-logical symbols are available at all and how many arguments each of these symbols takes. In this sense, a signature is the first building block of any predicate-logical language: it determines what one can talk about.

Definition 2.1 (Signature). *A signature of predicate logic is a triple*

$$\Sigma = \langle \mathrm{Pred}_\Sigma, \mathrm{Fun}_\Sigma, \mathrm{Ary}_\Sigma \rangle,$$

where

- Pred_Σ *is a set of predicate symbols,*
- Fun_Σ *is a set of function symbols, and*
- Ary_Σ *assigns to each symbol* $S \in \mathrm{Pred}_\Sigma \cup \mathrm{Fun}_\Sigma$ *its* arity $\mathrm{Ary}_\Sigma(S) \in \mathbb{N}_0$.

Symbols of arity 0 are called *constants*. They therefore behave like named objects.

Conventions. To improve readability, we usually adopt the following notation:

- predicate symbols are typically written as $P, Q, R, \ldots$ or $P_1, P_2, \ldots$,
- function symbols are typically written as $f, g, h, \ldots$ or $f_1, f_2, \ldots$,
- constants are typically written as $a, b, c, \ldots$.

As before, we use the letter Σ for signatures, as well as $\Lambda, \Sigma_1, \Sigma_2, \ldots$.

Set-theoretic notation. Often, we write a signature more compactly as a set together with arity indications, for example

$$\Sigma = \{P/2,\ f/1,\ c/0\}.$$

This means that Σ contains a binary predicate symbol P, a unary function symbol f, and a constant c. This notation is convenient, but it does not completely replace the triple-based description: without context, one cannot always tell directly from such a set which symbols are predicates and which are functions. For this reason, we adhere to the naming conventions given above.

Example 2.2. *We consider some typical signatures:*

$$\Sigma_{\text{Graph}} = \{E/2,\ a/0,\ b/0\},$$

where E is a binary relation for edges and a, b are two constants. Furthermore, let

$$\Sigma_{\text{Arith}} = \{< /2,\ 0/0,\ s/1,\ +/2\},$$

where $<$ is a binary predicate, 0 is a constant, s is a unary function symbol, and $+$ is a binary function symbol.

Why arity matters. The same symbol can occur meaningfully in different signatures only if it has the same arity and the same type in each of them. For example, a binary symbol $f/2$ is different from a unary symbol $f/1$. Thus, arity is not merely decorative extra information, but an essential part of the symbol's meaning.

Definition 2.3 (Subsignature and Intersection). *Let*

$$\Sigma = \langle \text{Pred}_\Sigma, \text{Fun}_\Sigma, \text{Ary}_\Sigma \rangle \quad \textit{and} \quad \Lambda = \langle \text{Pred}_\Lambda, \text{Fun}_\Lambda, \text{Ary}_\Lambda \rangle$$

be two signatures.

- *We write $\Sigma \subseteq \Lambda$ if*

  $$\text{Pred}_\Sigma \subseteq \text{Pred}_\Lambda, \qquad \text{Fun}_\Sigma \subseteq \text{Fun}_\Lambda,$$

  and for every symbol $S \in \text{Pred}_\Sigma \cup \text{Fun}_\Sigma$, we have:

  $$\text{Ary}_\Sigma(S) = \text{Ary}_\Lambda(S).$$

  A subsignature therefore contains all symbols of the larger signature with unchanged arity.

- *The intersection* $\Sigma \cap \Lambda$ *is the signature that contains exactly those symbols that occur in both signatures and have the same arity there:*

$$\begin{aligned}\mathrm{Pred}_{\Sigma\cap\Lambda} = \{\, P \in \mathrm{Pred}_\Sigma \cap \mathrm{Pred}_\Lambda \mid \\ \mathrm{Ary}_\Sigma(P) = \mathrm{Ary}_\Lambda(P)\,\}.\end{aligned}$$

$$\mathrm{Fun}_{\Sigma\cap\Lambda} = \{f \in \mathrm{Fun}_\Sigma \cap \mathrm{Fun}_\Lambda \mid \mathrm{Ary}_\Sigma(f) = \mathrm{Ary}_\Lambda(f)\}.$$

For all symbols in $\Sigma \cap \Lambda$*, the arity is then uniquely determined.*

Example 2.4. *We consider the signatures*

$$\begin{aligned}\Sigma_1 &= \{E/2,\ a/0,\ s/1\},\\ \Sigma_2 &= \{E/2,\ a/0,\ t/1\},\\ \Sigma_3 &= \{E/2,\ a/0,\ s/2\}.\end{aligned}$$

Then the following hold:

- $\Sigma_1 \subseteq \Sigma_1 \cup \Sigma_2$*, since all symbols occur with the same arity.*

- $\Sigma_1 \not\subseteq \Sigma_3$*, because* s *also occurs there, but with a different arity.*

- $\Sigma_1 \cap \Sigma_2 = \{E/2,\ a/0\}$.

- $\Sigma_1 \cap \Sigma_3 = \{E/2,\ a/0\}$*, since* s *is not included in the intersection because of its different arity.*

Summary. The signature determines the linguistic inventory of predicate logic. It specifies which relations, functions, and constants may be used at all in the theory under consideration. In the following subsections, we will build on this vocabulary and use it to develop terms, formulas, and ultimately the semantics of predicate logic.

2.2.2 Terms and Variables

Before we can construct formulas of predicate logic, we must first specify how we refer to objects in the domain under consideration. For this purpose, we use *terms*. Roughly speaking, a term is an expression that denotes an object: either directly by means of a variable or constant, or indirectly through the application of a function symbol to terms that have already been formed.

The role of variables is especially important here. A variable initially stands for an arbitrary object and is assigned a concrete value only later by an interpretation. Intuitively, variables can therefore be understood as placeholders. Terms, in turn, allow us to combine such placeholders or constants into new designations of objects.

Definition 2.5 (Terms). *Let Σ be a signature, and let V be a countably infinite set of variables. The set of Σ-terms over V, written* $\mathrm{Term}(\Sigma)$, *is defined inductively as follows:*

(T0) Every variable from V is a Σ-term.

(T1) If f is a function symbol in Σ with $\mathrm{Ary}(f) = n$ *and*

if $t_1, \ldots, t_n$ *are already* Σ*-terms, then*

$$f(t_1, \ldots, t_n)$$

is again a Σ*-term.*

Constants are nothing other than function symbols of arity 0. Strictly speaking, every constant could therefore also be written as a term of the form $c()$. For the sake of readability, however, we simply write c and treat constants as simple terms in their own right. Together with variables, we call such terms *simple terms.* All terms that arise through the application of a function symbol are called *compound* or *complex terms.*

Example 2.6. *Let*

$$\Sigma = \{P/1,\ R/2,\ f/1,\ g/2,\ h/3,\ a/0,\ b/0\}.$$

Then, for example, the following expressions are Σ*-terms:*

$$x, \qquad a, \qquad f(x), \qquad h(a, x, b), \qquad g(f(x), a).$$

The expression

$$R(g(x, a), b)$$

is not *a term, because here* R *is a predicate symbol rather than a function symbol. Likewise,*

$$h(x, a)$$

is not a term, because in this signature the function symbol h *expects three arguments, not two.*

Important Note. In propositional logic, we already used the word *term* in a different context, namely for conjunctions of literals in DNF. This is a different, historically established use of the same word. In each context, however, it is clear which meaning is intended.

Why We Keep Track of the Variable Set. Formally, the set of terms depends on the chosen variable set V. In the rest of this text, however, we always work with the same fixed countably infinite set of variables. For that reason, we will usually no longer mention V explicitly.

Remark 2.7. *Terms are the building blocks with which objects can be named and combined in predicate logic. On their basis, atomic formulas are later formed, and ultimately the full formal language emerges.*

2.2.3 Formulas

Terms designate objects. Formulas relate such objects to one another or ascribe properties to them. Only through the combination of atomic statements, logical connectives, and quantifiers does the actual language of predicate logic arise.

Intuitively, a formula can be understood as a logically structured statement about a world under consideration. An atomic component may, for example, state that an object has a certain property or that two objects stand in a certain relation. With the help of $\wedge$, $\vee$, $\neg$, $\forall$, and $\exists$, more complex statements can be constructed from these.

Definition 2.8 (Formulas). *Let Σ be a signature. The set of Σ-formulas, written* $\mathrm{FO}(\Sigma)$, *is defined inductively as follows:*

(FO0) *If P is a predicate symbol from Σ with* $\mathrm{Ary}(P) = n$ *and $t_1, \ldots, t_n$ are Σ-terms, then*

$$P(t_1, \ldots, t_n)$$

is a Σ-formula.

(FO1) *If φ and ψ are Σ-formulas, then the following are also Σ-formulas:*

$$(\varphi \land \psi) \qquad \textit{and} \qquad (\varphi \lor \psi)$$

(FO2) *If φ is a Σ-formula, then*

$$\neg\varphi$$

is also a Σ-formula.

(FO3) *If φ is a Σ-formula and x is a variable, then the following are also Σ-formulas:*

$$\forall x\, \varphi \qquad \textit{and} \qquad \exists x\, \varphi$$

Atomic formulas of the form $P(t_1, \ldots, t_n)$ are also called *atomic statements*. As in propositional logic, their negations are called *literals*. Formulas obtained by applying (FO1) to (FO3) are called *complex formulas*. Formulas of the form $\forall x\, \varphi$ or $\exists x\, \varphi$ are called *quantified formulas.* The symbols $\wedge, \vee, \neg, \forall, \exists$ are the logical building blocks that determine the structure of formulas.

As in propositional logic, we also use abbreviations here:

$$(\varphi \to \psi) := (\neg\varphi \vee \psi), \qquad (\varphi \leftrightarrow \psi) := (\varphi \to \psi) \wedge (\psi \to \varphi).$$

Depending on the context, we also count $\to$ and $\leftrightarrow$ among the connectives.

Example 2.9. *We consider the signature*

$$\Sigma = \{\,\text{Arbeitsstation}/1,\ \text{verbunden}/2,\ \text{Server}/0,\ \text{steht\_bei}/2,\ \text{root}/0\,\}.$$

Then, for example, the following expressions are Σ-formulas:

- $\forall x\,(\text{Arbeitsstation}(x) \to \exists y\,\text{verbunden}(x, y))$
- Server
- steht_bei(root, root)

Intuitively, the first formula could be read as follows: "Every workstation is connected to at least one other object."

The following, by contrast, are not Σ-formulas:

- $\exists x\,\text{verbunden}(x)$, *because* verbunden *is binary here,*
- $\forall x\,\text{Arbeitsstation}(x) \wedge \text{root}$, *if the parenthesization is missing and the structure is therefore not unambiguous,*

- verbunden(root), *because the predicate receives too few arguments.*

Parentheses and Readability. The formal definition requires complex expressions to be parenthesized unambiguously. In practice, however, this would quickly make formulas hard to read. We therefore adopt precedence conventions that allow formulas to be written more clearly.

Notation 2.10. *We use the following precedence conventions:*

- $\forall$, $\exists$, *and* $\neg$ *bind most strongly.*
- $\wedge$ *binds more strongly than* $\vee$.
- $\rightarrow$ *binds more weakly than* $\vee$.
- $\leftrightarrow$ *binds most weakly.*
- *We omit outer parentheses whenever this does not change the meaning.*
- *Parentheses may be omitted whenever the formula remains unambiguous under these precedence conventions.*
- *If several quantifiers occur in sequence, we may write the dot only once or omit it entirely, for example:*

$$\forall x \, \forall y \, \varphi \quad \textit{statt} \quad \forall x . \forall y . \, \varphi.$$

Accordingly, we can read the string

$$\forall x\, \text{Arbeitsstation}(x) \vee \neg \text{verbunden}(x, \text{root}) \wedge \text{Server}$$

unambiguously as

$$(\forall x\, \text{Arbeitsstation}(x)) \vee (\neg \text{verbunden}(x, \text{root}) \wedge \text{Server})$$

.

Free and Bound Variables. Within a formula, variables may either be bound by a quantifier or occur freely. This distinction is important: bound variables matter only within the scope of their quantifier, whereas free variables stand for objects that have not yet been specified.

Example 2.11. *In the formula*

$$\forall x\, (\text{Arbeitsstation}(x) \to \text{verbunden}(x, y))$$

x *is bound, whereas* y *occurs freely. By contrast, in*

$$\text{verbunden}(u, v) \to \forall u\, \text{Server}$$

the first occurrence of u *is free and the second is bound.*

Definition 2.12 (Free and Bound Variables, Subformulas). *For a term* t*, let* $\text{Var}(t)$ *denote the set of variables occurring in* t*. For a formula* φ*, we define the set of free variables* $\text{FV}(\varphi)$ *and the set of bound variables* $\text{BV}(\varphi)$ *recursively in the usual way:*

$$\text{FV}(P(t_1, \ldots, t_n)) = \text{Var}(t_1) \cup \cdots \cup \text{Var}(t_n),$$

$$\mathrm{FV}(\varphi \wedge \psi) = \mathrm{FV}(\varphi) \cup \mathrm{FV}(\psi), \qquad \mathrm{FV}(\neg\varphi) = \mathrm{FV}(\varphi),$$

$$\mathrm{FV}(\forall x\, \varphi) = \mathrm{FV}(\varphi) \setminus \{x\}, \qquad \mathrm{FV}(\exists x\, \varphi) = \mathrm{FV}(\varphi) \setminus \{x\},$$

and analogously for $\mathrm{BV}(\varphi)$. *We denote the set of subformulas of* φ *by* $\mathrm{TF}(\varphi)$.

Definition 2.13 (Sentences). *A formula* $\varphi \in \mathrm{FO}(\Sigma)$ *is called* closed *or a* sentence *if it has no free variables, that is, if*

$$\mathrm{FV}(\varphi) = \varnothing.$$

We denote the set of all Σ*-sentences by* $\mathrm{FOSatz}(\Sigma)$.

Signature of a Formula. Not every formula uses all symbols of a signature. We therefore write $\mathrm{Sig}(\varphi)$ for the set of all predicate and function symbols that actually occur in φ. A formula φ is compatible with a signature Σ if

$$\mathrm{Sig}(\varphi) \subseteq \Sigma.$$

This will become important later when we relate formulas and structures to one another.

Remark 2.14. *The concept of free variables is key to understanding first-order formulas: it shows which parts of a formula are already "fixed" and which still depend on a later interpretation. This is precisely why, in the next subsection, we distinguish carefully between formulas with free variables and closed sentences.*

2.2.4 Semantics

Syntax determines *how* first-order expressions are formed. Semantics answers the actual substantive question: *What do* these expressions mean in a concrete world? To do so, we need mathematical structures in which the symbols of a signature are assigned meanings. Only then can we meaningfully ask whether a formula is true.

Intuitively, a structure can be viewed as a model of a certain situation. In a database, for example, the elements of the domain may represent records or objects, predicates may represent tables, and function symbols may represent computations or assignments. In a graph-theoretic structure, the elements of the domain represent vertices, and a binary predicate represents the edge relation.

Definition 2.15 (Structure). *Let Σ be a signature. A Σ-structure is a tuple*

$$\mathcal{A} = \langle A, \{P^{\mathcal{A}}\}_{P \in \mathrm{Pred}_\Sigma}, \{f^{\mathcal{A}}\}_{f \in \mathrm{Fun}_\Sigma} \rangle,$$

such that:

- *$A \neq \varnothing$ is a nonempty set. It is called the* domain, underlying set, *or* universe *of the structure.*
- *For every predicate symbol $P \in \Sigma$ with* $\mathrm{Ary}(P) = n$, *$P^{\mathcal{A}} \subseteq A^n$ is an n-ary relation.*
- *For every function symbol $f \in \Sigma$ with* $\mathrm{Ary}(f) = n$, *$f^{\mathcal{A}} : A^n \to A$ is an n-ary function.*

We denote the set of all Σ-structures by $\mathrm{Struct}(\Sigma)$.

Constants are simply function symbols of arity 0. They are therefore interpreted as distinguished fixed elements of the domain.

Example 2.16. *Consider a signature for a small campus model:*

$$\Sigma_{\mathrm{Campus}} = \{\mathrm{Room}/1,\ \mathrm{connected}/2,\ \mathrm{cafe}/0,\ \mathrm{floor}/1\}.$$

Here, $\mathrm{Room}(x)$ *expresses that x is a room,* $\mathrm{connected}(x, y)$ *that two rooms are directly connected,* cafe *is a distinguished room, and* $\mathrm{floor}(x)$ *describes the floor level of a room.*

Assignments. The structure determines the meaning of the non-logical symbols. For variables, we additionally need an assignment that maps each variable symbol to an element of the domain.

Definition 2.17 (Assignment). *Let Σ be a signature and $\mathcal{A}$ be a Σ-structure. An assignment for $\mathcal{A}$ is a function*

$$\beta : V \to A,$$

that assigns to each variable an element of the domain.

Since only finitely many variables occur in a formula, in practice we often specify assignments only partially. For a variable x and an element $a \in A$, the notation

$$\beta[x \mapsto a]$$

denotes the assignment that agrees with β everywhere except that it maps x to a. For several variables, we write accordingly

$$\beta[x_1 \mapsto a_1, \ldots, x_n \mapsto a_n].$$

Definition 2.18 (Interpretation). *A Σ-interpretation is a tuple*

$$I = \langle \mathcal{A}, \beta \rangle,$$

where $\mathcal{A}$ is a Σ-structure and β is an assignment for $\mathcal{A}$.

Evaluation of Terms. Terms denote elements of the domain. Their meaning depends on the respective interpretation.

Definition 2.19 (Evaluation of Terms). *Let $I = \langle \mathcal{A}, \beta \rangle$ be a Σ-interpretation. Then the evaluation $[\![\cdot]\!]_I$ on terms is given by*

$$\begin{aligned} [\![x]\!]_I &= \beta(x) && \textit{for a variable } x, \\ [\![c]\!]_I &= c^{\mathcal{A}} && \textit{for a constant } c, \\ [\![f(t_1, \ldots, t_n)]\!]_I &= f^{\mathcal{A}}([\![t_1]\!]_I, \ldots, [\![t_n]\!]_I) \end{aligned}$$

for a compound term $f(t_1, \ldots, t_n)$.

Example 2.20. *Let $\mathcal{A}$ be a structure with domain*

$$A = \{\text{Office}, \text{Laboratory}, \text{Archive}, \text{Kitchen}\},$$

and let m be a constant with $m^{\mathcal{A}} = \text{Office}$. Furthermore, let $u/1$ be a function symbol with the meaning "the floor above." If $\beta(x) = \text{Laboratory}$, then we have

$$[\![m]\!]_I = \text{Office} \quad \textit{and} \quad [\![u(x)]\!]_I = u^{\mathcal{A}}(\text{Laboratory}).$$

Evaluation of Formulas. An atomic formula is true exactly when the tuple formed from its terms belongs to the corresponding relation. The logical connectives and quantifiers are then evaluated according to their meaning.

Definition 2.21 (Evaluation of Formulas). *Let $I = \langle \mathcal{A}, \beta \rangle$ be a Σ-interpretation. Then every Σ-formula φ is assigned a truth value from* $\{\text{true}, \text{false}\}$*:*

$$\llbracket P(t_1, \ldots, t_n) \rrbracket_I = \text{true} \iff (\llbracket t_1 \rrbracket_I, \ldots, \llbracket t_n \rrbracket_I) \in P^{\mathcal{A}},$$

$$\begin{aligned}
\llbracket \varphi \wedge \psi \rrbracket_I = \text{true} &\iff \llbracket \varphi \rrbracket_I = \text{true } \textit{and } \llbracket \psi \rrbracket_I = \text{true}, \\
\llbracket \varphi \vee \psi \rrbracket_I = \text{true} &\iff \llbracket \varphi \rrbracket_I = \text{true } \textit{or } \llbracket \psi \rrbracket_I = \text{true}, \\
\llbracket \neg \varphi \rrbracket_I = \text{true} &\iff \llbracket \varphi \rrbracket_I = \text{false}, \\
\llbracket \exists x\, \varphi \rrbracket_I = \text{true} &\iff \exists a \in A: \ \llbracket \varphi \rrbracket_{\langle \mathcal{A}, \beta[x \mapsto a] \rangle} = \text{true}, \\
\llbracket \forall x\, \varphi \rrbracket_I = \text{true} &\iff \forall a \in A: \ \llbracket \varphi \rrbracket_{\langle \mathcal{A}, \beta[x \mapsto a] \rangle} = \text{true}.
\end{aligned}$$

For the abbreviations $\rightarrow$ and $\leftrightarrow$, we correspondingly have:

$$\llbracket \varphi \rightarrow \psi \rrbracket_I = \text{true} \iff \llbracket \neg \varphi \vee \psi \rrbracket_I = \text{true},$$

$$\begin{aligned}
\llbracket \varphi \leftrightarrow \psi \rrbracket_I = \text{true}& \\
&\iff \llbracket \varphi \rightarrow \psi \rrbracket_I = \text{true} \\
&\quad\ \textit{and } \llbracket \psi \rightarrow \varphi \rrbracket_I = \text{true}.
\end{aligned}$$

Example 2.22. *We consider a structure for a small library scenario:*

$$A = \{\text{ReadingRoom}, \text{Shelf1}, \text{Shelf2}, \text{Entrance}\}.$$

Let z be a constant with $z^{\mathcal{A}} = \text{Entrance}$, let $B/1$ be the predicate "is an area containing books," and let $N/2$ be the relation "is adjacent to." We define

$$B^{\mathcal{A}} = \{\text{Shelf1}, \text{Shelf2}\},$$

$$N^{\mathcal{A}} = \{(\text{Entrance}, \text{ReadingRoom}), \\ (\text{ReadingRoom}, \text{Shelf1})\}.$$

Then the formula

$$\exists x\, (B(x) \land N(z, x))$$

is not true, because for $x = \text{ReadingRoom}$, we have $N(z, x)$, but $\text{ReadingRoom} \notin B^{\mathcal{A}}$. By contrast,

$$\forall x\, (B(x) \to N(z, x))$$

is false, because for $x = \text{Shelf1}$, $B(x)$ holds, but $N(z, x)$ does not.

The Notion of a Model. An interpretation is called a model of a formula if the formula is true in that interpretation. This allows the notions familiar from propositional logic to be transferred almost unchanged.

Definition 2.23 (Model, Satisfiability, and Validity). *Let Σ be a signature and φ a Σ-formula.*

- *A Σ-interpretation I is called a* model *of φ if $[\![\varphi]\!]_I = \text{true}$.*

- *φ is called* Σ-satisfiable *if φ has at least one Σ-model.*
- *φ is called* Σ-unsatisfiable *if φ has no Σ-model.*
- *φ is called* Σ-valid *if every Σ-interpretation is a model of φ.*

If a formula is satisfiable, we also call it consistent. *If it is unsatisfiable, we also call it* contradictory *or* inconsistent. *A valid formula is also called a* tautology.

Remark 2.24. *Thus, in predicate logic too, a clear distinction is made between syntax and semantics: syntax tells us how formulas are formed, whereas semantics tells us when they are true in a structure. The difference from propositional logic is that meaning is determined not only by truth values assigned to atomic formulas, but by entire structures, relations, functions, and assignments.*

2.2.5 Exercises and Solutions

Exercise 1. Consider the predicate-logical signature

$$\Sigma = \{P/1,\ i/1,\ e/0,\ b/0\}$$

and the formula

$$\varphi_4 = \neg(\exists x\, P(x) \lor P(b)).$$

For each of the following statements, decide whether it is true or false.

1. $\Sigma_1 = \{P/1, i/1\}$ is a sub-signature of Σ.
2. e is a Σ-term.
3. $P(y)$ is an atomic formula.
4. x occurs free in φ_4.

Solutions to Exercise 1.

1. **True.** All symbols in Σ_1 also occur in Σ, with the same arity.
2. **True.** e is a constant symbol, that is, a function symbol of arity 0; therefore e is a term.
3. **True.** Since P is a unary predicate symbol and y is a term, $P(y)$ is an atomic formula.
4. **False.** The occurrence of x lies within the scope of $\exists x$ and is therefore bound. No variables occur free in φ_4; hence $\mathrm{FV}(\varphi_4) = \varnothing$.

Exercise 2. Consider the predicate-logical signature

$$\Sigma = \{P/2,\ R/2,\ i/2,\ h/2,\ f/2,\ b/0\}$$

and the formula

$$\varphi_4 = \neg\forall z.(\neg\neg\exists v.\,(R(v, y) \wedge P(z, u))).$$

Decide for each of the following statements whether it is true or false.

1. We have $\text{Sig}(\varphi_1) = \Sigma_1$ for $\varphi_1 = \neg R(v, b)$ and
$$\Sigma_1 = \{P/2,\ R/2,\ i/2,\ b/0\}.$$
2. $\exists w\,(\neg P(v, w))$ is a Σ-formula.
3. $\exists z\, R(x, i(b, z))$ is a quantified formula.
4. z is free in φ_4.

Solutions to Exercise 2.

1. **False.** In $\varphi_1 = \neg R(v, b)$, only the predicate symbol $R/2$ and the constant $b/0$ occur. Hence the signature of the formula is
$$\text{Sig}(\varphi_1) = \{R/2,\ b/0\},$$
not Σ_1.
2. **True.** The expression is a valid Σ-formula. It is well-formed, and all non-logical symbols occurring in it, namely the predicate symbol P, belong to Σ. The variables v and w are used in the usual syntactic way as free and bound variables, respectively, which does not affect well-formedness.
3. **True.** The formula begins with an existential quantifier and is therefore a quantified formula.
4. **False.** The occurrence of z lies within the scope of the outer quantifier $\forall z$ and is therefore bound. The free variables in φ_4 are u and y, that is,
$$\text{FV}(\varphi_4) = \{u, y\}.$$

Brief feedback.

- A subsignature must contain all symbols together with their given arities.
- An expression is a Σ-formula exactly if it is well-formed and uses only symbols from Σ.
- An atomic expression has the form $P(t_1, \ldots, t_n)$.
- A variable is free exactly when it is not bound by a quantifier.
- For the signature of a formula, only the symbols that actually occur are counted.

2.3 Coincidence, Equivalence, and Substitution

In this section, we study several basic properties of predicate logic that are central for the later treatment of normal forms, substitutions, and model classes. One important guiding idea is the following: for the semantic evaluation of a formula, only those symbols are relevant that actually occur in it. Likewise, if a formula contains free variables, its truth depends only on the assignment of those variables; all other variables are irrelevant.

These observations are by no means merely intuitively plausible; they can be formulated precisely and proved rigorously. This is exactly what the coincidence lemma does. Building on it, we then consider logical equivalence, substitutions, and finally model classes. The following subsection first provides the technical foundation for these further steps.

2.3.1 Coincidence Lemma

Intuitively, the following is clear: if two interpretations agree on all symbols that are actually used in a formula, then they cannot evaluate that formula differently. The same holds for variable assignments: only the free variables of a formula are semantically relevant; bound variables are in any case "covered" by the quantifiers.

Example 2.25. *We consider the formula*

$$\varphi = \exists u \, \forall v \, R(u, v)$$

and the atomic formula

$$\psi = R(u, v).$$

Both formulas contain only the binary predicate symbol R and the variables u and v, while φ has no free variables.

Now let

$$\begin{aligned} A_1 &= \{1, 2, 3, 4\}, \\ R^{\mathcal{A}_1} &= \{(1,1), (1,2), (1,3), \\ &\qquad (1,4), (3,1)\}. \end{aligned}$$

and let β_1 be an assignment with $\beta_1(u) = 2$ and $\beta_1(v) = 4$. Then $\langle \mathcal{A}_1, \beta_1 \rangle$ is not a model of ψ, since $(2,4) \notin R^{\mathcal{A}_1}$. By *contrast, the same interpretation is a model of φ, because for u one may choose, for instance, the value 1, and then for every $v \in A_1$ we have $R(1, v)$.*

Now consider a second interpretation

$$A_2 = \{1, 2, 3, 4\}, \qquad R^{\mathcal{A}_2} = R^{\mathcal{A}_1}, \qquad S^{\mathcal{A}_2} = \{3, 4\},$$

with an assignment $\beta_2(u) = 3$, $\beta_2(v) = 1$. The additional symbol S plays no role for φ. Here again, φ is true, although the variable assignment has been chosen differently.

The following theorem summarizes this observation in general form.

Theorem 2.26 (Coincidence Lemma). *Let Σ and Λ be two signatures. Let $I_1 = \langle \mathcal{A}_1, \beta_1 \rangle$ be a Σ-interpretation and $I_2 = \langle \mathcal{A}_2, \beta_2 \rangle$ be a Λ-interpretation. Assume that both interpretations have the same domain and agree on all symbols in $\Sigma \cap \Lambda$. Then the following holds:*

1. *If t is a $\Sigma \cap \Lambda$-term and $\beta_1(x) = \beta_2(x)$ for all $x \in \mathrm{Var}(t)$, then*

$$[\![t]\!]_{I_1} = [\![t]\!]_{I_2}.$$

2. *If φ is a $\Sigma \cap \Lambda$-formula and $\beta_1(x) = \beta_2(x)$ for all $x \in \mathrm{FV}(\varphi)$, then*

$$I_1 \models^{\mathrm{FO}}_{\Sigma} \varphi \quad \textit{if and only if} \quad I_2 \models^{\mathrm{FO}}_{\Lambda} \varphi.$$

Proof. We prove both statements simultaneously by structural induction.

For (1): We proceed by induction on the structure of the term t.

- If $t = x$ is a variable, then by assumption we have
$$[\![t]\!]_{I_1} = \beta_1(x) = \beta_2(x) = [\![t]\!]_{I_2}.$$
- If $t = c$ is a constant, then c is interpreted in the same way in both interpretations. Hence,
$$[\![t]\!]_{I_1} = c^{\mathcal{A}_1} = c^{\mathcal{A}_2} = [\![t]\!]_{I_2}.$$
- If $t = f(t_1, \ldots, t_n)$, then the induction hypothesis yields
$$[\![t_i]\!]_{I_1} = [\![t_i]\!]_{I_2} \qquad \text{for all } i = 1, \ldots, n.$$
Since f is interpreted identically in both interpretations, it follows that
$$[\![f(t_1, \ldots, t_n)]\!]_{I_1} = [\![f(t_1, \ldots, t_n)]\!]_{I_2}.$$

This proves statement (1).

For (2): We proceed by induction on the structure of the formula φ.

- If $\varphi = P(t_1, \ldots, t_n)$ is atomic, then (1) shows that the evaluation of the terms coincides in both interpretations. Since, moreover, P is interpreted identically in both structures, it follows that
$$I_1 \models^{\text{FO}}_{\Sigma} \varphi \quad \text{if and only if} \quad I_2 \models^{\text{FO}}_{\Lambda} \varphi.$$

- If $\varphi = \psi_1 \wedge \psi_2$ or $\varphi = \psi_1 \vee \psi_2$, then the claim follows immediately from the induction hypothesis and the definition of the connectives.

- If $\varphi = \neg\psi$, then the statement follows directly from the induction hypothesis and the semantics of negation.

- If $\varphi = \forall x\, \psi$, then the truth of φ depends only on whether ψ is true under all possible assignments to x. Since the two interpretations agree on all free variables relevant to ψ, and since the domain is the same, the induction hypothesis yields the same truth value for every such assignment. Hence the claimed equivalence holds.

- The case $\varphi = \exists x\, \psi$ is analogous.

This also proves statement (2). □

Remark 2.27. *In words, the coincidence lemma says that for the semantic evaluation of a formula, only the symbols occurring in it and its free variables matter. Everything else may be changed arbitrarily without affecting the truth value of the formula.*

2.3.2 Equivalence

Not every formula expresses genuinely new content. Often, two formulas differ only in their outward form while describing exactly the same models. We do not want to treat such formulas as different, but rather as *logically equivalent.* This is particularly important because later we will systematically transform formulas by means of equivalence rules.

Definition 2.28 (Logical Equivalence). *Let Σ be a signature. Two Σ-formulas φ and ψ are called* logically equivalent, *written*

$$\varphi \equiv_\Sigma \psi,$$

if for every Σ-interpretation I we have:

$$[\![\varphi]\!]_I = [\![\psi]\!]_I.$$

Example 2.29. *The two formulas*

$$\forall x \, \mathrm{Raum}(x) \qquad \textit{and} \qquad \forall x \, (\mathrm{Raum}(x) \wedge \mathrm{Raum}(x))$$

are logically equivalent. Semantically, they express exactly the same thing: every object in the domain is a room. The second formula merely repeats this requirement unnecessarily.

Locality of equivalence. For the question whether two formulas are equivalent, it is irrelevant under which sufficiently large signature we consider them. What matters is only how they are evaluated with respect to the symbols relevant to them. We call this observation the *locality* of logical equivalence.

Proposition 2.30 (Locality). *Let Σ and Λ be two signatures, and let φ and ψ be two $\Sigma \cap \Lambda$-formulas. Then*

$$\varphi \equiv_\Sigma \psi \quad \Longleftrightarrow \quad \varphi \equiv_\Lambda \psi.$$

Proof. We first show the implication

$$\varphi \equiv_\Sigma \psi \implies \varphi \equiv_\Lambda \psi.$$

The converse direction then follows by exchanging the roles of Σ and Λ.

So assume that $\varphi \equiv_\Sigma \psi$ holds, and let $J = \langle B, \beta' \rangle$ be an arbitrary Λ-interpretation. We now consider a Σ-interpretation $I = \langle A, \beta \rangle$ with the same domain as J, such that both interpretations agree on all symbols from $\Sigma \cap \Lambda$. For the symbols that occur only in one of the two signatures, we may choose arbitrary interpretations.

Since φ and ψ use only symbols from $\Sigma \cap \Lambda$, the Coincidence Lemma immediately yields:

$$I \models_\Sigma \varphi \iff J \models_\Lambda \varphi,$$

and likewise

$$I \models_\Sigma \psi \iff J \models_\Lambda \psi.$$

Because $\varphi \equiv_\Sigma \psi$ holds by assumption, it follows that

$$I \models_\Sigma \varphi \iff I \models_\Sigma \psi.$$

Combining these, we obtain

$$J \models_\Lambda \varphi \iff J \models_\Lambda \psi.$$

Since J was arbitrary, it follows that

$$\varphi \equiv_\Lambda \psi.$$

The converse direction is proved in exactly the same way, simply with the two signatures exchanged. This completes the proof. □

Remark 2.31. *The statement means that as soon as both formulas use the same actually occurring symbols, their equivalence does not depend on whether we consider them in a smaller or in a larger signature.*

Classical Transformations. From propositional logic, we already know many equivalences, and these remain valid in predicate logic as well. They can be used like familiar calculation rules when transforming formulas.

Proposition 2.32 (Basic Equivalences). *Let φ, ψ, χ be predicate-logical formulas. Then the following hold:*

Idempotence:

$$\begin{aligned} \varphi \wedge \varphi &\equiv \varphi, \\ \varphi \vee \varphi &\equiv \varphi. \end{aligned}$$

Commutativity:

$$\begin{aligned} \varphi \wedge \psi &\equiv \psi \wedge \varphi, \\ \varphi \vee \psi &\equiv \psi \vee \varphi. \end{aligned}$$

Associativity:

$$\begin{aligned} (\varphi \wedge \psi) \wedge \chi &\equiv \varphi \wedge (\psi \wedge \chi), \\ (\varphi \vee \psi) \vee \chi &\equiv \varphi \vee (\psi \vee \chi). \end{aligned}$$

De Morgan's Laws:

$$\neg(\varphi \wedge \psi) \equiv \neg\varphi \vee \neg\psi,$$

$$\neg(\varphi \lor \psi) \equiv \neg\varphi \land \neg\psi.$$

Distributivity:

$$\varphi \lor (\psi \land \chi) \equiv (\varphi \lor \psi) \land (\varphi \lor \chi),$$
$$\varphi \land (\psi \lor \chi) \equiv (\varphi \land \psi) \lor (\varphi \land \chi).$$

Double-Negation Elimination:

$$\neg\neg\varphi \equiv \varphi.$$

Proof. We prove the claim by showing, for each of the listed formulas, that under an arbitrary interpretation it has the same truth value as the expression on the right-hand side of the stated equivalence. Let I be an arbitrary interpretation.

Idempotence. Since the connectives are evaluated semantically in a truth-functional way, we have

$$[\![\varphi \land \varphi]\!]_I = [\![\varphi]\!]_I \land [\![\varphi]\!]_I = [\![\varphi]\!]_I.$$

Hence $\varphi \land \varphi \equiv \varphi$. Analogously, we obtain

$$[\![\varphi \lor \varphi]\!]_I = [\![\varphi]\!]_I \lor [\![\varphi]\!]_I = [\![\varphi]\!]_I,$$

thus $\varphi \lor \varphi \equiv \varphi$.

Commutativity. We have

$$[\![\varphi \land \psi]\!]_I = [\![\varphi]\!]_I \land [\![\psi]\!]_I = [\![\psi]\!]_I \land [\![\varphi]\!]_I = [\![\psi \land \varphi]\!]_I.$$

This implies $\varphi \land \psi \equiv \psi \land \varphi$. The case for $\lor$ is analogous:

$$[\![\varphi \lor \psi]\!]_I = [\![\varphi]\!]_I \lor [\![\psi]\!]_I = [\![\psi]\!]_I \lor [\![\varphi]\!]_I = [\![\psi \lor \varphi]\!]_I.$$

Associativity. We obtain

$$\begin{aligned}[\![(\varphi \wedge \psi) \wedge \chi]\!]_I &= ([\![\varphi]\!]_I \wedge [\![\psi]\!]_I) \wedge [\![\chi]\!]_I \\ &= [\![\varphi]\!]_I \wedge ([\![\psi]\!]_I \wedge [\![\chi]\!]_I) \\ &= [\![\varphi \wedge (\psi \wedge \chi)]\!]_I.\end{aligned}$$

Hence $(\varphi \wedge \psi) \wedge \chi \equiv \varphi \wedge (\psi \wedge \chi)$. The case for $\vee$ follows in the same way.

De Morgan's Laws. We have

$$[\![\neg(\varphi\wedge\psi)]\!]_I = \neg([\![\varphi]\!]_I\wedge[\![\psi]\!]_I) = \neg[\![\varphi]\!]_I\vee\neg[\![\psi]\!]_I = [\![\neg\varphi\vee\neg\psi]\!]_I.$$

This yields $\neg(\varphi \wedge \psi) \equiv \neg\varphi \vee \neg\psi$. Analogously, we obtain

$$[\![\neg(\varphi\vee\psi)]\!]_I = \neg([\![\varphi]\!]_I\vee[\![\psi]\!]_I) = \neg[\![\varphi]\!]_I\wedge\neg[\![\psi]\!]_I = [\![\neg\varphi\wedge\neg\psi]\!]_I.$$

Distributivity. Here we use the classical distributivity of the Boolean operations:

$$\begin{aligned}[\![\varphi \vee (\psi \wedge \chi)]\!]_I &= [\![\varphi]\!]_I \vee ([\![\psi]\!]_I \wedge [\![\chi]\!]_I) \\ &= ([\![\varphi]\!]_I \vee [\![\psi]\!]_I) \wedge ([\![\varphi]\!]_I \vee [\![\chi]\!]_I) \\ &= [\![(\varphi \vee \psi) \wedge (\varphi \vee \chi)]\!]_I.\end{aligned}$$

Hence $\varphi \vee (\psi \wedge \chi) \equiv (\varphi \vee \psi) \wedge (\varphi \vee \chi)$. Analogously,

$$[\![\varphi \wedge (\psi \vee \chi)]\!]_I = [\![(\varphi \wedge \psi) \vee (\varphi \wedge \chi)]\!]_I.$$

Double Negation. Finally,

$$[\![\neg\neg\varphi]\!]_I = \neg(\neg[\![\varphi]\!]_I) = [\![\varphi]\!]_I.$$

Thus $\neg\neg\varphi \equiv \varphi$.

Since the truth values coincide under every arbitrary interpretation I, the stated logical equivalences follow. □

Example 2.33. *Using these rules, we can, for example, transform the formula*

$$\neg(\text{Raum}(x) \wedge \text{verbunden}(x, y))$$

into

$$\neg\text{Raum}(x) \vee \neg\text{verbunden}(x, y).$$

Such transformations will become especially important later in connection with normal forms and resolution.

Tautologies and Contradictions. The familiar special cases from propositional logic carry over directly.

Proposition 2.34 (Tautological and Contradictory Formulas). *Let φ be a predicate-logical formula, $\bot$ a contradictory formula, and $\top$ a valid formula. Then the following hold:*

$$\begin{aligned} \varphi \vee \neg\varphi &\equiv \top, & \varphi \wedge \neg\varphi &\equiv \bot, \\ \varphi \wedge \bot &\equiv \bot, & \varphi \vee \top &\equiv \top, \\ \varphi \wedge \top &\equiv \varphi, & \varphi \vee \bot &\equiv \varphi. \end{aligned}$$

Proof. We prove each statement by comparing the evaluations on both sides for an arbitrary interpretation I.

We write $[\![\psi]\!]_I$ for the truth value of the formula ψ under I.

So let I be arbitrary. By assumption, we have

$$[\![\top]\!]_I = \text{true} \qquad \text{and} \qquad [\![\bot]\!]_I = \text{false}.$$

(1) Tautology. We have

$$[\![\varphi \vee \neg\varphi]\!]_I = [\![\varphi]\!]_I \vee \neg[\![\varphi]\!]_I.$$

Since for every Boolean truth value a, the equation $a \vee \neg a = \text{true}$ holds, it follows that

$$[\![\varphi \vee \neg\varphi]\!]_I = \text{true} = [\![\top]\!]_I.$$

Hence $\varphi \vee \neg\varphi \equiv \top$.

(2) Elementary Contradiction. Analogously, we obtain

$$[\![\varphi \wedge \neg\varphi]\!]_I = [\![\varphi]\!]_I \wedge \neg[\![\varphi]\!]_I.$$

For every Boolean truth value a, we have $a \wedge \neg a = \text{false}$. Therefore,

$$[\![\varphi \wedge \neg\varphi]\!]_I = \text{false} = [\![\bot]\!]_I.$$

Hence $\varphi \wedge \neg\varphi \equiv \bot$.

(3) Absorbing Elements. First:

$$[\![\varphi \wedge \bot]\!]_I = [\![\varphi]\!]_I \wedge [\![\bot]\!]_I = [\![\varphi]\!]_I \wedge \text{false} = \text{false} = [\![\bot]\!]_I.$$

Thus $\varphi \wedge \bot \equiv \bot$.

Next:

$$[\![\varphi \vee \top]\!]_I = [\![\varphi]\!]_I \vee [\![\top]\!]_I = [\![\varphi]\!]_I \vee \text{true} = \text{true} = [\![\top]\!]_I.$$

Thus $\varphi \vee \top \equiv \top$.

(4) Identity Elements. We have

$$[\![\varphi \wedge \top]\!]_I = [\![\varphi]\!]_I \wedge [\![\top]\!]_I = [\![\varphi]\!]_I \wedge \text{true} = [\![\varphi]\!]_I.$$

This implies $\varphi \wedge \top \equiv \varphi$.

Finally:

$$[\![\varphi \vee \bot]\!]_I = [\![\varphi]\!]_I \vee [\![\bot]\!]_I = [\![\varphi]\!]_I \vee \text{false} = [\![\varphi]\!]_I.$$

Hence $\varphi \vee \bot \equiv \varphi$.

Since the evaluations agree for every arbitrary interpretation I, the claimed logical equivalences follow. □

Equivalences Involving Quantifiers. With quantifiers, one has to look a bit more carefully. Some transformations are always admissible, while others are admissible only when a variable does not occur free in the relevant subformula.

Proposition 2.35 (Equivalences Involving Quantifiers). *Let φ and ψ be predicate-logical formulas. Then the following hold:*

$$\varphi \equiv \forall x\, \varphi \qquad \text{if } x \notin \mathrm{FV}(\varphi), \tag{1}$$

$$\varphi \equiv \exists x\, \varphi \qquad \text{if } x \notin \mathrm{FV}(\varphi), \tag{2}$$

$$\forall x\, \varphi \equiv \neg \exists x\, \neg \varphi \tag{3}$$

$$\exists x\, \varphi \equiv \neg\forall x\, \neg\varphi \tag{4}$$

$$\forall x\, \forall y\, \varphi \equiv \forall y\, \forall x\, \varphi \tag{5}$$

$$\exists x\, \exists y\, \varphi \equiv \exists y\, \exists x\, \varphi \tag{6}$$

$$(\forall x\, \varphi) \wedge (\forall x\, \psi) \equiv \forall x\, (\varphi \wedge \psi) \tag{7}$$

$$(\exists x\, \varphi) \vee (\exists x\, \psi) \equiv \exists x\, (\varphi \vee \psi) \tag{8}$$

Proof. We prove the eight equivalences one by one. Throughout, let $I = \langle A, \beta \rangle$ be an arbitrary interpretation.

(1) and (2): First assume that $x \notin \mathrm{FV}(\varphi)$. Then the truth value of φ does not depend on the assignment of the variable x. Hence, for every $a \in A$,

$$[\![\varphi]\!]_{\langle A, \beta[x \mapsto a] \rangle} = [\![\varphi]\!]_I.$$

Therefore,

$$\begin{aligned} [\![\forall x\, \varphi]\!]_I &= \text{true} \\ &\iff \forall a \in A : \ [\![\varphi]\!]_{\langle A, \beta[x \mapsto a] \rangle} = \text{true} \\ &\iff [\![\varphi]\!]_I = \text{true}. \end{aligned}$$

Thus $\varphi \equiv \forall x\, \varphi$. Exactly the same argument also yields

$$\begin{aligned} [\![\exists x\, \varphi]\!]_I &= \text{true} \\ &\iff \exists a \in A : \ [\![\varphi]\!]_{\langle A, \beta[x \mapsto a] \rangle} = \text{true} \\ &\iff [\![\varphi]\!]_I = \text{true}. \end{aligned}$$

hence $\varphi \equiv \exists x\, \varphi$.

(3) and (4): For universal quantification, we have

$$[\![\forall x\, \varphi]\!]_I = \text{true} \iff \forall a \in A : \ [\![\varphi]\!]_{\langle A, \beta[x \mapsto a] \rangle} = \text{true}.$$

This is the case exactly when there is no $a \in A$ such that

$$[\![\varphi]\!]_{\langle A,\beta[x\mapsto a]\rangle} = \text{false}.$$

By the semantics of negation, this is equivalent to

$$\neg\exists a \in A : \; [\![\neg\varphi]\!]_{\langle A,\beta[x\mapsto a]\rangle} = \text{true},$$

that is,

$$[\![\neg\exists x\, \neg\varphi]\!]_I = \text{true}.$$

Therefore,

$$\forall x\, \varphi \equiv \neg\exists x\, \neg\varphi.$$

The second equivalence,

$$\exists x\, \varphi \equiv \neg\forall x\, \neg\varphi$$

follows analogously, or directly by interchanging $\forall$ and $\exists$.

(5) and (6): We first consider the universal quantifiers. We have

$$\begin{aligned} &[\![\forall x\, \forall y\, \varphi]\!]_I = \text{true} \\ &\qquad \iff \forall a \in A\; \forall b \in A : \; [\![\varphi]\!]_{\langle A,\beta[x\mapsto a,\, y\mapsto b]\rangle} = \text{true}. \end{aligned}$$

Since the order of the two universal quantifiers does not matter, this is equivalent to

$$\forall b \in A\; \forall a \in A : \; [\![\varphi]\!]_{\langle A,\beta[y\mapsto b, x\mapsto a]\rangle} = \text{true},$$

that is,

$$[\![\forall y\, \forall x\, \varphi]\!]_I = \text{true}.$$

Hence,

$$\forall x\, \forall y\, \varphi \equiv \forall y\, \forall x\, \varphi.$$

The same argument applies to existential quantifiers:

$$[\![\exists x\, \exists y\, \varphi]\!]_I = \text{true}$$
$$\iff \exists a \in A\ \exists b \in A :\ [\![\varphi]\!]_{\langle A, \beta[x \mapsto a,\, y \mapsto b]\rangle} = \text{true}.$$

and this holds exactly when

$$\exists b \in A\ \exists a \in A :\ [\![\varphi]\!]_{\langle A, \beta[y \mapsto b, x \mapsto a]\rangle} = \text{true},$$

that is,

$$\exists y\, \exists x\, \varphi \equiv \exists x\, \exists y\, \varphi.$$

(7): We have

$$[\![(\forall x\, \varphi) \wedge (\forall x\, \psi)]\!]_I = \text{true}$$

exactly when both

$$[\![\forall x\, \varphi]\!]_I = \text{true} \quad \text{as well as} \quad [\![\forall x\, \psi]\!]_I = \text{true}$$

hold. By the semantics of the universal quantifier, this is equivalent to saying that for every $a \in A$,

$$[\![\varphi]\!]_{\langle A, \beta[x \mapsto a]\rangle} = \text{true} \quad \text{and} \quad [\![\psi]\!]_{\langle A, \beta[x \mapsto a]\rangle} = \text{true}.$$

This, in turn, is exactly the case when for every $a \in A$,

$$[\![\varphi \wedge \psi]\!]_{\langle A, \beta[x \mapsto a]\rangle} = \text{true},$$

that is, exactly when

$$[\![\forall x\, (\varphi \wedge \psi)]\!]_I = \text{true}.$$

Therefore,

$$(\forall x\, \varphi) \wedge (\forall x\, \psi) \equiv \forall x\, (\varphi \wedge \psi).$$

(8): Analogously, we obtain

$$[\![(\exists x\, \varphi) \vee (\exists x\, \psi)]\!]_I = \text{true}$$

exactly when either $[\![\exists x\, \varphi]\!]_I = \text{true}$ or $[\![\exists x\, \psi]\!]_I = \text{true}$ holds. This is equivalent to saying that there exists an $a \in A$ such that

$$[\![\varphi \vee \psi]\!]_{\langle A, \beta[x \mapsto a]\rangle} = \text{true},$$

that is, exactly when

$$[\![\exists x\, (\varphi \vee \psi)]\!]_I = \text{true}.$$

Hence,

$$(\exists x\, \varphi) \vee (\exists x\, \psi) \equiv \exists x\, (\varphi \vee \psi).$$

Since I was arbitrary, all eight equivalences hold in general. □

Proposition 2.36 (Shifting Quantifiers across $\wedge$ and $\vee$)**.** *Let φ and ψ be predicate-logical formulas, and let x be a variable with $x \notin \mathrm{FV}(\psi)$. Then the following equivalences hold:*

$$\begin{aligned}
(\forall x\, \varphi) \wedge \psi &\equiv \forall x\, (\varphi \wedge \psi), \\
(\exists x\, \varphi) \wedge \psi &\equiv \exists x\, (\varphi \wedge \psi), \\
(\forall x\, \varphi) \vee \psi &\equiv \forall x\, (\varphi \vee \psi), \\
(\exists x\, \varphi) \vee \psi &\equiv \exists x\, (\varphi \vee \psi).
\end{aligned}$$

Proof. The first equivalence follows directly from (1) and (7) of the previous proposition:

$$(\forall x\, \varphi) \wedge \psi \equiv (\forall x\, \varphi) \wedge (\forall x\, \psi) \equiv \forall x\, (\varphi \wedge \psi),$$

since $x \notin \mathrm{FV}(\psi)$.

For the second equivalence, let $I = \langle \mathcal{A}, \beta \rangle$ be an arbitrary interpretation. Because $x \notin \mathrm{FV}(\psi)$, changing the assignment of x does not affect the truth value of ψ. Hence:

$$\begin{aligned} I \models (\exists x\, \varphi) \wedge \psi &\iff \exists a \in A:\ I[x \mapsto a] \models \varphi \text{ and} \\ &\qquad I[x \mapsto a] \models \psi \\ &\iff \exists a \in A:\ I[x \mapsto a] \models (\varphi \wedge \psi) \\ &\iff I \models \exists x\, (\varphi \wedge \psi). \end{aligned}$$

The two disjunction statements are proved analogously: since $x \notin \mathrm{FV}(\psi)$, the truth value of ψ remains unchanged when the assignment of x is varied. Therefore,

$$I \models (\forall x\, \varphi) \vee \psi \iff I \models \forall x\, (\varphi \vee \psi),$$

and likewise,

$$I \models (\exists x\, \varphi) \vee \psi \iff I \models \exists x\, (\varphi \vee \psi).$$

This proves all four equivalences. □

Example 2.37. *The formula*

$$\forall x\, \exists y\, \mathrm{connected}(x, y)$$

means: for every object, there exists a suitable object to which it is connected. Here, the order of the two quantifiers must not be exchanged, because

$$\exists y \, \forall x \, \text{connected}(x, y)$$

generally expresses something quite different: it would mean that there is a single object that stands in the chosen relation to all others.

Example 2.38. *The following two formulas are likewise not equivalent in general:*

$$(\forall x \, \text{Raum}(x)) \vee (\forall x \, \text{verbunden}(x, z))$$

and

$$\forall x \, (\text{Raum}(x) \vee \text{verbunden}(x, z)).$$

The first formula states that either all objects are rooms or all objects are connected to z. The second formula merely requires that every object has at least one of these two properties.

Remark 2.39. *The results of this section provide the formal tools that will later allow us to transform formulas systematically. Especially important is the distinction between what holds unconditionally and what is permissible only under an additional side condition involving free variables.*

Example 2.40. *The following equivalence does not hold in general:*

$$\forall x \, \exists y \, \varphi \;\not\equiv\; \exists y \, \forall x \, \varphi.$$

We show this by means of a counterexample structure. Let Σ be a signature with a binary predicate symbol R, and let

$$\mathcal{A} = \langle A, R^{\mathcal{A}} \rangle \quad \text{with} \quad A = \{0,1\}, \qquad R^{\mathcal{A}} = \{(0,0),(1,1)\}.$$

Consider the formula

$$\varphi(x,y) := R(x,y).$$

Then we have:

$$\mathcal{A} \models \forall x\, \exists y\, R(x,y),$$

because for $x = 0$ we may choose $y = 0$, and for $x = 1$ we may choose $y = 1$.

On the other hand,

$$\mathcal{A} \not\models \exists y\, \forall x\, R(x,y),$$

because there is no single $y \in A$ such that both

$$R(0,y) \quad \text{and} \quad R(1,y)$$

hold. For $y = 0$, $R(0,0)$ is true, but $R(1,0)$ is false; for $y = 1$, $R(1,1)$ is true, but $R(0,1)$ is false.

Hence,

$$\forall x\, \exists y\, R(x,y) \not\equiv \exists y\, \forall x\, R(x,y).$$

Thus, the claimed interchangeability of the universal and existential quantifiers is false in general.

Example 2.41 (Two Quantifier Equivalences That Fail in General). *The following two equivalences also do not hold in general:*

$$(\forall x\,\varphi) \vee (\forall x\,\psi) \not\equiv \forall x\,(\varphi \vee \psi),$$
$$(\exists x\,\varphi) \wedge (\exists x\,\psi) \not\equiv \exists x\,(\varphi \wedge \psi).$$

We show this by means of a single counterinterpretation. Let Σ be the signature with two unary predicate symbols P and Q, and let

$$\mathcal{A} = \langle A, P^{\mathcal{A}}, Q^{\mathcal{A}} \rangle, \quad \textit{with} \quad \begin{aligned} A &= \{0,1\}, \\ P^{\mathcal{A}} &= \{0\}, \\ Q^{\mathcal{A}} &= \{1\}. \end{aligned}$$

Consider now the formulas

$$\varphi(x) := P(x) \qquad \textit{and} \qquad \psi(x) := Q(x).$$

Then, for the first equivalence, we have:

$$\mathcal{A} \not\models \forall x\,P(x) \quad \textit{and} \quad \mathcal{A} \not\models \forall x\,Q(x),$$

thus

$$\mathcal{A} \not\models (\forall x\,P(x)) \vee (\forall x\,Q(x)).$$

On the other hand, for every $a \in A$, at least one of the two statements $P(a)$ or $Q(a)$ is true. Therefore,

$$\mathcal{A} \models \forall x\,(P(x) \vee Q(x)).$$

Hence,

$$(\forall x\, P(x)) \vee (\forall x\, Q(x)) \not\equiv \forall x\, (P(x) \vee Q(x)).$$

For the second equivalence, we obtain:

$$\mathcal{A} \models \exists x\, P(x) \quad \textit{and} \quad \mathcal{A} \models \exists x\, Q(x),$$

since $P^{\mathcal{A}} = \{0\}$ *and* $Q^{\mathcal{A}} = \{1\}$*. Therefore,*

$$\mathcal{A} \models (\exists x\, P(x)) \wedge (\exists x\, Q(x)).$$

But there is no $a \in A$ *such that both* $P(a)$ *and* $Q(a)$ *hold, so*

$$\mathcal{A} \not\models \exists x\, (P(x) \wedge Q(x)).$$

Thus,

$$(\exists x\, P(x)) \wedge (\exists x\, Q(x)) \not\equiv \exists x\, (P(x) \wedge Q(x)).$$

The two equivalences are therefore false in general.

2.3.3 Substitution

Substitutions are the predicate-logical counterpart of what is known in propositional logic as the targeted replacement of variables. The essential difference, however, is that in predicate logic we have not only variables but also compound terms as substitution objects. For this reason, special care is required with bound variables, so that no unintended binding of free variables arises.

Substitutions.

Definition 2.42 (Substitution). *Let* Σ *be a signature. A* Σ*-substitution is a function*

$$\sigma : V \to \mathrm{Term}(\Sigma)$$

with the property that $\sigma(x) = x$ *for all but finitely many variables* $x \in V$.

The set

$$\mathrm{Dom}(\sigma) := \{x \in V \mid \sigma(x) \neq x\}$$

is called the effective domain *of* σ. *We denote the set of variables occurring in* σ *by*

$$\mathrm{Var}(\sigma) := \{x \in V \mid x \in \mathrm{Var}(\sigma(y)) \text{ for some } y \in \mathrm{Dom}(\sigma)\}.$$

We usually write substitutions in the form

$$\sigma = \{x_1 \mapsto t_1, \ldots, x_n \mapsto t_n\}.$$

By this we mean the function that maps each x_i to t_i and leaves all other variables unchanged.

Example 2.43. *The substitution*

$$\sigma = \{x \mapsto g(z, a),\ y \mapsto f(z)\}$$

replaces the variable x *by the term* $g(z, a)$ *and the variable* y *by the term* $f(z)$. *All other variables remain unchanged. Here,* $\mathrm{Dom}(\sigma) = \{x, y\}$, *and* $\mathrm{Var}(\sigma) = \{z\}$.

Often, one modifies a substitution at only a single point. For this, we write:

$$\sigma[x \mapsto t],$$

where $\sigma[x \mapsto t]$ denotes the substitution that maps x to t and otherwise agrees with σ.

Application of Substitutions. A substitution acts on terms and formulas by simultaneously replacing all free occurrences of the variables concerned.

Definition 2.44 (Application of a Substitution). *Let Σ be a signature and let σ be a Σ-substitution.*

- *For a Σ-term t, we define $\sigma(t)$ inductively by*
  - $\sigma(x) = \sigma(x)$ *for* $x \in \mathrm{Dom}(\sigma)$,
  - $\sigma(x) = x$ *for* $x \notin \mathrm{Dom}(\sigma)$,
  - $\sigma(f(t_1, \ldots, t_n)) = f(\sigma(t_1), \ldots, \sigma(t_n))$.
- *For a Σ-formula φ, we define $\sigma(\varphi)$ inductively by*
  - $\sigma(P(t_1, \ldots, t_n)) = P(\sigma(t_1), \ldots, \sigma(t_n))$,
  - $\sigma(\psi \wedge \chi) = \sigma(\psi) \wedge \sigma(\chi)$,
  - $\sigma(\psi \vee \chi) = \sigma(\psi) \vee \sigma(\chi)$,
  - $\sigma(\neg\psi) = \neg\sigma(\psi)$.
- *For quantifiers, one must pay attention to bound variables:*

- *If* $\varphi = \forall x\, \psi$ *and* $x \notin \mathrm{Var}(\sigma)$, *we set*
$$\sigma(\varphi) = \forall x\, \sigma(\psi).$$
- *If* $\varphi = \exists x\, \psi$ *and* $x \notin \mathrm{Var}(\sigma)$, *we set*
$$\sigma(\varphi) = \exists x\, \sigma(\psi).$$
- *If, on the other hand,* $x \in \mathrm{Var}(\sigma)$, *then we choose a fresh variable* $u \notin \mathrm{Var}(\sigma) \cup \mathrm{FV}(\psi)$ *and define*
$$\sigma(\forall x\, \psi) = \forall u\, \sigma[x \mapsto u](\psi),$$
$$\sigma(\exists x\, \psi) = \exists u\, \sigma[x \mapsto u](\psi).$$

The last two cases prevent a free variable from the substitution image from suddenly being captured by a quantifier.

Example 2.45. *Consider the signature*
$$\Sigma = \{P/2,\ f/1,\ a/0\}$$
and the formula
$$\varphi = \exists y\, P(x, f(y)).$$
Let furthermore
$$\sigma = \{x \mapsto f(y)\}.$$
Then y *cannot simply remain inside* φ, *because* y *already occurs in* $\sigma(x) = f(y)$. *Therefore, the bound variable is first renamed to a fresh variable, and we obtain*
$$\sigma(\varphi) = \exists u\, P(f(y), f(u)).$$

Substitution on Interpretations. Substitutions do not change the structure itself, but only the assignment of variables.

Definition 2.46 (Substitution of an Interpretation). *Let Σ be a signature, σ a Σ-substitution, and $I = \langle A, \beta \rangle$ a Σ-interpretation. Then we define a new assignment β^σ by*

$$\beta^\sigma(x) := [\![\sigma(x)]\!]_I.$$

We write the corresponding interpretation as

$$\sigma(I) := \langle A, \beta^\sigma \rangle.$$

Example 2.47. *Let*

$$\begin{aligned} \Sigma &= \{P/1, c/0\}, \\ A &= \{0, 1\}, \\ P^{\mathcal{A}} &= \{1\}, \\ c^{\mathcal{A}} &= 1. \end{aligned}$$

and $\beta(x) = 0$. Then we have

$$[\![P(x)]\!]_{\langle \mathcal{A}, \beta \rangle} = \text{false}.$$

For the substitution $\sigma = \{x \mapsto c\}$, however, we obtain

$$\beta^\sigma(x) = [\![c]\!]_{\langle \mathcal{A}, \beta \rangle} = 1,$$

so $\sigma(\langle \mathcal{A}, \beta \rangle)$ is a model of $P(x)$. At the same time,

$$\sigma(P(x)) = P(c),$$

and this formula is also true in $\langle \mathcal{A}, \beta \rangle$.

The Substitution Lemma. The central claim is that syntactic substitution and semantic evaluation are compatible with one another.

Theorem 2.48 (Substitution Lemma). *Let Σ be a signature, I a Σ-interpretation, and σ a Σ-substitution. Then the following hold:*

1. *For every Σ-term t, we have*

$$[\![\sigma(t)]\!]_I = [\![t]\!]_{\sigma(I)}.$$

2. *For every Σ-formula φ, we have*

$$I \models_\Sigma \sigma(\varphi) \iff \sigma(I) \models_\Sigma \varphi.$$

Proof. Write $I = \langle A, \beta \rangle$, and let β^σ denote the assignment induced by σ, that is,

$$\beta^\sigma(x) := [\![\sigma(x)]\!]_I \qquad \text{for all } x \in V.$$

Thus $\sigma(I) = \langle A, \beta^\sigma \rangle$.

(1) Terms. We prove the claim by structural induction on the form of t.

Case $t = x$ (variable). By definition,

$$[\![\sigma(x)]\!]_I = \beta^\sigma(x) = [\![x]\!]_{\sigma(I)}.$$

Case $t = c$ (constant). Since substitutions affect only variables, we have $\sigma(c) = c$. Hence

$$[\![\sigma(c)]\!]_I = [\![c]\!]_I = c^A = [\![c]\!]_{\sigma(I)}.$$

Case $t = f(t_1, \ldots, t_n)$. By the induction hypothesis,

$$[\![\sigma(t_i)]\!]_I = [\![t_i]\!]_{\sigma(I)} \qquad \text{for all } i = 1, \ldots, n.$$

Since σ does not change the interpretation of function symbols, the symbol f is interpreted identically in I and $\sigma(I)$. Therefore,

$$\begin{aligned}
[\![\sigma(f(t_1, \ldots, t_n))]\!]_I &= [\![f(\sigma(t_1), \ldots, \sigma(t_n))]\!]_I \\
&= f^A([\![\sigma(t_1)]\!]_I, \ldots, [\![\sigma(t_n)]\!]_I) \\
&= f^A([\![t_1]\!]_{\sigma(I)}, \ldots, [\![t_n]\!]_{\sigma(I)}) \\
&= [\![f(t_1, \ldots, t_n)]\!]_{\sigma(I)}.
\end{aligned}$$

This proves (1).

(2) Formulas. Again, we proceed by structural induction on φ.

Case $\varphi = P(t_1, \ldots, t_n)$ *atomic.* Then part (1) yields:

$$\begin{aligned}
[\![\sigma(P(t_1, \ldots, t_n))]\!]_I &= \text{true} \\
&\iff ([\![\sigma(t_1)]\!]_I, \ldots, [\![\sigma(t_n)]\!]_I) \in P^A.
\end{aligned}$$

$$\begin{aligned}
&\iff ([\![t_1]\!]_{\sigma(I)}, \ldots, [\![t_n]\!]_{\sigma(I)}) \in P^A \\
&\iff [\![P(t_1, \ldots, t_n)]\!]_{\sigma(I)} = \text{true}.
\end{aligned}$$

Connectives. The cases $\wedge$, $\vee$, and $\neg$ are immediate, since their semantics is truth-functional. For example,

$$[\![\sigma(\psi \wedge \chi)]\!]_I = \text{true} \iff [\![\sigma(\psi)]\!]_I = \text{true and } [\![\sigma(\chi)]\!]_I = \text{true}.$$

By the induction hypothesis, this is equivalent to

$$[\![\psi]\!]_{\sigma(I)} = \text{true and } [\![\chi]\!]_{\sigma(I)} = \text{true},$$

hence to

$$[\![\psi \wedge \chi]\!]_{\sigma(I)} = \text{true}.$$

Quantifiers. We treat the existential quantifier as a representative case. The universal quantifier is handled analogously.

If $x \notin \text{Var}(\sigma)$, then

$$\sigma(\exists x\, \psi) = \exists x\, \sigma(\psi).$$

Therefore,

$$\begin{aligned} [\![\sigma(\exists x\, \psi)]\!]_I = \text{true} &\iff \exists a \in A:\ [\![\sigma(\psi)]\!]_{\langle A, \beta[x \mapsto a]\rangle} = \text{true} \\ &\quad \text{(semantics of the existential quantifier)} \\ &\iff \exists a \in A:\ [\![\psi]\!]_{\sigma(\langle A, \beta[x \mapsto a]\rangle)} = \text{true} \\ &\quad \text{(induction hypothesis for } \psi) \\ &\iff \exists a \in A:\ [\![\psi]\!]_{\langle A, \beta^\sigma[x \mapsto a]\rangle} = \text{true} \\ &\quad \text{(coincidence lemma)} \\ &\iff [\![\exists x\, \psi]\!]_{\sigma(I)} = \text{true}. \end{aligned}$$

If $x \in \text{Var}(\sigma)$, then

$$\sigma(\exists x\, \psi) = \exists u\, \sigma[x \mapsto u](\psi),$$

where $u \notin \text{Var}(\sigma) \cup \text{FV}(\psi)$ is chosen fresh. The same argument then applies, since the fresh variable u does not create any unwanted variable capture, and the induction hypothesis is applicable to $\sigma[x \mapsto u](\psi)$.

Thus, in all cases,

$$[\![\sigma(\varphi)]\!]_I = \text{true} \quad \Longleftrightarrow \quad [\![\varphi]\!]_{\sigma(I)} = \text{true}.$$

This is exactly

$$I \models_\Sigma \sigma(\varphi) \quad \Longleftrightarrow \quad \sigma(I) \models_\Sigma \varphi.$$

This proves both statements. □

Proposition 2.49. *Let Σ be a signature, let φ, ψ be two Σ-formulas, and let σ be a substitution. If*

$$\varphi \equiv_\Sigma \psi,$$

then also

$$\sigma(\varphi) \equiv_\Sigma \sigma(\psi).$$

Proof. Let I be an arbitrary Σ-interpretation. By the substitution lemma, we have

$$I \models_\Sigma \sigma(\varphi) \quad \Longleftrightarrow \quad \sigma(I) \models_\Sigma \varphi,$$

and likewise

$$I \models_\Sigma \sigma(\psi) \quad \Longleftrightarrow \quad \sigma(I) \models_\Sigma \psi.$$

Since $\varphi \equiv_\Sigma \psi$ is assumed, it follows that

$$\sigma(I) \models_\Sigma \varphi \quad \Longleftrightarrow \quad \sigma(I) \models_\Sigma \psi.$$

Combining these equivalences, we obtain

$$I \models_\Sigma \sigma(\varphi) \quad \Longleftrightarrow \quad I \models_\Sigma \sigma(\psi).$$

Thus $\sigma(\varphi)$ and $\sigma(\psi)$ are logically equivalent. □

Variable Renamings. An important special case is given by substitutions that merely rename variables.

Definition 2.50 (Variable Renaming). *A variable renaming is a function $\nu : V \to V$ such that $\nu(x) = x$ for all but finitely many variables $x \in V$.*

Such renamings do not change the semantics as long as only bound variables are affected.

Proposition 2.51. *Let Σ be a signature and φ a Σ-formula. Then:*

$$\forall x\, \varphi \equiv_\Sigma \forall y\, \nu(\varphi), \qquad \exists x\, \varphi \equiv_\Sigma \exists y\, \nu(\varphi),$$

provided that $y \notin \mathrm{FV}(\varphi)$ and ν maps the variable x to y.

Proof. We prove only the first equivalence; the second follows by the same argument. Thus let $\nu = \{x \mapsto y\}$ and assume that $y \notin \mathrm{FV}(\varphi)$.

Choose a fresh constant symbol $c \notin \Sigma$. For each $a \in A$, let $\mathcal{A}_a$ denote the expansion of $\mathcal{A}$ obtained by interpreting c as a, that is, $c^{\mathcal{A}_a} = a$, while all symbols from Σ are interpreted exactly as in $\mathcal{A}$.

Moreover, define

$$\sigma = \{x \mapsto c\} \qquad \text{and} \qquad \tau = \{y \mapsto c\}.$$

Then for every assignment β and every $a \in A$, we have:

$$\sigma(\langle \mathcal{A}_a, \beta \rangle) = \langle \mathcal{A}_a, \beta[x \mapsto a] \rangle,$$
$$\tau(\langle \mathcal{A}_a, \beta \rangle) = \langle \mathcal{A}_a, \beta[y \mapsto a] \rangle.$$

This follows immediately from the definition of substitution acting on interpretations.

We now first consider the left-hand expression:

$$
\begin{aligned}
\langle \mathcal{A}, \beta \rangle \models \forall x\, \varphi \iff & \forall a \in A : \\
& \langle \mathcal{A}, \beta[x \mapsto a] \rangle \models \varphi \\
& \text{(semantics of the universal quantifier)} \\
\iff & \forall a \in A : \\
& \sigma(\langle \mathcal{A}_a, \beta \rangle) \models \varphi \\
& \text{(coincidence lemma, since } c \notin \Sigma(\varphi)) \\
\iff & \forall a \in A : \\
& \langle \mathcal{A}_a, \beta \rangle \models \sigma(\varphi) \\
& \text{(substitution lemma).}
\end{aligned}
$$

Analogously, for the right-hand expression we obtain:

$$
\begin{aligned}
\langle \mathcal{A}, \beta \rangle \models \forall y\, \nu(\varphi) \iff & \forall a \in A : \\
& \langle \mathcal{A}, \beta[y \mapsto a] \rangle \models \nu(\varphi) \\
& \text{(semantics of the universal quantifier)} \\
\iff & \forall a \in A : \\
& \tau(\langle \mathcal{A}_a, \beta \rangle) \models \nu(\varphi) \\
& \text{(coincidence lemma, since } c \notin \Sigma(\varphi)) \\
\iff & \forall a \in A : \\
& \langle \mathcal{A}_a, \beta \rangle \models \tau(\nu(\varphi)) \\
& \text{(substitution lemma).}
\end{aligned}
$$

Now $\tau \circ \nu = \{x \mapsto c\} = \sigma$. Hence we also have

$$\tau(\nu(\varphi)) = \sigma(\varphi).$$

Therefore both formulas have exactly the same models, and consequently

$$\forall x\, \varphi \equiv_\Sigma \forall y\, \nu(\varphi).$$

The existential quantifier is handled in a completely analogous way; one only replaces the semantics of $\forall$ by those of $\exists$. □

Example 2.52. *The formulas*

$$\forall x\, P(x, z) \qquad \text{and} \qquad \forall u\, P(u, z)$$

are equivalent, provided that u does not already occur free in the formula. The situation is different for free variables: the formulas $P(x)$ and $P(y)$ are not equivalent in general, since their truth values depend on different assignments of the free variable.

Replacement of Subformulas. Not only variables, but also entire subformulas can be replaced by equivalent subexpressions.

Definition 2.53 (Replacement of Subformulas). *Let φ and κ be formulas, and let $H \in \mathrm{TF}(\varphi)$ be a subformula of φ. Then $\varphi[H \mapsto \kappa]$ denotes the formula obtained by replacing every occurrence of H in φ by κ.*

Proposition 2.54. *Let Σ be a signature, let φ be a Σ-formula, and let $H \in \mathrm{TF}(\varphi)$ be a subformula. If*

$$H \equiv_\Sigma \kappa,$$

then

$$\varphi \equiv_\Sigma \varphi[H \mapsto \kappa].$$

Proof. We first show that for every Σ-interpretation I,

$$[\![\varphi]\!]_I = [\![\varphi[H \mapsto \kappa]]\!]_I.$$

This immediately implies $\varphi \equiv_\Sigma \varphi[H \mapsto \kappa]$.

If $\varphi = H$, then

$$\varphi[H \mapsto \kappa] = \kappa,$$

and the claim follows directly from the assumption $H \equiv_\Sigma \kappa$.

Hence we may assume in the following that $\varphi \neq H$. We now prove the statement by structural induction on the form of φ.

(FO0) Atomic case. Let $\varphi = P(t_1, \ldots, t_n)$ be atomic. Then H can only be the entire formula φ itself, since atomic formulas have no proper subformulas. This contradicts the assumption $\varphi \neq H$. Thus this case cannot occur.

(FO1) Connective case. Let $\varphi = \psi \otimes \chi$ with $\otimes \in \{\wedge, \vee\}$. Then the induction hypothesis yields

$$\psi \equiv_\Sigma \psi[H \mapsto \kappa] \qquad \text{and} \qquad \chi \equiv_\Sigma \chi[H \mapsto \kappa].$$

Therefore, for every interpretation I,

$$[\![\psi]\!]_I = [\![\psi[H \mapsto \kappa]]\!]_I \qquad \text{and} \qquad [\![\chi]\!]_I = [\![\chi[H \mapsto \kappa]]\!]_I.$$

Since the truth functions for $\wedge$ and $\vee$ are defined componentwise, it follows that

$$[\![\varphi]\!]_I = [\![\varphi[H \mapsto \kappa]]\!]_I.$$

Hence $\varphi \equiv_\Sigma \varphi[H \mapsto \kappa]$.

(FO2) Negation. Let $\varphi = \neg\psi$. From the induction hypothesis we obtain

$$\psi \equiv_\Sigma \psi[H \mapsto \kappa].$$

Thus, for every interpretation I,

$$[\![\psi]\!]_I = [\![\psi[H \mapsto \kappa]]\!]_I.$$

By the definition of negation, it then follows that

$$[\![\varphi]\!]_I = [\![\varphi[H \mapsto \kappa]]\!]_I.$$

So in this case as well, $\varphi \equiv_\Sigma \varphi[H \mapsto \kappa]$.

(FO3) Quantifier case. Let $\varphi = Qx\,\psi$ with $Q \in \{\forall, \exists\}$. By the induction hypothesis,

$$\psi \equiv_\Sigma \psi[H \mapsto \kappa].$$

Now let $I = \langle \mathcal{A}, \beta \rangle$ be an arbitrary interpretation. Then, for every assignment $a \in A$,

$$[\![\psi]\!]_{\langle \mathcal{A}, \beta[x \mapsto a] \rangle} = [\![\psi[H \mapsto \kappa]]\!]_{\langle \mathcal{A}, \beta[x \mapsto a] \rangle}.$$

Therefore the formulas $Qx\,\psi$ and $Qx\,\psi[H \mapsto \kappa]$ have the same truth value under I. Hence,

$$[\![\varphi]\!]_I = [\![\varphi[H \mapsto \kappa]]\!]_I.$$

Altogether, it follows that for all interpretations I,

$$[\![\varphi]\!]_I = [\![\varphi[H \mapsto \kappa]]\!]_I.$$

Therefore,

$$\varphi \equiv_\Sigma \varphi[H \mapsto \kappa].$$

□

2.3.4 Model Classes

So far, we have considered formulas only relative to *interpretations*, that is, relative to a structure together with an assignment of variables. For many purposes, however, it is more natural to work directly with *structures*. The reason is simple: a structure fixes the meaning of the non-logical symbols, and for formulas without free variables, the truth value no longer depends on a particular assignment.

Semantically, the free variables of a formula behave like implicitly universally quantified variables. We now formulate this observation systematically.

Definition 2.55 (Structural Satisfaction and Model Class). *Let Σ be a signature and φ a Σ-formula.*

For a Σ-structure $\mathcal{A}$, we write

$$\mathcal{A} \models^{\mathrm{FO}}_{\Sigma} \varphi$$

if and only if

$$\langle \mathcal{A}, \beta \rangle \models_{\Sigma}^{\mathrm{FO}} \varphi$$

holds for every *assignment β on $\mathcal{A}$.*

We denote the set of all Σ-structures satisfying φ by

$$\mathrm{MK}_{\Sigma}^{\mathrm{FO}}(\varphi) = \{\mathcal{A} \in \mathrm{Struct}(\Sigma) \mid \mathcal{A} \models_{\Sigma}^{\mathrm{FO}} \varphi\}.$$

This set is called the Σ-model class *of φ.*

For comparison, the set of all Σ-interpretations satisfying φ is given by

$$\mathrm{Mod}_{\Sigma}^{\mathrm{FO}}(\varphi) = \{I \mid I \textit{ is a } \Sigma\textit{-interpretation and } I \models_{\Sigma}^{\mathrm{FO}} \varphi\}.$$

In what follows, however, we will mostly work with the model class $\mathrm{MK}_{\Sigma}^{\mathrm{FO}}(\varphi)$.

In words, the definition says: a structure satisfies a formula exactly when the formula is true under *every* possible assignment. For closed formulas, that is, formulas without free variables, this distinction is especially convenient, because then no assignment plays any role at all.

Proposition 2.56. *Let φ be a Σ-formula with*

$$\mathrm{FV}(\varphi) = \{x_1, \ldots, x_n\}.$$

Then for every Σ-structure $\mathcal{A}$, we have:

$$\mathcal{A} \models_{\Sigma}^{\mathrm{FO}} \varphi \iff \mathcal{A} \models_{\Sigma}^{\mathrm{FO}} \forall x_1 \cdots \forall x_n \, \varphi.$$

Proof. We prove both directions.

"$\Rightarrow$": Assume that $\mathcal{A} \models_{\Sigma}^{\mathrm{FO}} \varphi$. Then $\langle \mathcal{A}, \beta \rangle \models_{\Sigma}^{\mathrm{FO}} \varphi$ holds for every assignment β on $\mathcal{A}$. In particular, this remains true after any change in the values of $x_1, \ldots, x_n$. Hence

$$\langle \mathcal{A}, \beta \rangle \models_{\Sigma}^{\mathrm{FO}} \forall x_1 \cdots \forall x_n \, \varphi$$

for every assignment β, and therefore $\mathcal{A} \models_{\Sigma}^{\mathrm{FO}} \forall x_1 \cdots \forall x_n \, \varphi$.

"$\Leftarrow$": Assume that $\mathcal{A} \models_{\Sigma}^{\mathrm{FO}} \forall x_1 \cdots \forall x_n \, \varphi$. Let β be an arbitrary assignment on $\mathcal{A}$. Then

$$\langle \mathcal{A}, \beta \rangle \models_{\Sigma}^{\mathrm{FO}} \forall x_1 \cdots \forall x_n \, \varphi.$$

By the semantics of the universal quantifier, it follows that

$$\langle \mathcal{A}, \beta \rangle \models_{\Sigma}^{\mathrm{FO}} \varphi$$

holds. Since β was arbitrary, we conclude that $\mathcal{A} \models_{\Sigma}^{\mathrm{FO}} \varphi$.

This proves the claim. □

Example 2.57. *Consider the signature $\Sigma = \{E/2\}$, where $E(x, y)$ is read as "there is an edge from x to y." Then we have, for example, the two formulas*

$$\varphi := \forall x \, \exists y \, E(x, y) \qquad \text{and} \qquad \psi := \exists x \, \forall y \, \neg E(y, x).$$

The formula φ describes exactly those graphs in which every vertex has a successor. The formula ψ describes exactly those graphs in which there exists a vertex with no incoming edge.

Now consider the following two structures:

$$\mathcal{G}_1 = \langle G_1, E^{\mathcal{G}_1} \rangle$$

$$\text{with} \quad G_1 = \{a, b, c\},$$
$$E^{\mathcal{G}_1} = \{(a, b), (b, c), (c, a)\}.$$

and

$$\mathcal{G}_2 = \langle G_2, E^{\mathcal{G}_2} \rangle$$

$$\text{with} \quad G_2 = \{d, e, f, g\},$$
$$E^{\mathcal{G}_2} = \{(d, e), (d, f), (d, g)\}.$$

Then:

$$\mathcal{G}_1 \models_{\Sigma}^{\mathrm{FO}} \varphi, \qquad \mathcal{G}_1 \not\models_{\Sigma}^{\mathrm{FO}} \psi,$$

because in the cycle $a \to b \to c \to a$, every vertex has a successor, but no vertex is without an incoming edge.

For $\mathcal{G}_2$, the opposite holds:

$$\mathcal{G}_2 \not\models_{\Sigma}^{\mathrm{FO}} \varphi, \qquad \mathcal{G}_2 \models_{\Sigma}^{\mathrm{FO}} \psi,$$

because e, f, g have no successor, whereas d has no incoming edge.

Thus, the two formulas have different model classes.

Corollary 2.58 (Invariance under Variable Renaming). *Let Σ be a signature, φ a Σ-formula, and $\nu : V \to V$ a variable renaming. Then*

$$\mathrm{MK}_{\Sigma}^{\mathrm{FO}}(\varphi) = \mathrm{MK}_{\Sigma}^{\mathrm{FO}}(\nu(\varphi)).$$

Corollary 2.59 (Closed Formulas and Model Classes). *Let φ and ψ be closed Σ-formulas. Then*

$$\varphi \equiv_\Sigma \psi \quad \Longleftrightarrow \quad \mathrm{MK}_\Sigma^{\mathrm{FO}}(\varphi) = \mathrm{MK}_\Sigma^{\mathrm{FO}}(\psi).$$

Corollary 2.60 (Model Classes of Closed Formulas). *Let φ, ψ be closed Σ-sentences. Then:*

$$\mathrm{MK}_\Sigma^{\mathrm{FO}}(\varphi \wedge \psi) = \mathrm{MK}_\Sigma^{\mathrm{FO}}(\varphi) \cap \mathrm{MK}_\Sigma^{\mathrm{FO}}(\psi),$$

$$\mathrm{MK}_\Sigma^{\mathrm{FO}}(\varphi \vee \psi) = \mathrm{MK}_\Sigma^{\mathrm{FO}}(\varphi) \cup \mathrm{MK}_\Sigma^{\mathrm{FO}}(\psi),$$

$$\mathrm{MK}_\Sigma^{\mathrm{FO}}(\neg\varphi) = \mathtt{Struct}(\Sigma) \setminus \mathrm{MK}_\Sigma^{\mathrm{FO}}(\varphi),$$

$$\mathrm{MK}_\Sigma^{\mathrm{FO}}(\varphi \to \psi) = \mathrm{MK}_\Sigma^{\mathrm{FO}}(\neg\varphi) \cup \mathrm{MK}_\Sigma^{\mathrm{FO}}(\psi).$$

Proof. All statements follow immediately from the semantics of the logical connectives and from the definition of structural satisfaction. For closed formulas, the truth of a formula in a structure is independent of any assignment, so the semantic relationships transfer directly to the corresponding model classes. □

2.3.5 Exercises and Solutions

Exercise 2.61. *Consider the first-order signature*

$$\Sigma = \{P/2, R/2, Q/2, g/1, f/1, i/2, h/2, c/0\}.$$

Consider the following expressions:

$$\begin{aligned}
\varphi_1 &= \forall v\,(\neg\neg Q(g(h(v,c)),u)),\\
\psi_1 &= \forall v\,(\neg\neg\neg Q(g(h(v,c)),u)),\\
\varphi_2 &= \neg\neg\forall y\,(\neg Q(y,z)),\\
\psi_2 &= \neg\forall z\,(\neg\neg\neg\forall y\,(\neg Q(y,z))),\\
\sigma_3 &= \{u \mapsto c,\ z \mapsto h(c,w)\},\\
\varphi_4 &= \forall z\, R(i(h(c,z),c),v)\ \vee\ \neg\neg Q(x,i(c,c)),\\
\psi_4 &= \forall z\, R(i(h(c,z),c),c)\ \vee\ \neg\neg Q(x,i(c,c)),\\
\sigma_4 &= \{x \mapsto w,\ v \mapsto c\}.
\end{aligned}$$

Decide for each of the following statements whether it is true or false.

1. $\varphi_1 \equiv \psi_1$.

2. $\varphi_2 \equiv \psi_2$.

3. σ_3 *is a substitution in* Σ.

4. $\sigma_4(\varphi_4) = \psi_4$.

Solution 2.62. *1.* ***False.*** *The formulas are not equivalent. In* φ_1*, the statement under the universal quantifier contains a double negation, so the inner part is equivalent to*

$$Q(g(h(v,c)),u).$$

In ψ_1, by contrast, there is a triple negation, so the inner part is equivalent to

$$\neg Q(g(h(v,c)),u).$$

Therefore, the two formulas generally have different meanings.

2. ***False.*** *These two formulas are not equivalent either. The reason is that z occurs free in φ_2 and therefore cannot simply be moved into the scope of a quantifier without changing the semantics. The formula ψ_2 binds z, whereas in φ_2, z remains free.*

3. ***True.*** *σ_3 is a substitution over Σ. In particular,*

$$\mathrm{Dom}(\sigma_3) = \{u, z\}$$

is finite, and both terms $\sigma_3(u) = c$ and $\sigma_3(z) = h(c,w)$ are Σ-terms.

4. ***False.*** *The substitution $\sigma_4(\varphi_4)$ was not carried out correctly, because the occurrence of x in the right disjunct must also be replaced. The correct result is*

$$\sigma_4(\varphi_4) = \forall z\, R(i(h(c,z),c),c) \ \vee \ \neg\neg Q(w, i(c,c)).$$

In ψ_4, however, x remains unchanged. Hence, $\sigma_4(\varphi_4) \neq \psi_4$.

Brief feedback. *In (1) and (2), the decisive point is the semantics of negation and quantifiers: an additional negation inside a formula changes its meaning, and merely* *turning a free occurrence into a bound one changes the formula as well. For (3), it is enough to check that only finitely many variables are affected and that the replacement images are valid* Σ*-terms. In (4), every free variable from* $\mathrm{Dom}(\sigma_4)$ *must actually be replaced; bound variables remain untouched.*

2.4 Normal Forms and Ground Resolution

In the previous sections, we developed the syntax and semantics of first-order logic and also clarified how bound and free variables can be handled precisely. We now turn to a question that is particularly important for proof methods: How can first-order formulas be transformed in such a way that their logical structure becomes more transparent and can be treated mechanically by a resolution procedure?

The basic idea is the same as in propositional logic, but in first-order logic additional difficulties arise. Quantifiers, variable binding, and function symbols make it necessary to introduce suitable normal forms first. We will then show that formulas can be brought into a form in which their satisfiability can be studied very effectively by means of so-called Herbrand models. On this basis, we finally obtain a resolution procedure for first-order logic, which we call *ground resolution*.

The path through this section is divided into four steps: First, we extend the semantics from individual formulas to sets of formulas. Then we introduce normal forms such as prenex form and Skolem form and make precise the shape of the clauses with which we will work. In the third step, we consider Herbrand structures as canonical models for appropriate formulas in Skolem form. Finally, we use this preparation to derive a resolution procedure for first-order logic from Herbrand's theorem.

2.4.1 Semantics for Sets of Formulas

So far, we have considered satisfiability and validity only for individual formulas. For resolution, however, it is more natural to work not merely with a single formula, but with entire sets of formulas. Semantically, this is not a genuine new beginning: A set of formulas is satisfied exactly when each of its elements is satisfied.

Notation 2.63 (Semantics for Sets of Formulas). *Let Σ be a signature, A a Σ-structure, β an assignment for A, and F a set of Σ-formulas. Then we write*

$$\langle A, \beta \rangle \models^{\text{FO}}_{\Sigma} F \;\Leftrightarrow\; \forall \varphi \in F: \; \langle A, \beta \rangle \models^{\text{FO}}_{\Sigma} \varphi$$

and likewise

$$A \models^{\text{FO}}_{\Sigma} F \;\Leftrightarrow\; \forall \varphi \in F: \; A \models^{\text{FO}}_{\Sigma} \varphi$$

If

$$\langle A, \beta \rangle \models^{\text{FO}}_{\Sigma} F$$

holds, we call $\langle A, \beta \rangle$ a Σ-model *of F. Correspondingly, a structure A is called a model of F if*

$$A \models^{\text{FO}}_{\Sigma} F$$

holds.

We write the model class *of F as*

$$\mathcal{M}_\Sigma^{\mathrm{FO}}(F) = \{\, A \in \mathrm{Str}(\Sigma) \mid A \models_\Sigma^{\mathrm{FO}} F \,\}$$

Moreover, for two sets of Σ-formulas F and G, we define:

$$F \equiv_\Sigma G \;\;:\Longleftrightarrow\;\; \forall\ \Sigma\text{-interpretations } I : \\ I \models_\Sigma^{\mathrm{FO}} F \;\Leftrightarrow\; I \models_\Sigma^{\mathrm{FO}} G$$

For a single formula φ, we write $F \equiv_\Sigma \varphi$ as an abbreviation for $F \equiv_\Sigma \{\varphi\}$.

Definition 2.64. *Let Σ be a signature and F a set of Σ-formulas. Then we say that:*

- *F is* satisfiable *if there exists a Σ-model of F;*
- *F is* unsatisfiable *if F has no Σ-model;*
- *F is* valid *if every Σ-interpretation is a model of F.*

The next proposition summarizes some immediate but useful properties.

Proposition 2.65. *Let Σ be a signature. For all sets of Σ-sentences F, G and all Σ-sentences φ, ψ, the following hold:*

1. *$F \equiv_\Sigma G$ if and only if $\mathcal{M}_\Sigma^{\mathrm{FO}}(F) = \mathcal{M}_\Sigma^{\mathrm{FO}}(G)$.*
2. *$\mathcal{M}_\Sigma^{\mathrm{FO}}(\{\varphi\}) = \mathcal{M}_\Sigma^{\mathrm{FO}}(\varphi)$.*

3. $\mathcal{M}_{\Sigma}^{\text{FO}}(\{\varphi \wedge \psi\}) = \mathcal{M}_{\Sigma}^{\text{FO}}(\{\varphi, \psi\})$.

4. $\mathcal{M}_{\Sigma}^{\text{FO}}(F \cup G) = \mathcal{M}_{\Sigma}^{\text{FO}}(F) \cap \mathcal{M}_{\Sigma}^{\text{FO}}(G)$.

5. $\mathcal{M}_{\Sigma}^{\text{FO}}(F) = \bigcap_{\varphi \in F} \mathcal{M}_{\Sigma}^{\text{FO}}(\varphi)$.

Proof. All statements follow directly from the definition of the model class. In particular, (5) is simply the formulation that a structure satisfies F if and only if it satisfies each individual formula in F. The remaining statements are straightforward reformulations of this fact. □

An important difference from propositional logic is that, in predicate logic, a set of formulas cannot always be replaced by a single formula. In other words, there are sets F of Σ-formulas for which no formula φ with $F \equiv_{\Sigma} \varphi$ exists.

Example 2.66. *A classical example is the set of formulas*

$$F = \Big\{ \exists x_1 \ldots \exists x_n \bigwedge_{1 \leq i < j \leq n} x_i \neq x_j \;\Big|\; n \in \mathbb{N},\ n \geq 1 \Big\}.$$

This set expresses that there are arbitrarily many pairwise distinct elements, and hence that the structure is infinite. However, a single FO-formula cannot characterize this property exactly; this follows from the Compactness Theorem.

2.4.2 Normal Forms

As in propositional logic, normal forms are also a useful tool in predicate logic. They make formulas easier to read and bring their logical structure into a form that is especially well suited for further transformations and later proofs. In the quantifier-free case, we can adopt the familiar normal forms directly; in addition, for formulas with quantifiers we particularly need prenex form and Skolem form.

Let us first recall that a *clause* is a disjunction of literals. Thus, if

$$K = L_1 \vee \cdots \vee L_n$$

is such a disjunction, then we call K a clause. As usual, we also regard a clause as the set of its literals.

Definition 2.67. *Let Σ be a signature and let φ be a quantifier-free Σ-formula. Then φ is called*

- in negation normal form (NNF) *if negations occur only directly in front of atoms;*
- in conjunctive normal form (CNF) *if φ is a conjunction of clauses;*
- in disjunctive normal form (DNF) *if φ is a disjunction of conjunctions of literals.*

For quantifier-free formulas, we can use the same transformations as in propositional logic. Therefore, the following proposition is immediately plausible.

Proposition 2.68. *Let Σ be a signature and let φ be a quantifier-free Σ-formula. Then there exists, in each case, a Σ-formula equivalent to φ in* negation normal form, conjunctive normal form, *and* disjunctive normal form.

In the next steps, we are mainly interested in how quantifiers and free variables can be arranged. For this purpose, we first introduce two notions.

Definition 2.69. *Let Σ be a signature and let φ be a Σ-formula. Then φ is called*

- clean *if no variable occurs both free and bound and no variable is bound by two different quantifiers;*
- in prenex form *if there exist a quantifier-free formula ψ and quantifiers $Q_1, \ldots, Q_n \in \{\forall, \exists\}$ such that*

$$\varphi = Q_1 x_1 \cdots Q_n x_n . \psi.$$

The idea of prenex form is simple: all quantifiers are moved to the front, while the quantifier-free core of the formula remains inside. In this way, the binding structure becomes clearly visible.

Example 2.70. *Consider the following three formulas:*

$$\begin{aligned} &(1) \quad \forall x\,(R(x) \land \exists y\, S(y)),\\ &(2) \quad \forall x\, R(x) \land \exists x\, S(x),\\ &(3) \quad \exists u\, \forall v\,(R(u) \land S(v)). \end{aligned}$$

The formula in (1) is clean, but not in prenex form. The formula in (2) is neither clean nor in prenex form. The formula in (3) is both clean and in prenex form.

In general, every formula can be transformed into an equivalent clean formula in prenex form by renaming bound variables and moving the quantifiers.

Proposition 2.71. *Let Σ be a signature and let φ be a Σ-formula. Then the following hold:*

1. *There exists a clean Σ-formula equivalent to φ.*

2. *There exists a clean Σ-formula in prenex form equivalent to φ.*

Proof Sketch. For part (1), one renames bound variables in such a way that no collisions with free variables occur and no variable is bound twice. One only has to ensure that the new variable does not appear anywhere else in the formula.

For part (2), one starts with a clean formula. If the formula is not yet in prenex form, one looks for a subformula in which a quantifier still occurs inside a conjunction, disjunction, or negation, and then moves this quantifier outward using the usual equivalences. Repeating this step gradually shifts the quantifiers to the front of the formula. Since each round brings the formula closer to prenex form, the procedure terminates after finitely many steps. □

Example 2.72. *Consider the formula*

$$\varphi = \neg(\forall y\, T(x, f(y)) \vee U(x)) \;\vee\; \exists x\, \forall z\, V(g(x), z)$$

over the signature $\Sigma = \{T/2, U/1, V/2, f/1, g/1\}$. *We first clean up the formula: the free variable* x *on the left-hand side and the bound variable* x *on the right-hand side are separated by renaming. For instance, we obtain*

$$\varphi \equiv_\Sigma \neg(\forall u\, T(x, f(u)) \vee U(x)) \;\vee\; \exists w\, \forall z\, V(g(w), z).$$

We now move the quantifiers outward step by step. First, we transform the negated universal quantifier into an existential quantifier:

$$\neg(\forall u\, T(x, f(u)) \vee U(x)) \equiv_\Sigma \exists u\, \neg(T(x, f(u)) \vee U(x)).$$

After that, the quantifiers of the right-hand part are moved outward as well. In the end, we obtain an equivalent formula in clean prenex form, for example

$$\exists u\, \exists w\, \forall z\, \Big(\neg(T(x, f(u)) \vee U(x)) \;\vee\; V(g(w), z)\Big).$$

For resolution methods, we now need a special form in which no existential quantifiers occur any longer.

Definition 2.73. *Let* Σ *be a signature and let* φ *be a* Σ*-formula. We say that* φ *is* in Skolem form *if* φ *is clean, in prenex form, and contains only universal quantifiers.*

Not every formula is equivalent to a formula in Skolem form. However, the theorem behind Skolemization says that existential quantifiers can be eliminated in such a way that satisfiability is preserved.

Definition 2.74. *Let φ and ψ be two Σ-formulas. Then φ and ψ are called* satisfiability-equivalent *if the following holds:*

$$\varphi \text{ is satisfiable if and only if } \psi \text{ is satisfiable.}$$

The transformation into Skolem form consists of two steps: First, one brings the formula into clean prenex form. Then one replaces each existentially quantified variable by a new functional term that depends exactly on the preceding universally quantified variables.

Proposition 2.75. *For every Σ-formula, there exists a satisfiability-equivalent formula in Skolem form, where the signature may be extended by new function symbols.*

Proof. Let χ be a Σ-formula. By the previous proposition, there exists a formula φ in cleaned prenex form that is equivalent to χ. Let Λ be the signature containing Σ together with all function symbols required for the Skolemization of φ.

By the *Coincidence Lemma* (Theorem 2.26), we have:

$$\chi \text{ is } \Sigma\text{-satisfiable} \iff \chi \text{ is } \Lambda\text{-satisfiable.}$$

Moreover, locality (Proposition 2.30) implies that φ and χ are also satisfiability-equivalent over Λ.

It therefore suffices to show that φ has a satisfiability-equivalent Skolem form.

Induction on the number of existential quantifiers in φ:

Base case: If φ contains no existential quantifier, then φ is already in Skolem form. In that case, we set $\varphi^s := \varphi$, and satisfiability-equivalence is immediate.

Inductive step: Let φ be a formula with at least one existential quantifier. Then we can write φ as

$$\varphi \;=\; \forall x_1 \cdots \forall x_n \exists y \, \psi,$$

where ψ is again a formula (which may itself still contain quantifiers). For the Skolemization, we replace y by a new function symbol f of arity n:

$$\varphi' := \forall x_1 \cdots \forall x_n \, \psi[y \mapsto f(x_1, \ldots, x_n)].$$

If φ is satisfiable: Let $I = \langle A, \beta \rangle$ be a model of φ. Then for every tuple $(a_1, \ldots, a_n) \in A^n$:

$$\langle A, \beta[x_1 \mapsto a_1, \ldots, x_n \mapsto a_n] \rangle \models^{\mathrm{FO}}_{\Sigma} \exists y \, \psi.$$

That is, there exists a witness $b \in A$ such that

$$\langle A, \beta[x_1 \mapsto a_1, \ldots, x_n \mapsto a_n, \, y \mapsto b] \rangle \models^{\mathrm{FO}}_{\Sigma} \psi.$$

We now interpret f in such a way that $f(a_1, \ldots, a_n) := b$. By the *Substitution Lemma* (Theorem 2.48), we obtain

$$\begin{aligned} &\langle A, \beta[x_1 \mapsto a_1, \ldots, x_n \mapsto a_n] \rangle \\ &\qquad \models^{\mathrm{FO}}_{\mathcal{L}} \psi[y \mapsto f(x_1, \ldots, x_n)] = \varphi' \end{aligned}$$

Hence, after extending I to the enlarged signature Λ, it becomes a model of φ'. Therefore, φ' is satisfiable.

If φ' is satisfiable: Let $I' = \langle A, \beta\rangle$ be a model of φ'. Then for all $a_1, \ldots, a_n \in A$:

$$\langle A, \beta[x_1 \mapsto a_1, \ldots, x_n \mapsto a_n]\rangle \models_{\mathcal{L}}^{\text{FO}} \psi[y \mapsto f(x_1, \ldots, x_n)].$$

Applying the *Substitution Lemma* once again, it follows that:

$$\langle A, \beta[x_1 \mapsto a_1, \ldots, x_n \mapsto a_n, y \mapsto f(x_1, \ldots, x_n)]\rangle \models_{\mathcal{L}}^{\text{FO}} \psi.$$

It follows immediately that I' is also a model of φ, and thus φ is satisfiable.

Conclusion of the induction: The formula φ' contains one existential quantifier fewer than φ. By the induction hypothesis, there therefore exists a satisfiability-equivalent Skolem form φ^s of φ'. Hence, φ^s is also satisfiability-equivalent to φ and ultimately to χ.

Consequently, for every Σ-formula χ, there exists a satisfiability-equivalent formula in Skolem form, where the signature is extended by new function symbols if necessary. □

Example 2.76. *We consider the formula*

$$\exists y\, \forall x\, \exists z\, (P(y) \wedge Q(x, z)).$$

The first existential variable y is replaced by a new constant symbol c. The second existential variable z depends on the preceding universally quantified variable x

and is therefore replaced by a new unary function symbol $h(x)$*. In this way we obtain a satisfiability-equivalent formula in Skolem form:*

$$\forall x\,(P(c) \wedge Q(x, h(x))).$$

It is often also useful to eliminate free variables. This is done by binding them existentially.

Proposition 2.77. *Let* Σ *be a signature and let* φ *be a* Σ*-formula with*

$$FV(\varphi) = \{x_1, \ldots, x_n\}.$$

Then the formula

$$\exists x_1 \cdots \exists x_n\, \varphi$$

is closed and satisfiability-equivalent to φ*.*

Proof. Let φ be a Σ-formula with free variables

$$FV(\varphi) = \{x_1, \ldots, x_n\}.$$

We consider the closed formula

$$\exists x_1 \cdots \exists x_n\, \varphi.$$

We must show that this formula is satisfiability-equivalent to φ.

Idea: By introducing existential quantifiers for all free variables, we make φ closed without changing the satisfiable structures. Any assignment that satisfies φ can serve as a witness for the existential quantifiers, and conversely, any satisfiable closed formula automatically yields an assignment for φ.

Direction 1: φ *satisfiable* $\implies \exists x_1 \cdots \exists x_n \, \varphi$ *satisfiable*

Assume that φ is satisfiable. Then there exists a model $I = \langle A, \beta \rangle$ over the signature Σ such that

$$\langle A, \beta \rangle \models^{\text{FO}} \varphi.$$

Since β already assigns values in A to all free variables of φ, we may use these values directly as witnesses for the existential quantifiers. Concretely, this means:

$$\langle A, \beta[x_1 \mapsto \beta(x_1), \ldots, x_n \mapsto \beta(x_n)] \rangle \models^{\text{FO}} \varphi.$$

By the semantics of existential quantifiers, it follows immediately that:

$$\langle A, \beta \rangle \models^{\text{FO}} \exists x_1 \cdots \exists x_n \, \varphi.$$

Hence $\exists x_1 \cdots \exists x_n \, \varphi$ is satisfiable.

Direction 2: $\exists x_1 \cdots \exists x_n \, \varphi$ *satisfiable* $\implies \varphi$ *satisfiable*

Assume that the closed formula $\exists x_1 \cdots \exists x_n \, \varphi$ is satisfiable. Then there exists a model $I = \langle A, \beta \rangle$ over Σ such that

$$\langle A, \beta \rangle \models^{\text{FO}} \exists x_1 \cdots \exists x_n \, \varphi.$$

By the definition of the semantics of existential quantifiers, there exist elements $a_1, \ldots, a_n \in A$ such that

$$\langle A, \beta[x_1 \mapsto a_1, \ldots, x_n \mapsto a_n] \rangle \models^{\text{FO}} \varphi.$$

Thus the assignment that maps $x_i \mapsto a_i$ yields a model of φ. Therefore, φ is satisfiable.

Conclusion: Since both directions have been shown, we have

$$\varphi \text{ satisfiable} \iff \exists x_1 \cdots \exists x_n \, \varphi \text{ satisfiable.}$$

Hence the formula $\exists x_1 \cdots \exists x_n \, \varphi$ is closed and satisfiability-equivalent to φ. □

Finally, we arrive at the form needed for ground resolution.

2.4.3 Matrix Clause Form

Definition 2.78. *Let Σ be a signature and let φ be a Σ-formula. We say that φ is* in matrix clause form *if φ is closed, is in Skolem form, and its quantifier-free part is in conjunctive normal form.*

Thus, if

$$\varphi = \forall x_1 \cdots \forall x_n \, (C_1 \wedge \cdots \wedge C_m)$$

is in matrix clause form, we often write φ more briefly as

$$\{C_1, \ldots, C_m\}.$$

Here each C_i is a clause.

Theorem 2.79. *For every Σ-formula φ, there exists a satisfiability-equivalent formula in matrix clause form, where the signature may be extended by new function symbols if necessary.*

Proof. Let φ be a Σ-formula. We want to construct a satisfiability-equivalent formula in matrix clause form, possibly after extending the signature by new function symbols.

Step 1: Skolemization. First we apply Proposition 2.73. It states that every Σ-formula φ has a satisfiability-equivalent formula ψ in Skolem form. In doing so, the signature Σ may be expanded to a larger signature $\mathcal{L}$ in order to include the required Skolem functions. The advantage of the Skolem form ψ is that all existential quantifiers have already been eliminated, so that only universal quantifiers remain in front of the matrix.

Step 2: Closing the formula. In order to apply matrix clause form meaningfully, the formula must be closed. By Proposition 2.75, we can transform ψ into a closed formula without changing its satisfiability. Concretely, all free variables are bound by existential quantifiers (cf. Proposition 2.76), which ensures that we can later work with an actual matrix.

Step 3: Conjunctive normal form of the matrix. The matrix of ψ is the quantifier-free part of the formula. Let this matrix be χ. By Proposition 2.26, every quantifier-free formula χ can be transformed into a satisfiability-equivalent conjunctive normal form χ'. Conjunctive normal form is a conjunction of disjunctions of literals, which is exactly the form required for matrix clause form.

Step 4: Replacing the matrix by its conjunctive normal form. We replace the original matrix χ in ψ

by its conjunctive normal form χ'. Since χ and χ' are satisfiability-equivalent, the whole formula remains satisfiability-equivalent to φ. The resulting formula now has the form

$$\forall x_1 \dots \forall x_n \, \chi',$$

where χ' is in conjunctive normal form. Hence we have obtained a formula in matrix clause form.

Conclusion: Thus, in four steps we have constructed from φ a formula in matrix clause form:

1. Skolemization of all existential quantifiers (Proposition 2.73),
2. closing the formula (Proposition 2.75),
3. transforming the matrix into conjunctive normal form (Proposition 2.26),
4. inserting the normal-form matrix back into the formula.

Since each of these transformations preserves satisfiability, the resulting formula in matrix clause form is satisfiability-equivalent to φ.

Therefore, for every Σ-formula there exists a satisfiability-equivalent formula in matrix clause form, where the signature may be extended by new function symbols if necessary. □

Example 2.80. *The formula*

$$\forall x \, \exists y \, (R(x) \to S(x, y))$$

is initially not in matrix clause form. After transformation into Skolem form, we obtain

$$\forall x\,(\neg R(x) \vee S(x, h(x))),$$

and this is already a formula in matrix clause form, since the quantifier-free part is a clause. The corresponding clause representation is simply

$$\{\{\neg R(x),\ S(x, h(x))\}\}.$$

2.4.4 Herbrand Models

For ground resolution, a special kind of structure is crucial, one in which terms no longer appear merely as labels, but themselves constitute the elements of the domain. The underlying idea is quite natural: instead of evaluating a formula in an abstract structure, we first consider all terms that can be formed from the signature at all. These terms then become the "building blocks" of the model.

Definition 2.81. *A Σ-term is called a* Σ-ground term *if it contains no variables. Thus, a ground term consists only of constants and function symbols.*

The set of all Σ-ground terms is called the Herbrand universe *of Σ and is denoted by*

$$\mathrm{GT}(\Sigma) = \{\, t \in \mathrm{Term}(\Sigma) \mid t \textit{ is a } \Sigma\textit{-ground term} \,\}.$$

Example 2.82. *Consider the signatures*

$$\Sigma_1 = \{a/0, f/1, g/2\} \qquad \textit{and} \qquad \Sigma_2 = \{c/0\}.$$

Then the Herbrand universe of Σ_1 *is already infinite:*

$$\mathrm{GT}(\Sigma_1) = \{a,\ f(a),\ f(f(a)),\ f(f(f(a))),\ \ldots,\\ g(a,a),\ g(f(a),a),\ g(a,f(a)),\ \ldots\}.$$

By contrast,

$$\mathrm{GT}(\Sigma_2) = \{c\}.$$

Thus, the more function symbols a signature has, the larger its associated Herbrand universe will typically be.

If a signature contains no constant symbol, then $\mathrm{GT}(\Sigma)$ is empty. For our later applications, however, this is not a serious restriction: by the coincidence lemma, we may extend the signature by a fresh constant if necessary, without essentially changing the satisfiability of the formulas under consideration.

We now turn to Herbrand structures themselves. The central idea is that the interpretation of the function symbols is already fixed: a compound term is interpreted simply by syntactically assembling the corresponding ground terms.

Definition 2.83. *Let* Σ *be a signature containing at least one constant symbol. A* Σ*-structure*

$$H = \langle H, P^H, \ldots, f^H, \ldots \rangle$$

is called a Herbrand structure *if the following holds:*

- *The domain* H *is exactly the Herbrand universe* $\mathrm{GT}(\Sigma)$.

- *Every constant symbol c is interpreted by the corresponding ground term, that is, by $c^H = c$.*
- *Every n-ary function symbol f is interpreted in the Herbrand structure in such a way that for every assignment $\beta\colon \mathrm{Var} \to H$,*

$$[\![f(t_1,\dots,t_n)]\!]_{\langle H,\beta\rangle} = f\big([\![t_1]\!]_{\langle H,\beta\rangle},\dots,[\![t_n]\!]_{\langle H,\beta\rangle}\big).$$

Thus, the only degrees of freedom in a Herbrand structure lie in the interpretation of the predicate symbols.

Example 2.84. *Let again $\Sigma_1 = \{a/0, f/1, g/2\}$. One possible Herbrand structure H over Σ_1 is given by*

$$\begin{aligned} H &= \mathrm{GT}(\Sigma_1), \\ f^H(t) &= f(t), \\ g^H(s,t) &= g(s,t), \\ a^H &= a. \end{aligned}$$

and, for example,

$$P^H = \{\, a,\ f(a),\ g(a,a)\,\}$$

for a unary predicate symbol $P/1$. If one were to choose a different Herbrand structure over the same signature, then the domain as well as the interpretation of the function symbols and constants would remain the same; only the set P^H could differ.

A *Herbrand interpretation* is an interpretation $\langle H, \beta \rangle$ in which H is a Herbrand structure. A Herbrand structure H is called a *Herbrand model* of a formula φ if $H \models^{\text{FO}} \varphi$.

For formulas in Skolem form, the existence of a Herbrand model is especially important. In fact, for such formulas, satisfiability is exactly the same as the existence of a Herbrand model.

Proposition 2.85. *Let Σ be a signature containing at least one constant symbol, and let φ be a closed Σ-sentence in Skolem form. Then:*

$$\varphi \textit{ is satisfiable} \quad \Longleftrightarrow \quad \varphi \textit{ has a Herbrand model.}$$

Proof. Let

$$\varphi = \forall x_1 \cdots \forall x_n \, \psi$$

with quantifier-free part ψ.

The implication from right to left is immediate: if φ has a Herbrand model, then φ has a model at all, and hence φ is satisfiable.

For the converse, assume that φ is satisfiable. Then there exists a Σ-structure A such that

$$A \models^{\text{FO}} \varphi.$$

From this we construct a Herbrand structure H.

The domain of H is the Herbrand universe $\text{GT}(\Sigma)$. The constant and function symbols are interpreted syntactically in H, so that ground terms in H are evaluated as themselves. For each k-ary predicate symbol P, we set

$$P^H = \{(t_1, \ldots, t_k) \in \text{GT}(\Sigma)^k \mid A \models^{\text{FO}} P(t_1, \ldots, t_k)\}.$$

Now define the map

$$\pi : \mathrm{GT}(\Sigma) \to A, \qquad \pi(t) = [\![t]\!]_A.$$

For each assignment β in H, we define

$$\beta^A := \pi \circ \beta.$$

We first prove the following auxiliary claim:

Auxiliary claim. For every quantifier-free formula θ and every assignment β in H, we have

$$\langle H, \beta \rangle \models^{\mathrm{FO}} \theta \iff \langle A, \beta^A \rangle \models^{\mathrm{FO}} \theta.$$

We prove this claim by induction on the structure of θ.

Case 1: θ is atomic, that is, $\theta = P(t_1, \ldots, t_k)$. It suffices to show that for every term t,

$$\pi\Big([\![t]\!]_{\langle H,\beta \rangle}\Big) = [\![t]\!]_{\langle A,\beta^A \rangle}.$$

This follows by induction on the structure of t. For variables, the statement is immediate from the definition of β^A; the same holds for constants. In the case of a function expression, one uses the syntactic interpretation of the function symbols in H. Hence, for atomic formulas,

$$\langle H, \beta \rangle \models^{\mathrm{FO}} P(t_1, \ldots, t_k),$$

if and only if

$$\langle A, \beta^A \rangle \models^{\mathrm{FO}} P(t_1, \ldots, t_k).$$

Case 2: $\theta = \neg\theta_1$. Then the claim follows directly from the induction hypothesis and the semantics of negation.

Case 3: $\theta = \theta_1 \wedge \theta_2$. Then

$$\langle H, \beta\rangle \models^{\mathrm{FO}} \theta_1 \wedge \theta_2$$

holds if and only if both formulas θ_1 and θ_2 are satisfied in $\langle H, \beta\rangle$. By the induction hypothesis, this is the case if and only if both formulas are also satisfied in $\langle A, \beta^A\rangle$. Thus the claim also holds for conjunctions.

Case 4: $\theta = \theta_1 \vee \theta_2$. The proof is analogous to the conjunction case.

This proves the auxiliary claim.

We now show that H satisfies the formula φ. Let β be any assignment in H. Since $A \models^{\mathrm{FO}} \forall x_1 \cdots \forall x_n\, \psi$, it follows in particular that

$$\langle A, \beta^A\rangle \models^{\mathrm{FO}} \psi.$$

By the auxiliary claim, this yields

$$\langle H, \beta\rangle \models^{\mathrm{FO}} \psi.$$

Since β was arbitrary, it follows that

$$H \models^{\mathrm{FO}} \forall x_1 \cdots \forall x_n\, \psi.$$

Hence H is a Herbrand model of φ.

Therefore we have shown:

$$\varphi \text{ is satisfiable} \iff \varphi \text{ has a Herbrand model.}$$

□

Example 2.86. *Consider the signature*

$$\Sigma = \{a/0, f/1, R/1\}$$

and the formula

$$\forall x\,(R(x) \lor R(f(x))).$$

A Herbrand structure over Σ *has the domain*

$$\{a, f(a), f(f(a)), f(f(f(a))), \ldots\}.$$

If, for example,

$$R^H = \{a, f(f(a)), f(f(f(f(a))))\}$$

is chosen, then H *is not a model of the formula, because for* $x = f(a)$ *neither* $R(f(a))$ *nor* $R(f(f(a)))$ *is true. If, on the other hand,* R^H *is chosen in such a way that for every ground term at least one of the term itself or its* f*-successor belongs to* R^H*, then the formula is satisfied. This example shows how strongly satisfaction in a Herbrand structure depends on the interpretation of the predicates.*

2.4.5 Herbrand's Theorem

The Herbrand models from the previous subsection are particularly useful because they reduce the semantic question of satisfiability to a purely syntactic level. The next step is no longer to consider a formula in Skolem form directly, but instead to collect all of its possible ground instances. This is precisely how the Herbrand expansion arises.

Definition 2.87. *Let Σ be a signature containing at least one constant, and let*

$$\varphi = \forall x_1 \cdots \forall x_n \, \psi$$

be a closed Σ-sentence in Skolem form, where ψ is quantifier-free. Then

$$\mathrm{HE}(\varphi) = \{\psi[x_1 \mapsto t_1, \ldots, x_n \mapsto t_n] \mid t_1, \ldots, t_n \in \mathrm{GT}(\Sigma)\}$$

is called the Herbrand expansion *of φ.*

The replacement of variables by ground terms is also called ground instantiation, *or simply* grounding.

The Herbrand expansion therefore contains all formulas obtained from the matrix of φ by independently replacing the universally bound variables with arbitrary ground terms.

Example 2.88. *We consider the signature*

$$\Sigma = \{a/0, f/1, R/2\}$$

and the sentence

$$\varphi = \forall x \, \forall y \, R(x, f(y)).$$

Then the Herbrand expansion begins with the instances

$$\begin{array}{rr} R(a, f(a)), & R(a, f(f(a))), \\ R(f(a), f(a)), & R(f(a), f(f(a))), \\ & \ldots \end{array}$$

More precisely,

$$\mathrm{HE}(\varphi) = \{\, R(t_1, f(t_2)) \mid t_1, t_2 \in \mathrm{GT}(\Sigma) \,\}.$$

For later use, it is often helpful to translate the Herbrand expansion into a purely propositional form. To do so, each occurring atom is replaced by a new 0-ary predicate symbol.

Definition 2.89. *Let ψ be a quantifier-free formula. We write*

$$\mathrm{Atom}(\psi) = \{\mu \in \mathrm{TF}(\psi) \mid \mu \textit{ is an atom}\}$$

for the set of its atomic subformulas.

The propositional Herbrand expansion *of φ is the set*

$$\begin{aligned}\mathrm{AHE}(\varphi) = \{\psi[\mu_1 \mapsto A_{\mu_1}, \ldots, \mu_m \mapsto A_{\mu_m}] \mid \\ \psi \in \mathrm{HE}(\varphi),\ \mathrm{Atom}(\psi) = \{\mu_1, \ldots, \mu_m\}\}.\end{aligned}$$

where, for each atom μ, a new 0-ary predicate symbol A_μ is introduced.

The term *propositional* is justified because the resulting formulas no longer contain variables or function symbols. All occurring predicates are 0-ary and behave semantically like propositional variables: they are either true or false.

Example 2.90. *For the formula from the previous example, we obtain*

$$\mathrm{AHE}(\varphi) = \{\, A_{R(a,f(a))},\ A_{R(a,f(f(a)))},\ A_{R(f(a),f(a))}, \ldots\}.$$

This becomes even more transparent with a slightly richer matrix, for example

$$\psi = P(a) \vee \neg Q(f(a)).$$

which yields the propositional formula

$$A_{P(a)} \vee \neg A_{Q(f(a))}.$$

The connection between satisfiability and the Herbrand expansion is close.

Proposition 2.91. *Let Σ be a signature containing at least one constant, and let φ be a closed Σ-sentence in Skolem form. Then:*

$$\begin{aligned} \varphi \textit{ is satisfiable} &\iff \mathrm{HE}(\varphi) \textit{ is satisfiable} \\ &\iff \mathrm{AHE}(\varphi) \textit{ is satisfiable.} \end{aligned}$$

Proof. We write

$$\varphi = \forall x_1 \cdots \forall x_n \, \psi$$

with quantifier-free part ψ.

By the previous subsection, φ is satisfiable if and only if φ has a Herbrand model. For a Herbrand structure H, however, we have:

$$H \models^{\mathrm{FO}} \forall x_1 \cdots \forall x_n \, \psi$$

if and only if for all ground terms $t_1, \ldots, t_n \in \mathrm{GT}(\Sigma)$ we have:

$$H \models^{\mathrm{FO}} \psi[x_1 \mapsto t_1, \ldots, x_n \mapsto t_n].$$

This is exactly the statement that H is a model of $\mathrm{HE}(\varphi)$. Hence,

$$\varphi \text{ is satisfiable} \iff \mathrm{HE}(\varphi) \text{ is satisfiable.}$$

For the propositional Herbrand expansion, one simply replaces each atom by a new 0-ary predicate symbol. In this way, the truth of a ground atom is encoded directly as the truth of the corresponding 0-ary symbol. Therefore, a Herbrand structure is a model of HE(φ) if and only if the corresponding propositional interpretation is a model of AHE(φ). Consequently,

$$\text{HE}(\varphi) \text{ is satisfiable} \iff \text{AHE}(\varphi) \text{ is satisfiable.}$$

□

We can now formulate Herbrand's theorem proper.

Theorem 2.92 (Herbrand's Theorem). *Let Σ be a signature containing at least one constant, and let φ be a closed Σ-sentence in Skolem form. Then:*

1. *φ is satisfiable if and only if every finite subset of* AHE(φ) *is satisfiable.*

2. *φ is unsatisfiable if and only if there exists a finite unsatisfiable subset of* AHE(φ).

Theorem 2.93 (Herbrand's Theorem). *Let Σ be a signature containing at least one constant, and let φ be a closed Σ-sentence in Skolem form. Then:*

1. *φ is satisfiable if and only if every finite subset of* AHE(φ) *is satisfiable.*

2. *φ is unsatisfiable if and only if there exists a finite unsatisfiable subset of* AHE(φ).

Proof. By the preceding proposition, φ is satisfiable if and only if $\mathrm{AHE}(\varphi)$ is satisfiable. Thus it suffices to show:

A set of quantifier-free formulas over 0-ary predicates is satisfiable if and only if every finite subset is satisfiable.

The implication from left to right is immediate: if a set is satisfiable, then every subset of it is satisfiable, and in particular every finite subset.

For the converse, assume that every finite subset of $\mathrm{AHE}(\varphi)$ is satisfiable. List the 0-ary predicate symbols occurring in $\mathrm{AHE}(\varphi)$ as

$$A_1, A_2, A_3, \ldots$$

in some fixed order.

We now construct inductively an increasing sequence of sets

$$\Phi_0 \subseteq \Phi_1 \subseteq \Phi_2 \subseteq \cdots$$

with $\Phi_0 := \mathrm{AHE}(\varphi)$ and the property that every finite subset of Φ_i remains satisfiable.

Suppose Φ_{i-1} has already been defined, and consider the symbol A_i. Then at least one of the following statements holds:

- every finite subset of $\Phi_{i-1} \cup \{A_i\}$ is satisfiable,
- every finite subset of $\Phi_{i-1} \cup \{\neg A_i\}$ is satisfiable.

For if both sets had unsatisfiable finite subsets, then their union would already yield a finite unsatisfiable subset of Φ_{i-1}, contradicting the induction hypothesis.

We therefore define

$$\Phi_i := \begin{cases} \Phi_{i-1} \cup \{A_i\}, & \text{if every finite subset of} \\ & \Phi_{i-1} \cup \{A_i\} \text{ is satisfiable,} \\ \Phi_{i-1} \cup \{\neg A_i\}, & \text{otherwise.} \end{cases}$$

This preserves at each step the property that every finite subset of Φ_i is satisfiable.

Now let

$$\Phi := \bigcup_{i \in \mathbb{N}} \Phi_i.$$

Then every finite subset of Φ is already contained in one of the sets Φ_i and is therefore satisfiable. Moreover, for each A_i, exactly one of the two formulas A_i or $\neg A_i$ belongs to Φ.

We now define a structure B with domain $\{b\}$ by stipulating for each 0-ary predicate symbol A_i that

$$A_i^B = \begin{cases} \{()\}, & \text{if } A_i \in \Phi, \\ \emptyset, & \text{if } \neg A_i \in \Phi. \end{cases}$$

We now show that $B \models^{\mathrm{FO}} \Phi$.

Assume this were not the case. Then there exists a formula $\alpha \in \Phi$ with $B \not\models^{\mathrm{FO}} \alpha$. Let

$$\{A_{i_1}, \ldots, A_{i_m}\}$$

be the set of 0-ary predicate symbols occurring in α. Then

$$\Delta := \{\alpha\} \cup \{A_{i_j} \mid A_{i_j} \in \Phi\} \cup \{\neg A_{i_j} \mid \neg A_{i_j} \in \Phi\}$$

is a finite subset of Φ and hence satisfiable. Let I be a model of Δ. Then I and B agree on all symbols occurring in α. By the coincidence lemma, proved earlier, it follows that

$$I \models^{\text{FO}} \alpha \quad \Longleftrightarrow \quad B \models^{\text{FO}} \alpha.$$

Since $B \not\models^{\text{FO}} \alpha$, it follows that $I \not\models^{\text{FO}} \alpha$ as well, contradicting the fact that I is a model of Δ. Hence,

$$B \models^{\text{FO}} \Phi$$

must hold.

Thus Φ is satisfiable, and therefore $\text{AHE}(\varphi)$ is satisfiable as well. Together with the preceding proposition, we obtain

$$\begin{aligned} \varphi \text{ is satisfiable} \quad &\Longleftrightarrow \quad \text{AHE}(\varphi) \text{ is satisfiable} \\ &\Longleftrightarrow \quad \text{every finite subset of AHE}(\varphi) \\ &\text{is satisfiable.} \end{aligned}$$

Statement (2) is the contrapositive form of (1). □

Example 2.94. *Consider the formula*

$$\varphi = \forall x \left(P(x) \vee \neg P(f(x))\right)$$

over the signature $\Sigma = \{a/0, f/1, P/1\}$. *Then* $\text{HE}(\varphi)$ *contains, among others, the formulas*

$$\begin{aligned} &P(a) \vee \neg P(f(a)), \\ &P(f(a)) \vee \neg P(f(f(a))), \\ &P(f(f(a))) \vee \neg P(f(f(f(a)))), \\ &\ldots \end{aligned}$$

Even a small finite subset of this Herbrand expansion may suffice to derive a contradiction. This is precisely the strength of Herbrand's theorem: to show that a formula is unsatisfiable, it is enough to find a suitable finite set of ground instances.

2.4.6 Ground Resolution

Herbrand's theorem shows that the unsatisfiability of a formula in Skolem form can already be witnessed by a finite set of ground instances. This is precisely the idea we use for a resolution procedure in predicate logic: we first transform the formula into matrix clause form, then instantiate all universally bound variables by ground terms, and finally apply propositional resolution to the clauses obtained in this way.

The goal of the procedure is the following decision problem:

Unsat(FO) Input: Σ-formula φ,
Output: Yes if and only if φ is unsatisfiable.

Definition 2.95. *Let $M = \{C_1, \ldots, C_m\}$ be a clause set in matrix clause representation. We then denote by*

$$\mathrm{HE}(M) = \mathrm{HE}(C_1) \cup \cdots \cup \mathrm{HE}(C_m)$$

the Herbrand expansion of M, and by

$$\mathrm{AHE}(M) = \mathrm{AHE}(C_1) \cup \cdots \cup \mathrm{AHE}(C_m)$$

the propositional Herbrand expansion of M.

Here, each clause set in HE(M) *is a set of ground instances of the original clauses, and in* AHE(M) *all atomic formulas are replaced by new* 0*-ary predicate symbols.*

Example 2.96. *We consider the formula*

$$\varphi = \forall x\, (\neg P(x) \vee P(f(x))) \ \wedge \ P(a) \ \wedge \ \neg P(f(f(a)))$$

over the signature $\Sigma = \{a/0, f/1, P/1\}$. *It is already in matrix clause form, and the corresponding clause representation is*

$$M = \{\{\neg P(x), P(f(x))\}, \{P(a)\}, \{\neg P(f(f(a)))\}\}.$$

Then HE(M) *contains, among others, the clauses*

$$\{\neg P(a), P(f(a))\}, \qquad \{\neg P(f(a)), P(f(f(a)))\},$$
$$\{P(a)\}, \qquad \{\neg P(f(f(a)))\}.$$

The propositional Herbrand expansion AHE(M) *is obtained from this by replacing the atoms, for instance, with* $A_{P(a)}$, $A_{P(f(a))}$, $A_{P(f(f(a)))}$, *and so on.*

The following proposition explains why, in ground resolution, we may handle arbitrary finite subsets of the propositional Herbrand expansion by means of propositional resolution.

Proposition 2.97. *Let* Σ *be a signature containing at least one constant, and let* M *be a clause set in matrix clause form. Then every finite set*

$$K \subseteq \mathrm{AHE}(M)$$

viewed as a first-order set of formulas is unsatisfiable if and only if K, regarded as a propositional clause set, is unsatisfiable.

Proof. Let $K \subseteq \mathrm{AHE}(M)$ be finite. Then K contains only finitely many 0-ary predicate symbols. We denote them by

$$A_1, \ldots, A_n.$$

Since no variables and no function symbols occur in K, every formula in K is already a propositional clause over the signature $\mathrm{Sig}(K) = \{A_1, \ldots, A_n\}$.

We first show that if K is satisfiable in first-order logic, then it is also satisfiable in propositional logic. To this end, let $I = \langle A, \beta \rangle$ be a model of K. By the coincidence lemma, we may restrict the signature to $\mathrm{Sig}(K)$ without changing the satisfiability of K. We now define a propositional valuation ω_I by

$$\omega_I(A_\mu) = \begin{cases} \text{true}, & \text{if } I \models^{\text{FO}} \mu, \\ \text{false}, & \text{otherwise.} \end{cases}$$

Since every formula in K consists only of 0-ary predicate symbols, its truth value in I coincides exactly with the truth value of the corresponding propositional clause under ω_I. Hence ω_I is a propositional model of K.

For the converse, assume that K is satisfiable in propositional logic. Let ω be a propositional valuation with $\omega \models K$. We define a Σ-structure A_ω with one-element domain

$$A = \{a\}$$

and assign, for every 0-ary predicate symbol A_μ occurring in K,

$$A_\mu^{A_\omega} = \begin{cases} \{()\}, & \text{if } \omega(A_\mu) = \text{true}, \\ \emptyset, & \text{if } \omega(A_\mu) = \text{false}. \end{cases}$$

Then, for every clause in K, its truth value in A_ω is exactly the same as its propositional truth value under ω. Consequently, $A_\omega \models^{\text{FO}} K$.

Thus first-order satisfiability and propositional satisfiability of K are equivalent. □

Together with Theorem 1.68, it now follows that a finite set $K \subseteq \text{AHE}(M)$ is unsatisfiable if and only if the empty clause can be derived from K by propositional resolution.

On this basis, we can now formulate the ground resolution procedure.

Definition 2.98 (Ground Resolution Procedure). *The input is a Σ-formula φ.*

1. *First transform φ into a satisfiability-equivalent formula in matrix clause form and denote its clause representation by M.*

2. *Initialize $K := \emptyset$.*

3. *Go through the clauses of* AHE(M) *one by one and in each step set*

$$K := K \cup \{C\}.$$

4. *After each step, check whether K is propositionally unsatisfiable, that is, whether the empty clause can be derived from K by propositional resolution.*

5. *If so, return* **Yes**. *Otherwise continue with the next clause.*

Theorem 2.99. *Given a Σ-formula φ as input, the ground resolution procedure returns* **Yes** *if and only if φ is unsatisfiable.*

Proof. Let M be the clause representation of a formula in matrix clause form that is satisfiability-equivalent to φ.

If the procedure returns **Yes**, *then φ is unsatisfiable.* Then there exists a finite set $K \subseteq \mathrm{AHE}(M)$ that is propositionally unsatisfiable. By the preceding proposition, K is then also unsatisfiable in predicate logic. Hence $\mathrm{AHE}(M)$ is unsatisfiable. By Herbrand's theorem, it follows that M is unsatisfiable and therefore φ is also unsatisfiable.

If φ is unsatisfiable, then the procedure returns **Yes**. Then, by Herbrand's theorem, some finite subset of $\mathrm{AHE}(M)$ is already unsatisfiable. Since the procedure adds all clauses of $\mathrm{AHE}(M)$ step by step, this finite subset will eventually be contained entirely in K. For this finite K, the propositional completeness of resolution yields the empty clause. Therefore the procedure returns **Yes**.

This shows that the ground resolution procedure returns **Yes** if and only if φ is unsatisfiable. □

Example 2.100. *We consider once again*

$$\varphi = \forall x\,(\neg P(x) \vee P(f(x))) \;\wedge\; P(a) \;\wedge\; \neg P(f(f(a))).$$

The relevant ground clauses are

$$\{\neg P(a), P(f(a))\}, \qquad \{\neg P(f(a)), P(f(f(a)))\},$$
$$\{P(a)\}, \qquad \{\neg P(f(f(a)))\}.$$

By propositional resolution, we obtain successively

$$\{P(a)\},\ \{P(f(a))\},\ \{P(f(f(a)))\},\ \emptyset.$$

Thus, a finite subset of AHE(M) *is already unsatisfiable, and the procedure detects the unsatisfiability of the original formula.*

2.4.7 Exercises and Solutions

Exercise 2.101. *Consider the first-order signature*

$$\Sigma = \{Q/2, S/1, f/2, i/1, b/0, c/0\}$$

and the following expressions:

$$\varphi_1 = Q(c,u) \wedge \neg Q(f(c,x),z),$$
$$\varphi_2 = \exists x\,(\neg Q(c,y) \vee \exists z\,\forall u\,Q(z,u)),$$
$$\varphi_3 = \forall y\,\exists w\,\forall u\,\neg(\neg\neg Q(u,i(w)) \wedge S(y)),$$
$$\psi_3 = \forall y\,\forall u\,\neg(\neg\neg Q(u,i(s(y))) \wedge S(y)),$$
$$\varphi_4 = \forall z\,\forall u\,(S(i(u)) \wedge S(i(i(z)))).$$

Determine for each of the following statements whether it is true or false:

1. *φ_1 is in NNF.*

2. *φ_2 is in rectified prenex form.*

3. *ψ_3 is the Skolemization of φ_3.*

4. *φ_4 is in matrix form.*

Solution 2.102. 1. ***True.*** *In φ_1, negations occur only directly in front of atomic formulas. Therefore, φ_1 is in negation normal form.*

2. ***False.*** *The formula is not in rectified prenex form, since not all quantifiers have been moved to the front.*

3. ***True.*** *ψ_3 is the Skolemization of φ_3. In particular, ψ_3 is in rectified prenex form and contains no existential quantifier.*

4. ***True.*** *φ_4 is closed, in Skolem form, and its quantifier-free part is in conjunctive normal form. Hence, φ_4 is in matrix form.*

2.5 Consequence and Resolution

In this section, we begin with resolution in first-order logic. The basic idea is the same as in propositional resolution: from two clauses with complementary literals, a new clause, the so-called resolvent, is formed. The difference is that in first-order logic we systematically handle variables and terms by means of substitutions.

As a result, proofs are often significantly shorter, since clauses do not have to be replaced completely by ground instances, but only specialized as far as necessary.

To develop this procedure rigorously, we first need a precise formulation of logical consequence in first-order logic. We then consider substitutions and general properties of consequence. After that, we introduce unification, from which first-order resolution is then built. We will later prove the soundness and completeness of the procedure via logical consequence and unsatisfiability.

2.5.1 Logical Consequence

As in propositional logic, we define logical consequence in first-order logic as a relation between a set of formulas and a single formula.

Definition 2.103. *Let Σ be a signature. The logical consequence relation for Σ is the relation*

$$\models_{\Sigma}^{\mathrm{FO}} \subseteq \mathcal{P}(\mathrm{FO}(\Sigma)) \times \mathrm{FO}(\Sigma)$$

with the following meaning: for every set $F \subseteq \mathrm{FO}(\Sigma)$ and every Σ-formula φ, we have

$$F \models_{\Sigma}^{\mathrm{FO}} \varphi$$

if and only if every Σ-model of F is also a Σ-model of φ.

If $F \models_{\Sigma}^{\mathrm{FO}} \varphi$ holds, we say that φ is a logical consequence of F. For two Σ-formulas φ and ψ, we write, by abbreviation,

$$\varphi \models_{\Sigma}^{\mathrm{FO}} \psi$$

for

$$\{\varphi\} \models^{\mathrm{FO}}_{\Sigma} \psi.$$

Because of the coincidence lemma, the signature is not essential in many situations, and we then simply write $\models^{\mathrm{FO}}$.

Example 2.104. *We consider the following consequences:*

$$\{P(a),\, P(a) \rightarrow Q(a)\} \models^{\mathrm{FO}} Q(a)$$

and

$$\forall x\, P(x) \models^{\mathrm{FO}} \exists x\, P(x).$$

In the first case, the consequence is a direct application of modus ponens. The second case shows that in first-order logic, an existential statement may follow from a universally valid statement, provided that the domain of the structures is non-empty.

For sentences, logical consequence behaves similarly to the way it does in propositional logic. Two simple but very useful observations concern substitutions and implications.

We begin by considering the effect of substitutions on sets of sentences.

Proposition 2.105. *Let* σ *be a substitution, and let* F *be a set of* Σ*-sentences. Then*

$$F \models^{\mathrm{FO}} \psi \quad \textit{for every } \psi \in \sigma(F).$$

Proof. Let $I = \langle A, \beta \rangle$ be an arbitrary model of F. Then I is a model of every formula $\varphi \in F$. Since φ is a sentence, its truth value does not depend on the assignment. Hence, in particular, for every assignment β' we have

$$\langle A, \beta' \rangle \models^{\mathrm{FO}} \varphi.$$

Applying this to the assignment induced by σ and using the substitution lemma, we obtain

$$I \models^{\mathrm{FO}} \sigma(\varphi).$$

This holds for every $\varphi \in F$. Therefore, I is also a model of $\sigma(F)$. It follows that

$$F \models^{\mathrm{FO}} \psi \quad \text{for every } \psi \in \sigma(F).$$

□

The next theorem is the predicate-logical form of the deduction theorem.

Theorem 2.106. *Let F be a set of Σ-sentences, and let φ, ψ be two Σ-sentences. Then*

$$F \cup \{\varphi\} \models^{\mathrm{FO}} \psi \iff F \models^{\mathrm{FO}} (\varphi \rightarrow \psi).$$

Proof. We argue in terms of model classes. First,

$$F \cup \{\varphi\} \models^{\mathrm{FO}} \psi$$

holds if and only if every structure in

$$\mathrm{MK}^{\mathrm{FO}}_{\Sigma}(F \cup \{\varphi\})$$

is also a model of ψ. Now,

$$\mathrm{MK}_{\Sigma}^{\mathrm{FO}}(F \cup \{\varphi\}) = \mathrm{MK}_{\Sigma}^{\mathrm{FO}}(F) \cap \mathrm{MK}_{\Sigma}^{\mathrm{FO}}(\varphi).$$

Hence the above statement is equivalent to saying that for every structure A with $A \models^{\mathrm{FO}} F$ and $A \models^{\mathrm{FO}} \varphi$, we also have $A \models^{\mathrm{FO}} \psi$. This, in turn, is exactly the case if for every structure A with $A \models^{\mathrm{FO}} F$ we have:

$$A \models^{\mathrm{FO}} \varphi \to \psi.$$

This proves that

$$F \cup \{\varphi\} \models^{\mathrm{FO}} \psi \quad \iff \quad F \models^{\mathrm{FO}} (\varphi \to \psi).$$

□

The connection between logical consequence and unsatisfiability is especially important for resolution.

Theorem 2.107. *Let F be a set of Σ-sentences, and let φ be a Σ-sentence. Then*

$$F \models^{\mathrm{FO}} \varphi \quad \iff \quad F \cup \{\neg\varphi\} \textit{ is unsatisfiable.}$$

Proof. First, assume that $F \cup \{\neg\varphi\}$ is unsatisfiable. Then no structure is a model of both F and $\neg\varphi$. Therefore, every structure that satisfies F must automatically be a model of φ. This is precisely what it means that

$$F \models^{\mathrm{FO}} \varphi.$$

Conversely, assume that $F \models^{\mathrm{FO}} \varphi$. Then every structure that satisfies F is also a model of φ. Consequently, there can be no structure that satisfies both F and $\neg\varphi$. Hence,

$$F \cup \{\neg\varphi\}$$

is unsatisfiable.

Thus both directions have been established. □

Example 2.108. *A useful application example is the entailment*

$$\forall x\,(P(x) \to Q(x)),\ P(a) \models^{\mathrm{FO}} Q(a).$$

By the previous theorem, this is equivalent to saying that the set

$$\{\forall x\,(P(x) \to Q(x)),\ P(a),\ \neg Q(a)\}$$

is unsatisfiable. This is precisely the point at which first-order resolution enters later on: it attempts to derive a contradiction from such sets systematically in the form of the empty clause.

2.5.2 Unification

In ground resolution, we work with fully instantiated clauses. This is very intuitive for a first introduction, but it is often unnecessarily coarse. In first-order resolution, we therefore want to leave variables in clauses and instantiate only as much as is actually needed for a single resolution step. This is exactly why we need *unification.*

The basic idea is simple: Two expressions are to be transformed by a substitution in such a way that they become syntactically identical afterwards. If this succeeds, the expressions are called *unifiable*. Of particular importance is the case in which one finds a substitution that is as general as possible, that is, a substitution that fixes only what is truly necessary while still permitting later specializations.

Definition 2.109. *Let A and B be expressions, that is, terms or atomic formulas. A substitution σ is called a* unifier *of A and B if*

$$\sigma(A) = \sigma(B)$$

holds. If such a unifier exists, then A and B are called unifiable.

We also transfer this notion to sets of terms: A substitution σ is a unifier of a set F of expressions if all expressions in $\sigma(F)$ are syntactically identical.

Example 2.110. *Consider the two atomic formulas*

$$A_1 = P(f(x), g(y), a) \qquad \textit{and} \qquad A_2 = P(f(b), g(c), a).$$

Then

$$\sigma = \{x \mapsto b,\ y \mapsto c\}$$

is a unifier, because

$$\sigma(A_1) = P(f(b), g(c), a) = \sigma(A_2).$$

By contrast,

$$B_1 = P(x, f(x)) \qquad \textit{and} \qquad B_2 = P(f(y), y)$$

are not unifiable. Indeed, any substitution that maps x to the term $f(y)$ would at the same time require, in the second argument, that $f(x)$ and y become equal. This would lead to an equation of the form

$$y = f(f(y))$$

and no finite term can satisfy such an equation.

Unifiers are not unique. We are therefore often interested in a unifier that commits as little as possible.

Definition 2.111. *A unifier σ of two expressions A and B is called a* most general unifier, *or* MGU, *if for every unifier τ of A and B there exists a substitution ρ such that*

$$\tau = \rho \circ \sigma.$$

In other words, every other unifier is obtained from σ by further specialization.

Example 2.112. *For the expressions*

$$C_1 = P(h(x), f(y, z)) \qquad \text{and} \qquad C_2 = P(h(f(a)), f(b, z))$$

we have

$$\sigma = \{x \mapsto f(a),\ y \mapsto b\}$$

as an MGU. The variable z need not be instantiated at all. By contrast,

$$\tau = \{x \mapsto f(a),\ y \mapsto b,\ z \mapsto c\}$$

is also a unifier, but not a most general one, because it is obtained from σ by further specialization.

For first-order resolution, we need a procedure that computes an MGU whenever one exists. Robinson's classical algorithm proceeds step by step from left to right and eliminates, at each stage, the first remaining visible syntactic discrepancy.

1. Start with the empty substitution $\sigma := \emptyset$.
2. Compare $\sigma(A)$ and $\sigma(B)$.
3. If the two expressions are identical, return σ.
4. Find the first position at which $\sigma(A)$ and $\sigma(B)$ differ.
5. If two different function symbols or constants occur at that position, then A and B are not unifiable.
6. Otherwise, one side contains a variable v and the other side a term t.
7. If v occurs in t, abort: in that case no unifier exists (*occurs check*).
8. Otherwise, extend the substitution by $v \mapsto t$ and set
$$\sigma := \{v \mapsto t\} \circ \sigma.$$
Then return to step 2.

Proposition 2.113. *Robinson's unification algorithm is correct and complete. That is: if two expressions A_1, A_2 are unifiable, then the algorithm produces a most general* *unifier (MGU); if they are not unifiable, it returns "not unifiable".*

Proof. Let A_1, A_2 be the inputs of the algorithm. We write

$$\sigma_0 = \mathrm{id}$$

for the empty substitution, and we denote by

$$\sigma_{k+1} = \gamma_k \circ \sigma_k$$

the substitution obtained at the kth step from a substitution of the form

$$\gamma_k = \{v_k \mapsto t_k\}$$

. We also set

$$B_k := \sigma_k(A_1), \qquad C_k := \sigma_k(A_2).$$

Termination. As long as the algorithm has not yet terminated, B_k and C_k differ at a first position when read from left to right. At the kth step, precisely this first point of difference is eliminated by the choice of γ_k. All preceding positions remain unchanged, because there B_k and C_k were already equal and the same substitution is applied to both expressions. Hence, after each successful step, the first differing symbol is shifted strictly to the right. Since A_1 and A_2 are finite expressions, this can happen only finitely many times. Therefore, the algorithm terminates unless it aborts earlier with the outcome "not unifiable".

Correctness in the success case. If the algorithm terminates with a substitution σ_n, then by the stopping condition

$$\sigma_n(A_1) = \sigma_n(A_2).$$

Thus, by definition, σ_n is a unifier of A_1 and A_2.

Generality of the substitution found. We now show that σ_n is in fact a most general unifier. To this end, we prove by induction on the executed steps the following stronger statement:

For every unifier θ of A_1 and A_2, there exists a substitution τ_k such that

$$\theta = \tau_k \circ \sigma_k.$$

Base case: For $k = 0$, we have $\sigma_0 = \mathrm{id}$. Then trivially

$$\theta = \theta \circ \mathrm{id},$$

so the claim holds with $\tau_0 = \theta$.

Inductive step: Assume the claim holds for σ_k. We consider the next step with

$$\sigma_{k+1} = \gamma_k \circ \sigma_k, \qquad \gamma_k = \{v_k \mapsto t_k\}.$$

Let θ be any unifier of A_1 and A_2. By the induction hypothesis, there exists a substitution τ_k such that

$$\theta = \tau_k \circ \sigma_k.$$

Then τ_k is a unifier of B_k and C_k, since

$$\tau_k(B_k) = \theta(A_1) = \theta(A_2) = \tau_k(C_k).$$

Since γ_k removes the first point of difference between B_k and C_k, we have there a variable v_k on one side and a term t_k on the other. Because τ_k is a unifier, we have

$$\tau_k(v_k) = \tau_k(t_k).$$

From $v_k \notin \mathrm{Var}(t_k)$ it follows that τ_k factors through a substitution τ_{k+1} that agrees with τ_k on all variables except v_k and in addition sets $v_k \mapsto \tau_k(t_k)$. Then

$$\tau_k = \tau_{k+1} \circ \gamma_k.$$

Consequently, we obtain

$$\theta = \tau_k \circ \sigma_k = (\tau_{k+1} \circ \gamma_k) \circ \sigma_k = \tau_{k+1} \circ (\gamma_k \circ \sigma_k) = \tau_{k+1} \circ \sigma_{k+1}.$$

This proves the inductive step.

After finitely many steps, therefore, for every unifier θ there exists a substitution τ_n such that

$$\theta = \tau_n \circ \sigma_n.$$

Hence σ_n is a most general unifier.

Non-unifiability. The algorithm can terminate with "not unifiable" only in the following two cases:

1. At the first point of difference, the two sides contain different constants or different function symbols. Then no substitution can make the two sides agree at that position, because substitutions alter only variables. Hence no unifier exists.

2. The occurs check fails: a variable v_k is to be replaced by a term t_k in which v_k itself occurs. Suppose, nevertheless, that a unifier θ existed. Then we would have

$$\theta(v_k) = \theta(t_k)$$

. Since v_k occurs in t_k, the term $\theta(t_k)$ would contain $\theta(v_k)$ as a proper subterm. This is impossible for finite terms. Therefore, in this case as well, no unifier can exist.

It follows that if A_1 and A_2 are unifiable, then the algorithm terminates with a most general unifier; if they are not unifiable, then it correctly returns "not unifiable".

□

Example 2.114. *We apply the algorithm to*

$$A_1 = P(f(x),\, g(y,y),\, h(z)),$$
$$A_2 = P(f(g(a)),\, g(b,b),\, h(c))$$

First, we compare the two expressions at the first position. There, A_1 contains the variable x, whereas A_2 contains the term $g(a)$. Hence we set

$$\sigma_1 = \{x \mapsto g(a)\}.$$

Then we obtain

$$\sigma_1(A_1) = P(f(g(a)),\, g(y,y),\, h(z))$$

and

$$\sigma_1(A_2) = P(f(g(a)),\, g(b,b),\, h(c)).$$

Next, we compare the second position. There, $\sigma_1(A_1)$ *contains the variable* y*, while* $\sigma_1(A_2)$ *contains the term* b*. Therefore, we extend the substitution to*

$$\sigma_2 = \{y \mapsto b\} \circ \sigma_1.$$

Then

$$\sigma_2(A_1) = P(f(g(a)),\ g(b, b),\ h(z))$$

and

$$\sigma_2(A_2) = P(f(g(a)),\ g(b, b),\ h(c)).$$

Finally, we compare the third position. There, $\sigma_2(A_1)$ *contains the variable* z*, while* $\sigma_2(A_2)$ *contains the term* c*. Thus we set*

$$\sigma_3 = \{z \mapsto c\} \circ \sigma_2.$$

This yields

$$\sigma_3(A_1) = \sigma_3(A_2) = P(f(g(a)),\ g(b, b),\ h(c)).$$

Hence,

$$\sigma_3 = \{x \mapsto g(a),\ y \mapsto b,\ z \mapsto c\}$$

is a unifier of A_1 *and* A_2*. By the preceding proposition,* σ_3 *is in fact a most general unifier.*

Example 2.115. *The notion can also be extended to sets of expressions. A substitution* σ *unifies a set* F *if* $\sigma(F)$ *contains exactly one element. For example,*

$$\sigma = \{x \mapsto a,\ z \mapsto f(a)\}$$

is a unifier of the set

$$\{\, h(x,z),\ h(a,f(x)),\ h(a,f(a))\,\},$$

because under σ all three expressions become

$$h(a,f(a)).$$

In fact, σ is again a most general unifier here.

2.5.3 First-Order Resolution

Having introduced the Herbrand expansion and Herbrand's theorem in the preceding sections, we can now turn to first-order resolution proper. The basic idea is the same as in propositional logic: two clauses are combined via complementary literals. The decisive difference is that in first-order logic we do not need to ground all terms in advance. Instead, we keep variables and instantiate them only where this is necessary for the individual resolution step. As a result, proofs become significantly shorter and more structured.

The central technical notion here is unification. Two literals may be resolved only if their atomic parts can be made syntactically identical by means of a substitution. We choose this substitution as generally as possible, that is, as a most general unifier.

Definition 2.116. *A clause is a finite set of literals, read in the usual way as the disjunction of its literals. Two clauses C_1 and C_2 are called* variable-disjoint *if* *no variable occurs in both clauses. By renaming bound variables, we can always achieve this.*

Definition 2.117. *Let C_1 and C_2 be two variable-disjoint clauses. Further, let*

$$H_1 = \nu_1(C_1) \qquad \textit{and} \qquad H_2 = \nu_2(C_2)$$

be two variants of C_1 and C_2 obtained by variable renaming and still variable- disjoint. Let $L_1 \in H_1$ and $\neg L_2 \in H_2$ be two literals whose atomic parts are unifiable, and let σ be a most general unifier of these atomic parts.

Then

$$R = \sigma(H_1 \setminus \{L_1\}) \cup \sigma(H_2 \setminus \{\neg L_2\})$$

is called a first-order resolvent *of C_1 and C_2.*

In substance, the definition says the following: we first make the two clauses variable- disjoint, then match the selected complementary literals by means of an MGU, and finally carry over all remaining literals after applying that same substitution.

Example 2.118. *Consider the clauses*

$$C_1 = \{\neg P(x),\, Q(x),\, T(a)\},$$
$$C_2 = \{P(f(y)),\, S(y)\}.$$

The literals $\neg P(x)$ and $P(f(y))$ are unifiable with the most general unifier

$$\sigma = \{x \mapsto f(y)\}.$$

The corresponding resolvent is therefore

$$R = \sigma(\{Q(x), T(a)\}) \cup \sigma(\{S(y)\}) = \{Q(f(y)), T(a), S(y)\}.$$

The next result shows that a resolution step is logically sound.

Proposition 2.119 (Resolution Lemma). *Let C_1 and C_2 be two clauses, and let R be a first-order resolvent of C_1 and C_2. Then*

$$\mathrm{MK}_{\Sigma}^{\mathrm{FO}}(\{C_1, C_2\}) \subseteq \mathrm{MK}_{\Sigma}^{\mathrm{FO}}(R).$$

In particular, every common model of C_1 and C_2 is also a model of the resolvent.

Proof. Let $H_1 = \nu_1(C_1)$ and $H_2 = \nu_2(C_2)$ be two variable-disjoint variants of C_1 and C_2. Further, let $L_1 \in H_1$ and $\neg L_2 \in H_2$ be the complementary literals used in the resolution step, and let σ be a most general unifier of their atomic parts. Then

$$R = \sigma(H_1 \setminus \{L_1\}) \cup \sigma(H_2 \setminus \{\neg L_2\}).$$

Now let A be a Σ-structure such that

$$A \models^{\mathrm{FO}} C_1 \qquad \text{and} \qquad A \models^{\mathrm{FO}} C_2.$$

Since renaming variables does not affect satisfiability, it follows that

$$A \models^{\mathrm{FO}} H_1 \qquad \text{and} \qquad A \models^{\mathrm{FO}} H_2.$$

We now show that $A \models^{\mathrm{FO}} R$.

To this end, let β be an arbitrary assignment. From $A \models^{\mathrm{FO}} H_1$ and $A \models^{\mathrm{FO}} H_2$ it follows in particular that

$$\langle A, \beta^\sigma \rangle \models^{\mathrm{FO}} H_1 \qquad \text{and} \qquad \langle A, \beta^\sigma \rangle \models^{\mathrm{FO}} H_2,$$

where β^σ denotes the assignment induced by σ. By the substitution lemma, we obtain

$$\langle A, \beta \rangle \models^{\mathrm{FO}} \sigma(H_1) \qquad \text{and} \qquad \langle A, \beta \rangle \models^{\mathrm{FO}} \sigma(H_2).$$

Assume, for contradiction, that $\langle A, \beta \rangle \not\models^{\mathrm{FO}} R$. Then all literals in

$$\sigma(H_1 \setminus \{L_1\}) \cup \sigma(H_2 \setminus \{\neg L_2\})$$

are false under β. Since $\sigma(H_1)$ and $\sigma(H_2)$ are each disjunctions of literals, the two selected literals $\sigma(L_1)$ and $\sigma(\neg L_2) = \neg\sigma(L_2)$ would therefore both have to be true. But because σ is a unifier, we have $\sigma(L_1) = \sigma(L_2)$. Hence we would have both

$$\langle A, \beta \rangle \models^{\mathrm{FO}} \sigma(L_1) \qquad \text{and} \qquad \langle A, \beta \rangle \models^{\mathrm{FO}} \neg\sigma(L_1),$$

which is a contradiction.

Therefore $\langle A, \beta \rangle \models^{\mathrm{FO}} R$ must hold. Since β was arbitrary, it follows that $A \models^{\mathrm{FO}} R$. This proves that

$$\mathrm{MK}_\Sigma^{\mathrm{FO}}(\{C_1, C_2\}) \subseteq \mathrm{MK}_\Sigma^{\mathrm{FO}}(R).$$

□

The resolution lemma is the correctness theorem for a single resolution step. For completeness, however, we also need a so-called lifting lemma. Informally, it says that every propositional resolution step on ground instances can be lifted to a first-order resolution step.

Proposition 2.120 (Lifting Lemma). *Let C_1 and C_2 be two clauses, and let σ_1, σ_2 be substitutions such that $\sigma_1(C_1)$ and $\sigma_2(C_2)$ are ground. If R' is a propositional resolvent of $\sigma_1(C_1)$ and $\sigma_2(C_2)$, then there exist a substitution τ and a first-order resolvent R of C_1 and C_2 such that*

$$R' = \tau(R).$$

Proof. Let

$$\sigma_1(C_1) = C_1' = \{A_0, \ldots, A_n\},$$
$$\sigma_2(C_2) = C_2' = \{B_0, \ldots, B_m\}.$$

be the two ground clauses. Without loss of generality, assume that $B_0 = \neg A_0$ and that

$$R' = \{A_1, \ldots, A_n, B_1, \ldots, B_m\}$$

is the propositional resolvent.

We may also assume that C_1 and C_2 are variable-disjoint; otherwise, we first rename variables. Since C_1 and C_2 are variable-disjoint, there exists a substitution σ' such that

$$\sigma'(C_1) = C_1' \qquad \text{and} \qquad \sigma'(C_2) = C_2'.$$

Now let $A \in C_1$ and $\neg B \in C_2$ be the literals used in forming the resolvent, so that $\sigma'(A) = A_0$ and $\sigma'(B) = B_0$. Since R' is a propositional resolvent, we have

$$\sigma'(A) = \sigma'(B).$$

Hence the atomic parts of A and B are unifiable. By the proposition on the Robinson algorithm, this unification has a most general unifier σ. Therefore,

$$R = \sigma(C_1 \setminus \{A\}) \cup \sigma(C_2 \setminus \{\neg B\})$$

is a first-order resolvent of C_1 and C_2.

Since σ' is a unifier of A and B and σ is a most general unifier, there exists a substitution τ such that

$$\sigma' = \tau \circ \sigma.$$

It follows immediately that

$$R' = \tau(R).$$

□

Example 2.121. *Consider the clauses*

$$C_1 = \{\neg R(x), T(x), U(a)\},$$
$$C_2 = \{R(g(y)), V(y)\}.$$

The two complementary literals $\neg R(x)$ and $R(g(y))$ are unifiable. A most general unifier is

$$\sigma = \{x \mapsto g(y)\}.$$

Hence

$$R = \sigma(\{T(x), U(a)\}) \cup \sigma(\{V(y)\}) = \{T(g(y)),\, U(a),\, V(y)\}$$

is a first-order resolvent of C_1 *and* C_2.

To illustrate the lifting lemma, let us now instantiate y, *for example, by the constant* a. *This yields the ground clauses*

$$\{\,\neg R(g(a)),\, T(g(a)),\, U(a)\,\} \qquad \textit{and} \qquad \{\,R(g(a)),\, V(a)\,\}.$$

On the propositional level, the resolvent

$$\{\,T(g(a)),\, U(a),\, V(a)\,\}$$

can be derived from these clauses. This is exactly the clause obtained by instantiating the first-order resolvent $R = \{T(g(y)), U(a), V(y)\}$ *with* $y \mapsto a$.

We can now iteratively collect all clauses obtainable by resolution. For a set of clauses K, we define: The set $\mathrm{Res}(K)$ consists of all first-order resolvents of pairs of clauses from K.

$$\mathrm{Res}_0(K) = K, \qquad \mathrm{Res}_{k+1}(K) = \mathrm{Res}_k(K) \cup \mathrm{Res}(\mathrm{Res}_k(K)),$$

and

$$\mathrm{Res}_\infty(K) = \bigcup_{k \in \mathbb{N}} \mathrm{Res}_k(K).$$

Thus, the set $\mathrm{Res}_\infty(K)$ contains all clauses that can be obtained from K in finitely many resolution steps.

Example 2.122. *Consider the clause set*

$$K = \{\{\neg S(x), T(x)\}, \{S(g(a))\}, \{\neg T(g(a))\}\}.$$

First, we resolve

$$\{\neg S(x), T(x)\} \quad \textit{and} \quad \{S(g(a))\}.$$

The complementary literals $\neg S(x)$ and $S(g(a))$ are unifiable with the unifier $x \mapsto g(a)$. Hence the resolvent is

$$\{T(g(a))\}.$$

Next, we resolve this clause with $\{\neg T(g(a))\}$. The literals $T(g(a))$ and $\neg T(g(a))$ are complementary, and their resolvent is the empty clause

$$\emptyset.$$

Therefore, K is first-order inconsistent.

We can now state the central completeness result.

Theorem 2.123 (Resolution Theorem)**.** *Let Σ be a signature and let φ be a Σ-formula. Then φ is unsatisfiable if and only if, for the clause representation M of the matrix clause form of φ, we have:*

$$\emptyset \in \mathrm{Res}_\infty(M).$$

Proof. We prove both directions.

If $\emptyset \in \mathrm{Res}_\infty(M)$, then φ is unsatisfiable. By the resolution lemma, each individual resolution step preserves the property that the class of models does not become larger. Hence, for all $k \in \mathbb{N}$, we have:

$$\mathrm{MK}_{\Sigma}^{\mathrm{FO}}(M) = \mathrm{MK}_{\Sigma}^{\mathrm{FO}}(\mathrm{Res}_k(M)).$$

Therefore,

$$\mathrm{MK}_{\Sigma}^{\mathrm{FO}}(M) = \mathrm{MK}_{\Sigma}^{\mathrm{FO}}(\mathrm{Res}_\infty(M)).$$

Now if $\emptyset \in \mathrm{Res}_\infty(M)$, then

$$\mathrm{MK}_{\Sigma}^{\mathrm{FO}}(\emptyset) = \emptyset.$$

Thus $\mathrm{MK}_{\Sigma}^{\mathrm{FO}}(M) = \emptyset$ as well, and so M is unsatisfiable. Since M is equisatisfiable with φ, it follows that φ is unsatisfiable.

If φ is unsatisfiable, then $\emptyset \in \mathrm{Res}_\infty(M)$. By Herbrand's theorem, there exists a finite set $K \subseteq \mathrm{AHE}(M)$ that is unsatisfiable. This set is a finite propositional clause set. Hence, by the propositional resolution theorem, there is a propositional resolution proof deriving the empty clause from K.

By the lifting lemma, each propositional resolution step can be lifted to a first-order resolution step. Therefore, the entire propositional proof can be translated step by step into a first-order proof. Thus, for a suitable k, we obtain:

$$\emptyset \in \mathrm{Res}_k(M),$$

and therefore, a fortiori,

$$\emptyset \in \mathrm{Res}_\infty(M).$$

□

Finally, we formulate the actual resolution calculus for formulas.

Definition 2.124. *For Σ-formulas φ and ψ, we write*

$$\varphi \vdash_{\mathrm{Res}} \psi$$

if and only if, for the clause representation M of the matrix clause form of

$$\varphi \wedge \neg\psi$$

the following holds:

$$\emptyset \in \mathrm{Res}_{\infty}(M).$$

Proposition 2.125. *For all Σ-formulas φ and ψ, the following equivalence holds:*

$$\varphi \models^{\mathrm{FO}} \psi \quad \Longleftrightarrow \quad \varphi \vdash_{\mathrm{Res}} \psi.$$

Proof. By the theorem on unsatisfiability and logical consequence (Theorem 2.103), we have

$$\varphi \models^{\mathrm{FO}} \psi \quad \Longleftrightarrow \quad \varphi \wedge \neg\psi \text{ is unsatisfiable.}$$

By the resolution theorem, this is the case if and only if the clause representation M of the matrix clause form of $\varphi \wedge \neg\psi$ satisfies

$$\emptyset \in \mathrm{Res}_{\infty}(M).$$

But by definition, this is exactly the statement $\varphi \vdash_{\mathrm{Res}} \psi$. □

2.5.4 Gödel's Completeness Theorem

In the previous section, we saw that first-order resolution is correct and complete for individual formulas: from every semantic consequence, a resolution proof can be obtained. In applications, however, one is usually given not just a single formula, but an entire set of assumptions. We therefore now extend the notion of resolution to sets of formulas.

The decisive idea is simple: a resolution proof is always finite. Hence, when working with a set of formulas, we may only rely on finite subsets. The essential question is therefore whether the restriction to finite subsets is too strong. The answer is no; this is precisely what the following compactness theorem provides.

Definition 2.126. *Let Σ be a signature, and let F be a set of Σ-formulas. We write*

$$F \vdash_{\text{Res}} \varphi$$

if and only if there exists a finite subset $F' \subseteq F$ such that

$$\bigwedge_{\psi \in F'} \psi \vdash_{\text{Res}} \varphi.$$

In other words, F proves φ in the resolution calculus if a finite subset of F already suffices together with φ.

That only finite subsets are allowed is of course necessary: on the one hand, every formula has finite length; on the other hand, successful resolution proofs always consist of only finitely many steps.

Theorem 2.127 (Compactness Theorem / Finiteness Theorem). *Let Σ be a finite signature, and let F be a countable set of Σ-formulas. Then the following hold:*

1. *F is satisfiable if and only if every finite subset of F is satisfiable.*
2. *F is unsatisfiable if and only if F has an unsatisfiable finite subset.*

Proof. Let

$$F = \{\varphi_1, \varphi_2, \varphi_3, \dots\}.$$

Statement (2) is the contrapositive form of (1); it therefore suffices to prove (1).

The direction from left to right is immediate: if F is satisfiable, then of course every subset of F is satisfiable, in particular every finite subset.

We now prove the converse direction. So assume that every finite subset of F is satisfiable. For each formula φ_i, by Theorem 2.73 we choose a formula $\varphi_i^{\text{Skolem}}$ in Skolem form that is equisatisfiable with φ_i. In doing so, we use fresh function symbols for different i, so that the individual Skolem forms do not interfere with one another.

For every finite subset

$$G = \{\psi_1, \dots, \psi_n\} \subseteq F$$

we consider the conjunction

$$\chi_G := \bigwedge_{j=1}^{n} \psi_j^{\text{Skolem}}.$$

Since each ψ_j^{Skolem} is equisatisfiable with ψ_j, and since the additional function symbols were chosen freshly for different formulas, χ_G is satisfiable.

Now consider the set of all propositional Herbrand expansions of these finite Skolem conjunctions:

$$\Psi_0 := \bigcup_{\substack{G \subseteq F \\ G \text{ finite}}} \text{AHE}(\chi_G).$$

Every finite subset of Ψ_0 is already contained in one of the sets $\text{AHE}(\chi_G)$ and is therefore satisfiable. By the same idea as in the proof of Herbrand's theorem (Theorem 2.90), it follows that Ψ_0 is satisfiable.

Now let I be a model of Ψ_0. From I, we construct a Herbrand structure H, exactly as in the proof of the proposition on the propositional Herbrand expansion (Proposition 2.89). Then we have:

$$H \models^{\text{FO}} \chi_G$$

for every finite subset $G \subseteq F$. For Ψ_0 contains the propositional Herbrand expansion of every such formula χ_G.

Thus H satisfies every finite subset of F. Since every φ_i is contained in some suitable finite subset of F, it even follows that

$$H \models^{\text{FO}} \varphi_i \qquad \text{for all } i \in \mathbb{N}.$$

Hence H is a model of F. Therefore F is satisfiable.

This proves (1), and (2) follows immediately by contraposition. □

Theorem 2.128 (Gödel's Completeness Theorem). *Let Σ be a finite signature. Then for every countable set F of Σ-formulas and every Σ-formula φ, the following holds:*

$$F \models^{\mathrm{FO}} \varphi \iff F \vdash_{\mathrm{Res}} \varphi.$$

Proof. We prove the two implications separately.

Direction "⇒". Assume that $F \models^{\mathrm{FO}} \varphi$. By Theorem 2.103, this is equivalent to the statement that

$$F \cup \{\neg\varphi\}$$

is unsatisfiable. By the Compactness Theorem, there is therefore already a finite subset $F' \subseteq F$ such that

$$F' \cup \{\neg\varphi\}$$

is unsatisfiable. Again by Theorem 2.103, it follows that

$$\bigwedge_{\psi \in F'} \psi \models^{\mathrm{FO}} \varphi.$$

For finite sets of formulas, however, we already showed in Proposition 2.121 that semantic consequence coincides exactly with derivability by resolution. Hence,

$$\bigwedge_{\psi \in F'} \psi \vdash_{\mathrm{Res}} \varphi.$$

By the definition of $\vdash_{\mathrm{Res}}$ for sets, this immediately implies

$$F \vdash_{\mathrm{Res}} \varphi.$$

Direction "$\Leftarrow$". Assume that $F \vdash_{\text{Res}} \varphi$. Then there exists a finite subset $F' \subseteq F$ such that

$$\bigwedge_{\psi \in F'} \psi \vdash_{\text{Res}} \varphi.$$

By Proposition 2.121, it follows that

$$\bigwedge_{\psi \in F'} \psi \models^{\text{FO}} \varphi.$$

Since $\bigwedge_{\psi \in F'} \psi$ is logically equivalent to F', we obtain

$$F' \models^{\text{FO}} \varphi.$$

Because $F' \subseteq F$, it follows a fortiori that

$$F \models^{\text{FO}} \varphi.$$

This proves the equivalence. □

In general, the Compactness Theorem and Gödel's Completeness Theorem also hold for uncountable sets of formulas. However, the corresponding proof requires additional set-theoretic tools, which we will discuss later.

2.5.5 Deductive Closures and Theories

Having introduced logical consequence for sets of formulas, it is natural to consider not only individual consequences, but also entire sets that are closed under logical consequence. This leads directly to the notions of deductive closure and theory. These concepts are particularly important for the later discussion of completeness and for the transition to Gödel's results.

In what follows, we write

$$\mathrm{FOSent}(\Sigma)$$

for the set of all closed Σ-formulas, that is, all sentences over Σ.

Definition 2.129. *Let Σ be a signature. The* deductive closure *of a set $F \subseteq \mathrm{FO}(\Sigma)$ is the set*

$$\mathrm{Cn}_\Sigma(F) = F \cup \{\varphi \in \mathrm{FOSent}(\Sigma) \mid F \models^{\mathrm{FO}} \varphi\}.$$

Thus, all sentences that follow logically from F are added to F.

A set $F \subseteq \mathrm{FO}(\Sigma)$ is called deductively closed *if*

$$F = \mathrm{Cn}_\Sigma(F).$$

A Σ-theory *is a deductively closed set of Σ-sentences.*

Accordingly, a theory is precisely a set of sentences that already contains all of its logical consequences. In this sense, a theory is a "closed" set of formulas.

One often writes Cn instead of Cn_Σ when the signature is clear from the context. Likewise, one usually speaks simply of a theory when a Σ-theory is meant.

Proposition 2.130 (Properties of the Consequence Operator)**.** *For every signature Σ, the operator Cn_Σ satisfies the following properties for all $F, F_1, F_2 \in \mathcal{P}(\mathrm{FO}(\Sigma))$:*

1. $F \subseteq \mathrm{Cn}_\Sigma(F)$ *(Extensivity),*
2. $\mathrm{Cn}_\Sigma(F) = \mathrm{Cn}_\Sigma(\mathrm{Cn}_\Sigma(F))$ *(Idempotence),*
3. *if* $F_1 \subseteq F_2$*, then* $\mathrm{Cn}_\Sigma(F_1) \subseteq \mathrm{Cn}_\Sigma(F_2)$ *(Monotonicity).*

Proof. We prove the three statements separately.

(1) Extensivity. By the definition of the consequence operator, we have

$$\mathrm{Cn}_\Sigma(F) = F \cup \{\varphi \in \mathrm{FOSent}(\Sigma) \mid F \models^{\mathrm{FO}} \varphi\}.$$

Thus, F is already contained in $\mathrm{Cn}_\Sigma(F)$ as the first component of the union. Therefore,

$$F \subseteq \mathrm{Cn}_\Sigma(F).$$

(2) Idempotence. First, we show that

$$\mathrm{Cn}_\Sigma(F) \subseteq \mathrm{Cn}_\Sigma(\mathrm{Cn}_\Sigma(F)).$$

This is immediate, because every set is contained in the collection of its own logical consequences: if $\mathrm{Cn}_\Sigma(F) \models^{\mathrm{FO}} \psi$, then by definition $\psi \in \mathrm{Cn}_\Sigma(\mathrm{Cn}_\Sigma(F))$.

For the reverse inclusion, let

$$\varphi \in \mathrm{Cn}_\Sigma(\mathrm{Cn}_\Sigma(F)).$$

Then either $\varphi \in \mathrm{Cn}_\Sigma(F)$, in which case we are done, or

$$\mathrm{Cn}_\Sigma(F) \models^{\mathrm{FO}} \varphi.$$

It therefore suffices to show that every model of F is also a model of $\mathrm{Cn}_\Sigma(F)$. So let A be a Σ-structure with $A \models^{\mathrm{FO}} F$. We must show that A satisfies every element of $\mathrm{Cn}_\Sigma(F)$.

If $\psi \in \mathrm{Cn}_\Sigma(F)$, then by definition there are two possibilities: either $\psi \in F$, in which case $A \models^{\mathrm{FO}} \psi$ holds immediately, or $F \models^{\mathrm{FO}} \psi$. In the second case, every structure satisfying F, and in particular A, satisfies the formula ψ. Hence,

$$A \models^{\mathrm{FO}} \mathrm{Cn}_\Sigma(F).$$

Therefore, from $\mathrm{Cn}_\Sigma(F) \models^{\mathrm{FO}} \varphi$ we immediately obtain

$$F \models^{\mathrm{FO}} \varphi,$$

hence

$$\varphi \in \mathrm{Cn}_\Sigma(F).$$

Thus,

$$\mathrm{Cn}_\Sigma(\mathrm{Cn}_\Sigma(F)) \subseteq \mathrm{Cn}_\Sigma(F)$$

is proved. Together with the first inclusion, this yields idempotence.

(3) Monotonicity. Let $F_1 \subseteq F_2$ and suppose that

$$\varphi \in \mathrm{Cn}_\Sigma(F_1).$$

Again, either $\varphi \in F_1$ or $F_1 \models^{\mathrm{FO}} \varphi$.

In the first case, since $F_1 \subseteq F_2$, we immediately have $\varphi \in F_2$ and therefore

$$\varphi \in \mathrm{Cn}_\Sigma(F_2).$$

In the second case, let A be an arbitrary Σ-structure with $A \models^{\mathrm{FO}} F_2$. Since $F_1 \subseteq F_2$, it follows that $A \models^{\mathrm{FO}} F_1$. From $F_1 \models^{\mathrm{FO}} \varphi$ we then obtain $A \models^{\mathrm{FO}} \varphi$. Since A was arbitrary, we have

$$F_2 \models^{\mathrm{FO}} \varphi,$$

hence $\varphi \in \mathrm{Cn}_\Sigma(F_2)$.

In both cases, it follows that

$$\mathrm{Cn}_\Sigma(F_1) \subseteq \mathrm{Cn}_\Sigma(F_2).$$

Thus, extensivity, idempotence, and monotonicity are proved. □

Theories are usually infinite. It is therefore often useful to consider theories that, in the strongest possible sense, are already completely decided.

Definition 2.131. *A Σ-theory T is called* complete *if, for every Σ-formula φ, the following holds:*

$$T \models^{\mathrm{FO}} \varphi \quad \textit{or} \quad T \models^{\mathrm{FO}} \neg\varphi.$$

Intuitively, this means that the theory already decides, for every formula, whether it holds or does not hold. For complete theories, no further semantic distinction is possible.

A central result now states that every consistent set of sentences can be extended to such a complete theory.

Theorem 2.132 (Lindenbaum-Lemma). *Every consistent set of sentences can be extended, by adding suitable sentences, to a consistent complete theory.*

The completeness of a theory should not be confused with the completeness of a resolution calculus. In the latter case, the question is whether every semantic consequence is also syntactically derivable. Here, by contrast, "complete" means that the theory itself contains a decision for every formula.

Besides the closure operator on sets of formulas, one also considers the associated *theory operator* on classes of structures.

Definition 2.133. *Let* Σ *be a signature. By*

$$\mathrm{Th}_\Sigma : \mathcal{P}(\mathrm{Struct}(\Sigma)) \to \mathcal{P}(\mathrm{FOSent}(\Sigma))$$

we denote the function

$$\mathrm{Th}_\Sigma(M) = \{\varphi \in \mathrm{FOSent}(\Sigma) \mid \textit{for all } I \in M,\ I \models^{\mathrm{FO}} \varphi\}.$$

The set $\mathrm{Th}_\Sigma(M)$ *is called the* theory of M.

Intuitively, $\mathrm{Th}_\Sigma(M)$ collects exactly those sentences that are true in all structures belonging to M. Thus, $\mathrm{Th}_\Sigma(M)$ is the semantic description of M at the level of sentences.

The relationship between Cn_Σ and Th_Σ reflects the same basic idea as in propositional logic: logical consequences on the one hand and the common truths of a model class on the other are tightly connected through a Galois correspondence. For our purposes, the key point is that starting from a set of sentences, its theory is again a deductively closed set. Likewise, a theory can be understood as the set of all sentences supported by its model class.

Theorem 2.134 (Relationship between Cn_Σ and Th_Σ). *Let Σ be a signature. The closure operator Cn_Σ and the theory operator Th_Σ are related in exactly the same close way as in propositional logic: from a set of sentences one obtains precisely its logical consequences, and from a class of structures one obtains precisely the sentences that hold in all of those structures.*

Formally, for all sets of Σ-sentences $F \in \mathcal{P}(\mathrm{FO}(\Sigma))$ and all sets of Σ-structures $M \in \mathcal{P}(\mathrm{Struct}(\Sigma))$, the following hold:

1. $\mathrm{Th}_\Sigma(\mathrm{MK}^{\mathrm{FO}}_\Sigma(F)) = \mathrm{Cn}_\Sigma(F)$,
2. $\mathrm{MK}^{\mathrm{FO}}_\Sigma(\mathrm{Th}_\Sigma(M)) = M$,
3. $\mathrm{Cn}_\Sigma(\mathrm{Th}_\Sigma(M)) = \mathrm{Th}_\Sigma(M)$.

In particular, $\mathrm{Th}_\Sigma(M)$ is deductively closed.

Proof. We use the definitions

$$\mathrm{MK}^{\mathrm{FO}}_\Sigma(F) = \{\, \mathcal{A} \in \mathrm{Struct}(\Sigma) \mid \mathcal{A} \models^{\mathrm{FO}} F \,\},$$

$$\mathrm{Th}_\Sigma(M) = \{\, \varphi \in \mathrm{FOSent}(\Sigma) \mid \forall \mathcal{A} \in M : \; \mathcal{A} \models^{\mathrm{FO}} \varphi \,\}.$$

as well as

$$\mathrm{Cn}_\Sigma(F) = F \cup \{\, \varphi \in \mathrm{FOSent}(\Sigma) \mid F \models^{\mathrm{FO}} \varphi \,\}.$$

(1) We first show that

$$\mathrm{Th}_\Sigma(\mathrm{MK}^{\mathrm{FO}}_\Sigma(F)) \subseteq \mathrm{Cn}_\Sigma(F).$$

So let

$$\varphi \in \mathrm{Th}_\Sigma(\mathrm{MK}_\Sigma^{\mathrm{FO}}(F)).$$

By definition, every structure in $\mathrm{MK}_\Sigma^{\mathrm{FO}}(F)$ satisfies φ. Hence,

$$\mathrm{MK}_\Sigma^{\mathrm{FO}}(F) \models^{\mathrm{FO}} \varphi.$$

But the structures in $\mathrm{MK}_\Sigma^{\mathrm{FO}}(F)$ are exactly those that satisfy F, so this is equivalent to

$$F \models^{\mathrm{FO}} \varphi.$$

Therefore $\varphi \in \mathrm{Cn}_\Sigma(F)$.

For the converse inclusion, let

$$\varphi \in \mathrm{Cn}_\Sigma(F).$$

Then either $\varphi \in F$ or $F \models^{\mathrm{FO}} \varphi$. In the first case, every structure in $\mathrm{MK}_\Sigma^{\mathrm{FO}}(F)$ satisfies φ immediately; in the second case, the same conclusion follows directly from $F \models^{\mathrm{FO}} \varphi$. Thus,

$$\varphi \in \mathrm{Th}_\Sigma(\mathrm{MK}_\Sigma^{\mathrm{FO}}(F)).$$

Hence we obtain the equality

$$\mathrm{Th}_\Sigma(\mathrm{MK}_\Sigma^{\mathrm{FO}}(F)) = \mathrm{Cn}_\Sigma(F).$$

(2) First, we always have

$$M \subseteq \mathrm{MK}_\Sigma^{\mathrm{FO}}(\mathrm{Th}_\Sigma(M)).$$

Indeed, let $\mathcal{A} \in M$. By the definition of $\mathrm{Th}_\Sigma(M)$, $\mathcal{A}$ satisfies every formula in $\mathrm{Th}_\Sigma(M)$. Hence,

$$\mathcal{A} \models^{\mathrm{FO}} \mathrm{Th}_\Sigma(M),$$

and therefore

$$\mathcal{A} \in \mathrm{MK}_\Sigma^{\mathrm{FO}}(\mathrm{Th}_\Sigma(M)).$$

So the inclusion is proved.

The fact that strict inclusion may occur in general is no accident: $\mathrm{Th}_\Sigma(M)$ does not describe the concrete class M itself, but only the sentences that hold in all structures from M. Hence, additional structures may satisfy exactly the same sentences and therefore also belong to $\mathrm{MK}_\Sigma^{\mathrm{FO}}(\mathrm{Th}_\Sigma(M))$.

Equality holds precisely when M is an elementary class, that is, when it is of the form

$$M = \mathrm{MK}_\Sigma^{\mathrm{FO}}(F)$$

for some set of sentences F.

If $M = \mathrm{MK}_\Sigma^{\mathrm{FO}}(F)$ for some set of sentences F, then part (1) yields

$$\begin{aligned}\mathrm{MK}_\Sigma^{\mathrm{FO}}(\mathrm{Th}_\Sigma(M)) &= \mathrm{MK}_\Sigma^{\mathrm{FO}}(\mathrm{Th}_\Sigma(\mathrm{MK}_\Sigma^{\mathrm{FO}}(F))) \\ &= \mathrm{MK}_\Sigma^{\mathrm{FO}}(\mathrm{Cn}_\Sigma(F)).\end{aligned}$$

Since F and $\mathrm{Cn}_\Sigma(F)$ have the same models, it follows that

$$\mathrm{MK}_\Sigma^{\mathrm{FO}}(\mathrm{Cn}_\Sigma(F)) = \mathrm{MK}_\Sigma^{\mathrm{FO}}(F) = M.$$

Conversely, every set of the form

$$\mathrm{MK}_\Sigma^{\mathrm{FO}}(\mathrm{Th}_\Sigma(M))$$

is, by definition, again a model class, hence an elementary class. Therefore, if

$$\mathrm{MK}_\Sigma^{\mathrm{FO}}(\mathrm{Th}_\Sigma(M)) = M,$$

then M itself is an elementary class.

(3) First, by extensivity of the closure operator, we have

$$\mathrm{Th}_\Sigma(M) \subseteq \mathrm{Cn}_\Sigma(\mathrm{Th}_\Sigma(M)).$$

It remains to prove the converse inclusion. So let

$$\varphi \in \mathrm{Cn}_\Sigma(\mathrm{Th}_\Sigma(M)).$$

Then either $\varphi \in \mathrm{Th}_\Sigma(M)$ or

$$\mathrm{Th}_\Sigma(M) \models^{\mathrm{FO}} \varphi.$$

In the second case, every structure satisfying all sentences from $\mathrm{Th}_\Sigma(M)$ also satisfies φ. In particular, this holds for every structure in M, and thus

$$\varphi \in \mathrm{Th}_\Sigma(M).$$

Hence,

$$\mathrm{Cn}_\Sigma(\mathrm{Th}_\Sigma(M)) \subseteq \mathrm{Th}_\Sigma(M).$$

Together with the first inclusion, we obtain

$$\mathrm{Cn}_\Sigma(\mathrm{Th}_\Sigma(M)) = \mathrm{Th}_\Sigma(M).$$

Therefore $\mathrm{Th}_\Sigma(M)$ is deductively closed.

□

The especially important point is this: if one starts with a class of structures and collects all sentences valid in it, one again obtains a deductively closed theory. This is precisely what makes theories a central object of model-theoretic study.

In what follows, we will need these notions in order to prepare carefully for the transition from logical consequence via the completeness theorem to Gödel's incompleteness theorems.

The Axiom of Choice and Zorn's Lemma: Foundations of Maximal Extensions

To prove the *Lindenbaum lemma* in full generality, we need a set-theoretic tool that goes beyond the logical notions used so far. The crucial issue is not the syntax of formulas, but the existence of *maximal extensions*—a concept that is central in many areas of mathematics. This is where the *axiom of choice* and *Zorn's lemma* enter the picture: two principles that are not constructively provable, yet have become indispensable in modern mathematics.

Basic Concepts: Order, Chain, and Maximality

Definition 2.135 (Partial Order, Chain, Maximal Element)**.** *Let* $(P, \leq)$ *be a* partially ordered set*, that is, a set* P *together with a relation* $\leq$ *that is reflexive, transitive, and antisymmetric.*

- *A subset $C \subseteq P$ is called a* chain *if for all $x, y \in C$, either $x \leq y$ or $y \leq x$ holds. Thus, a chain is a* totally ordered *(or* linearly ordered*) subset of P.*

- *An element $m \in P$ is called* maximal *if there is no element $p \in P$ with $m < p$ (that is, there is no p such that $m \leq p$ and $m \neq p$).*

Example 2.136. *Consider the set $P = \mathcal{P}(\mathbb{N})$ of all subsets of the natural numbers, ordered by inclusion $\subseteq$. A chain in P is a family of sets that are pairwise comparable (for example $\{1\}, \{1, 2\}, \{1, 2, 3\}, \dots$). A maximal element in P would be, for instance, $\mathbb{N}$ itself, provided it belongs to the family under consideration.*

The Central Principles

Axiom 2.137 (Axiom of Choice (AC))**.** *For every family $\mathcal{F} = \{A_i \mid i \in I\}$ of nonempty sets, there exists a* choice function *$f : I \to \bigcup_{i \in I} A_i$ such that for every $i \in I$, we have $f(i) \in A_i$.*

Remark 2.138. *The axiom of choice is a* principle of existence, *not of construction. It guarantees the existence of a choice function without providing an explicit rule for how to construct such a function. This is a central difference from constructive approaches in mathematics.*

Theorem 2.139 (Zorn's Lemma (ZL))**.** *Let $(P, \leq)$ be a nonempty partially ordered set. If every chain in P has an* upper bound *in P, then P contains a maximal element.*

Remark 2.140. *Zorn's Lemma is not a theorem in the ordinary sense, but rather a* maximality principle *in set theory. It is equivalent to the axiom of choice and makes it possible, in many contexts, to prove the existence of maximal objects without having to describe them explicitly.*

Theorem 2.141 (Well-Ordering Theorem)**.** *Every set M can be* well-ordered, *that is, there exists a linear order $\leq$ on M such that every nonempty subset of M has a* least element.

Example 2.142. *The natural numbers $\mathbb{N}$ are already well-ordered under their usual order. The real numbers $\mathbb{R}$ can also be well-ordered, but not by their usual order — this requires the axiom of choice.*

Equivalence of the Principles

In *Zermelo–Fraenkel set theory (ZF)*, the following statements are equivalent:

1. the *Axiom of Choice (AC)*,
2. *Zorn's Lemma (ZL)*,
3. the *Well-Ordering Theorem.*

Proof sketch. The equivalence of these principles is one of the central results of modern set theory. We briefly outline the main implications:

- **AC $\Rightarrow$ ZL:** Let $(P, \leq)$ be a partially ordered set in which every chain has an upper bound. One then uses the Axiom of Choice to carry out a transfinite construction that selects compatible upper bounds along chains. This yields a maximal chain, and its upper bound is a maximal element of P.

- **ZL $\Rightarrow$ AC:** Let $\mathcal{F} = \{A_i \mid i \in I\}$ be a family of nonempty sets. Consider the set $\mathcal{P}$ of all *partial choice functions* on $\mathcal{F}$, ordered by inclusion. Every chain in $\mathcal{P}$ has an upper bound, namely its union. By Zorn's Lemma, there exists a maximal partial choice function. One shows that it must in fact be defined on all of I; otherwise, it could be extended further, contradicting maximality. Hence a choice function on the whole family exists.

- **ZL $\Rightarrow$ Well-Ordering Theorem:** Let M be any set. Consider the collection $\mathcal{W}$ of all well-orderings on subsets of M, ordered by extension. Every chain in $\mathcal{W}$ has an upper bound given by its union. By Zorn's Lemma, there is a maximal well-ordering on some subset $N \subseteq M$. If $N \neq M$, one can extend this well-ordering by adjoining another element of M, which contradicts maximality. Therefore $N = M$, and M is well-orderable.

□

Significance for Logic: Lindenbaum's Lemma and Completeness

The equivalence of these principles is of fundamental importance for logic, since it guarantees the existence of *maximally consistent extensions* – a central tool in the proof of *Lindenbaum's Lemma*:

Theorem 2.143 (Idea of Lindenbaum's Lemma). *Every consistent set of sentences* Γ *in a language* $\mathcal{L}$ *can be extended to a* consistent and complete *theory* Γ^*.

Proof.

- Consider the set $\mathcal{T}$ of all consistent supersets of Γ, partially ordered by inclusion.
- Every chain in $\mathcal{T}$ has an upper bound, namely its union.
- By Zorn's Lemma (or the Axiom of Choice), there exists a maximal consistent set $\Gamma^* \in \mathcal{T}$.
- Γ^* is complete, because for every formula φ, either $\varphi \in \Gamma^*$ or $\neg\varphi \in \Gamma^*$ must hold; otherwise, Γ^* could be extended by φ or $\neg\varphi$, contradicting maximality.

□

The existence of such maximally consistent theories is the key to the *Henkin construction* and thus to the proof of *Gödel's Completeness Theorem*:

- Lindenbaum's Lemma provides maximally consistent theories.

- In the Henkin construction, these are extended by new constants to form a *Henkin theory.*
- From the Henkin theory, one constructs a *canonical model.*
- Gödel's Completeness Theorem follows: a theory Γ is satisfiable if and only if it is consistent.

Summary: Why These Principles Are Indispensable

The Axiom of Choice, Zorn's Lemma, and the Well-Ordering Theorem are not merely technical tools, but *cornerstones of modern mathematics.* They make it possible to establish the existence of maximal objects, even when these cannot be constructed explicitly. In logic, they are indispensable for:

- the proof of *Lindenbaum's Lemma* and thus the completeness of first-order logic,
- the construction of models for theories,
- the treatment of infinite sets in model theory.

Without these principles, the foundation of classical mathematics and logic would not be possible – a fact that underscores their central role.

2.5.6 Lindenbaum's Lemma: Existence of Maximal Consistent Theories

Theorem 2.144 (Lindenbaum's Lemma). *Let* $\Gamma \subseteq \mathrm{FOSatz}(\mathcal{L})$ *be a consistent set of sentences. Then there exists a maximal consistent and complete theory* Γ^* *such that*

$$\Gamma \subseteq \Gamma^*.$$

Proof. Consider the set

$$\mathcal{T} = \{\Delta \subseteq \mathrm{FOSatz}(\mathcal{L}) \mid \Gamma \subseteq \Delta \text{ and } \Delta \text{ is consistent}\},$$

that is, the set of all consistent supersets of Γ, ordered by inclusion. Since Γ itself is consistent, we have $\Gamma \in \mathcal{T}$; hence $\mathcal{T}$ is nonempty.

Now let $C \subseteq \mathcal{T}$ be a chain. We show that

$$\Sigma := \bigcup_{\Delta \in C} \Delta$$

is again an element of $\mathcal{T}$, and thus an upper bound for C. Clearly, $\Gamma \subseteq \Sigma$, because every $\Delta \in C$ already contains Γ. It remains to show that Σ is consistent.

Assume, for contradiction, that Σ were inconsistent. Then there would exist a finite subset

$$\{\varphi_1, \ldots, \varphi_n\} \subseteq \Sigma$$

that is already inconsistent. Since Σ is the union of a chain, all of $\varphi_1, \ldots, \varphi_n$ already belong to a common $\Delta \in C$. But since $\Delta \in \mathcal{T}$, the set Δ is consistent, and therefore every finite subset of Δ is consistent as well. This is a contradiction. Hence Σ is consistent, and therefore

$$\Sigma \in \mathcal{T}.$$

Consequently, every chain in $\mathcal{T}$ has an upper bound.

By Zorn's lemma, $\mathcal{T}$ therefore has a maximal element Γ^*. By definition, Γ^* is consistent and contains Γ.

It remains to show that Γ^* is complete. Let φ be an arbitrary $\mathcal{L}$-formula. We show that

$$\varphi \in \Gamma^* \quad \text{or} \quad \neg\varphi \in \Gamma^*.$$

Assume this were not the case. Then both

$$\Gamma^* \cup \{\varphi\} \qquad \text{and} \qquad \Gamma^* \cup \{\neg\varphi\}$$

are proper supersets of Γ^*. By the maximality of Γ^*, both sets must be inconsistent.

From the inconsistency of $\Gamma^* \cup \{\varphi\}$, the deduction theorem yields

$$\Gamma^* \vdash \neg\varphi.$$

Likewise, from the inconsistency of $\Gamma^* \cup \{\neg\varphi\}$, we obtain

$$\Gamma^* \vdash \varphi.$$

Hence Γ^* itself would be inconsistent, contradicting the choice of Γ^* as an element of $\mathcal{T}$. Therefore, for every formula φ, we must have:

$$\varphi \in \Gamma^* \quad \text{or} \quad \neg\varphi \in \Gamma^*.$$

Thus Γ^* is complete.

Altogether, Γ^* is a maximal consistent and complete theory with $\Gamma \subseteq \Gamma^*$. □

Significance for Logic

The Lindenbaum lemma is a central starting point for the subsequent completeness results. In particular, it provides the foundation for

- the *Henkin construction*, which yields a model for a consistent theory,
- the proof of *Gödel's completeness theorem*,
- and the treatment of infinite sets of formulas in model theory.

It shows that consistent theories can be extended to maximally consistent and complete theories, regardless of the cardinality of the language. In this way, it provides an essential building block for the classical semantics of first-order logic.

2.5.7 Henkin Construction

The Lindenbaum lemma gives us, for every consistent set of sentences, a maximally consistent and complete theory. However, this alone is not quite sufficient for model construction. The crucial additional step is to introduce *witnesses* for existential statements: if a theory claims that there exists an object with a certain property, then the language should also contain a term that names this object. This is exactly what the Henkin construction accomplishes.

The basic idea is simple but very powerful: for each existential statement, we introduce a fresh function symbol that provides a specific witness. This makes the later term interpretation a natural candidate for a model.

Definition 2.145. *Let $\mathcal{L}$ be a first-order language. For every $\mathcal{L}$-formula*

$$\exists x\, \varphi(x, \bar{y})$$

with possibly additional free variables $\bar{y} = y_1, \ldots, y_n$, we introduce a fresh n-ary function symbol f_φ that does not yet occur in $\mathcal{L}$. If $n = 0$, then f_φ is therefore a new constant symbol.

We denote the resulting expanded language by $\mathcal{L}^{\mathrm{H}}$.

Associated with each such existential formula is the corresponding *Henkin axiom*

$$\forall \bar{y}\, (\exists x\, \varphi(x, \bar{y}) \to \varphi(f_\varphi(\bar{y}), \bar{y})).$$

This axiom says: if there exists any witness for $\exists x\, \varphi(x, \bar{y})$ corresponding to the parameters $\bar{y}$, then the term named by $f_\varphi(\bar{y})$ is such a witness.

Definition 2.146. *Let $\Gamma \subseteq \mathrm{FOSent}(\mathcal{L})$ be a theory. The* Henkin extension *of Γ is the theory*

$$\Gamma^{\mathrm{H}} = \Gamma \cup \Big\{\forall \bar{y}\, (\exists x\, \varphi(x, \bar{y}) \to \varphi(f_\varphi(\bar{y}), \bar{y})) \\ \mid \exists x\, \varphi(x, \bar{y}) \in \mathrm{FO}(\mathcal{L})\Big\}.$$

where for each existential formula a fresh function symbol f_φ is chosen as above.

Intuitively, this means that in the expanded language every existential statement is equipped with a syntactic witness. This is exactly the property we will later need for term interpretation. In the term structure, the new symbols are read directly as constructions from their parameters, so existential statements can be witnessed by actually available terms.

Example 2.147. *Consider the formula*

$$\forall y\, \exists x\, R(x, y).$$

For this formula, we introduce a new unary function symbol f_R *and extend the theory by the Henkin axiom*

$$\forall y\, R(f_R(y), y).$$

Indeed, the statement

$$\forall y\, \exists x\, R(x, y)$$

is now witnessed by the fact that for each y*, the term* $f_R(y)$ *is available as a concrete candidate for a suitable* x*.*

If the formula is already closed, for example

$$\exists x\, P(x),$$

then a new constant symbol c_P *and the Henkin axiom*

$$\exists x\, P(x) \to P(c_P).$$

Thus, the Henkin construction extends a theory not only by complete decisions, but also by the linguistic means required for the later model construction. In the next section, we will combine this idea with the Lindenbaum lemma and derive Henkin's theorem from it.

2.5.8 Henkin's Theorem

The Henkin construction from the previous section provides suitable witnesses for existential statements in the form of new function or constant symbols. Together with the Lindenbaum lemma, this yields a model from a consistent theory. For languages with equality, an additional quotient construction is required in order to interpret equality correctly. To avoid obscuring the essential ideas, we first consider the case without equality.

Theorem 2.148 (Henkin's Theorem). *Let $\mathcal{L}$ be a first-order language, and let*

$$\Gamma \subseteq \mathrm{FOSent}(\mathcal{L})$$

be a consistent set of sentences. Then, after a Henkin extension of the language, there exists a maximally consistent and complete Henkin theory

$$\Delta \supseteq \Gamma,$$

whose term interpretation is a model of Δ. In particular, Γ is satisfiable.

Proof. First, as described in the Henkin construction, we extend the language $\mathcal{L}$ by fresh witness function symbols and, if necessary, by a fresh constant symbol, so that every existential statement in the extended language has an appropriate syntactic witness. We denote the resulting Henkin theory by Γ^{H}. It is consistent and contains Γ.

By the Lindenbaum lemma, there exists a maximally consistent and complete theory

$$\Delta \supseteq \Gamma^{\mathrm{H}}.$$

Since Δ is maximally consistent, it is in particular deductively closed.

We now define the term interpretation $\mathcal{T}^{\Delta}$ of Δ:

- The domain of $\mathcal{T}^{\Delta}$ is the set of all closed terms of the extended language.
- Each constant symbol is interpreted by the corresponding closed term.
- Each function symbol is evaluated syntactically, that is, by composing terms.
- A predicate symbol P holds for a tuple of closed terms $(t_1, \ldots, t_n)$ if and only if the atomic formula $P(t_1, \ldots, t_n)$ belongs to Δ.

We now prove the truth lemma: for every formula $\varphi(\bar{x})$ and every assignment s of the free variables to closed terms, we have

$$\varphi[s] \in \Delta \quad \Longrightarrow \quad \mathcal{T}^{\Delta}, s \models^{\mathrm{FO}} \varphi.$$

The proof proceeds by induction on the structure of φ.

Atomic case. If φ is atomic, the claim follows immediately from the definition of the predicate interpretation of $\mathcal{T}^{\Delta}$.

Negation. Assume that $\neg\psi[s] \in \Delta$. Since Δ is complete, it follows that $\psi[s] \notin \Delta$. If we had $\mathcal{T}^{\Delta}, s \models^{\mathrm{FO}} \psi$, then by the induction hypothesis $\psi[s] \in \Delta$ would hold, contradicting the previous observation. Hence $\mathcal{T}^{\Delta}, s \models^{\mathrm{FO}} \neg\psi$.

Conjunction. Assume that $(\psi \wedge \chi)[s] \in \Delta$. Since Δ is deductively closed, it follows that $\psi[s] \in \Delta$ and $\chi[s] \in \Delta$. By the induction hypothesis, therefore,

$$\mathcal{T}^{\Delta}, s \models^{\mathrm{FO}} \psi \qquad \text{and} \qquad \mathcal{T}^{\Delta}, s \models^{\mathrm{FO}} \chi,$$

so $\mathcal{T}^{\Delta}, s \models^{\mathrm{FO}} \psi \wedge \chi$.

Disjunction. Assume that $(\psi \vee \chi)[s] \in \Delta$. If neither $\psi[s]$ nor $\chi[s]$ were in Δ, then, by completeness of Δ, both negations $\neg\psi[s]$ and $\neg\chi[s]$ would belong to Δ. But then $\neg(\psi \vee \chi)[s] \in \Delta$ would also follow, contradicting the consistency of Δ. Thus at least one of the two formulas $\psi[s]$ or $\chi[s]$ belongs to Δ. By the induction hypothesis, either ψ or χ is therefore true in $\mathcal{T}^{\Delta}$, and hence so is $\psi \vee \chi$.

Universal quantifier. Assume that $(\forall x\, \psi)[s] \in \Delta$. Then deductive closure of Δ implies that for every closed term t,

$$\psi[s[x \mapsto t]] \in \Delta$$

holds. By the induction hypothesis, for every closed term t we also have

$$\mathcal{T}^{\Delta}, s[x \mapsto t] \models^{\mathrm{FO}} \psi.$$

Thus $\mathcal{T}^{\Delta}, s$ satisfies the universal statement $\forall x\, \psi$.

Existential quantifier. Assume that $(\exists x\, \psi)[s] \in \Delta$. By the Henkin construction, Δ contains the corresponding Henkin axiom, in particular the formula

$$\forall \bar{y} (\exists x\, \psi(x, \bar{y}) \rightarrow \psi(f_{\psi}(\bar{y}), \bar{y})),$$

where f_{ψ} is the fresh witness function symbol for $\exists x\, \psi$. Applying this to the assignment s and using modus ponens, we obtain

$$\psi[s(x \mapsto f_{\psi}(\bar{y}))] \in \Delta.$$

The induction hypothesis therefore yields

$$\mathcal{T}^{\Delta}, s[x \mapsto f_{\psi}(\bar{y})] \models^{\mathrm{FO}} \psi.$$

Hence,

$$\mathcal{T}^{\Delta}, s \models^{\mathrm{FO}} \exists x\, \psi.$$

This proves the truth lemma. In particular, every formula in Δ is true in $\mathcal{T}^{\Delta}$. Hence $\mathcal{T}^{\Delta}$ is a model of Δ.

Since $\Gamma \subseteq \Delta$, $\mathcal{T}^{\Delta}$ is in particular also a model of Γ. Therefore Γ is satisfiable.

Together with the soundness theorem for predicate logic proved earlier, we now obtain Gödel's completeness theorem in its general form as an immediate consequence: a set of sentences is semantically unsatisfiable if and only if it syntactically leads to a contradiction. □

Remark 2.149 (Equality). *In the proof above, we assumed that the language does not contain an equality symbol. If the language does contain equality, the interpretation of terms must be adapted accordingly.*

In that case, one defines an equivalence relation on the set of closed terms by

$$t_1 \sim t_2 \iff t_1 = t_2 \in \Delta.$$

The domain of the model then consists of the equivalence classes of closed terms. Functions and relations are interpreted in a well-defined way on these classes.

With this construction, one also obtains a model of the theory in the presence of equality. Hence, the completeness theorem remains fully valid for languages with equality as well.

2.5.9 Gödel's Completeness Theorem

In what follows, we use the Henkin construction developed earlier, together with Henkin's theorem, to prove Gödel's completeness theorem directly. Unlike the previous version, we do not proceed via a reduction to the compactness theorem. Instead, the unsatisfiability of $\Gamma \cup \{\neg\varphi\}$ is turned directly into a syntactic refutation, which can be formulated as a resolution derivation by means of Proposition 2.121.

Theorem 2.150 (Gödel's Completeness Theorem). *Let $\mathcal{L}$ be a first-order language and let $\Gamma \subseteq \mathrm{FOSent}(\mathcal{L})$ be a* *set of sentences. Then, for every sentence φ, the following holds:*

$$\Gamma \models^{\mathrm{FO}} \varphi \quad \Longleftrightarrow \quad \Gamma \vdash_{\mathrm{Res}} \varphi.$$

Proof. We prove the two directions separately.

Direction "$\Rightarrow$". Assume that $\Gamma \models^{\mathrm{FO}} \varphi$. Then there can be no model that satisfies all sentences in Γ and also satisfies $\neg\varphi$. Therefore, the set

$$\Gamma \cup \{\neg\varphi\}$$

is unsatisfiable.

By Henkin's theorem, every consistent set of sentences has a model. Hence, $\Gamma \cup \{\neg\varphi\}$ must be inconsistent.

Now let M be the matrix clause form associated with $\Gamma \cup \{\neg\varphi\}$. Then Proposition 2.121, or equivalently the resolution theorem used there, implies that the empty clause is derivable:

$$\emptyset \in \mathrm{Res}^{\infty}(M).$$

By definition, this means exactly that

$$\Gamma \vdash_{\mathrm{Res}} \varphi.$$

Direction "$\Leftarrow$". Assume that $\Gamma \vdash_{\mathrm{Res}} \varphi$. Then the matrix clause form M of $\Gamma \cup \{\neg\varphi\}$ is refutable. By the soundness of the resolution calculus, it follows that

$$\Gamma \cup \{\neg\varphi\}$$

is unsatisfiable. Hence, no model of Γ can satisfy $\neg\varphi$ at the same time, and therefore

$$\Gamma \models^{\mathrm{FO}} \varphi.$$

This proves the equivalence. ∎

The Compactness Theorem

The completeness theorem proved above allows us to derive the compactness theorem of first-order logic elegantly as a direct consequence.

Theorem 2.151 (Compactness Theorem). *Let $\mathcal{L}$ be a first-order language and let $\Gamma \subseteq \mathrm{FOSatz}(\mathcal{L})$ be a set of sentences. Then the following holds:*

Γ is satisfiable if and only if every finite subset $\Gamma_0 \subseteq \Gamma$ is satisfiable.

Proof. The implication from left to right is immediate: if Γ is satisfiable, then Γ has a model, and every such model also satisfies every subset $\Gamma_0 \subseteq \Gamma$.

For the converse, assume that every finite subset $\Gamma_0 \subseteq \Gamma$ is satisfiable, and show that Γ itself is satisfiable.

Suppose, for contradiction, that Γ were unsatisfiable. Then in particular,

$$\Gamma \models^{\mathrm{FO}} \bot.$$

By Gödel's completeness theorem, it follows that

$$\Gamma \vdash_{\mathrm{Res}} \bot.$$

By the definition of a resolution derivation, there then exists a finite subset $\Gamma_0 \subseteq \Gamma$ such that already

$$\Gamma_0 \vdash_{\text{Res}} \bot$$

holds.

By the soundness of the resolution calculus, Γ_0 is therefore unsatisfiable. This contradicts the assumption that every finite subset of Γ is satisfiable.

Hence Γ must be satisfiable. □

2.5.10 From the Completeness Theorem to the Incompleteness Theorems

With Gödel's completeness theorem, we established a central connection between syntax and semantics in predicate logic: a formula is logically valid if and only if it is provable. In this sense, predicate logic is complete – every semantically true statement can also be derived syntactically.

It is crucial, however, to understand that this claim refers exclusively to logic itself, not to specific mathematical theories. This is precisely where Gödel's incompleteness theorems enter the picture. Whereas the completeness theorem highlights the expressive power and effectiveness of formal proof systems, the incompleteness theorems reveal fundamental limitations of such systems.

To avoid misunderstandings, we therefore distinguish between two different notions of completeness:

- **Semantic completeness (Gödel, completeness theorem):** A calculus is complete if every formula that is true in all models is also provable, i.e.

$$\models \varphi \Rightarrow \vdash \varphi.$$

- **Syntactic completeness (theories):** A theory T is called complete if, for every formula φ, either $T \vdash \varphi$ or $T \vdash \neg\varphi$ holds.

Gödel's completeness theorem guarantees the first property for predicate logic. Gödel's incompleteness theorems, however, show that sufficiently expressive, recursively enumerable theories (such as formalized arithmetic) cannot have the second property – provided they are consistent.

In particular, the first incompleteness theorem states that every such theory contains true but unprovable statements. The second incompleteness theorem sharpens this result by showing that such a theory cannot prove its own consistency.

To formulate and prove these results precisely, we need a number of new concepts that go beyond the purely logical perspective developed so far. Central to this is the ability to encode syntactic objects – such as formulas and proofs – within arithmetic itself and to reason about them there. This leads to the following fundamental notions:

- **Formalized Theories and Arithmetic:** We consider theories that are sufficiently strong to express elementary statements about the natural numbers.

- **Recursive and Recursively Enumerable Sets:** These notions make precise when a set of formulas (for example, the axioms or the provable sentences of a theory) can be described algorithmically.

- **Gödel Numbering:** A systematic encoding of formulas, terms, and proofs by natural numbers, so that syntactic statements can be formulated arithmetically.

- **Representability of Relations:** The possibility of expressing within arithmetic properties such as "x is the Gödel code of a proof of φ."

- **Self-Reference and the Fixed-Point Theorem:** Techniques by which formulas are constructed that speak about their own provability.

- **Consistency and ω-Consistency:** More refined notions of consistency that are relevant for the precise formulation of the incompleteness theorems.

The transition from completeness to incompleteness thus marks a shift in perspective: whereas up to this point we have investigated what is in principle provable in logic, we now turn to the question of which statements must remain unprovable within concrete formal theories.

In what follows, we develop the necessary technical tools and then present the proofs of Gödel's two incompleteness theorems.

Formalized Theories and Arithmetic

We must first make precise the framework in which the following results are formulated. Gödel's incompleteness theorems do not concern arbitrary theories, but rather those that are sufficiently strong to express the elementary arithmetic of the natural numbers. This means that the theory is able to formalize basic statements about natural numbers and their simple arithmetic operations.

Accordingly, in what follows we consider formalized theories T in a suitable signature that contains, in particular, constants and functions for describing arithmetic, such as 0, a successor function, and operations like addition and multiplication. Such theories make it possible to formulate and investigate elementary arithmetical relations in a fully formal way.

This embedding of arithmetic is essential. It later allows us to encode syntactic objects such as formulas, terms, and proofs by natural numbers, and to express their properties within the theory itself. In this way, the theory can, in a certain sense, speak about its own syntax.

In the next step, we make precise what it means for such a theory to be described effectively. For this purpose, we introduce the notions of recursiveness and recursive enumerability.

Recursiveness and Recursive Enumerability

Having fixed the arithmetical framework in which we work, we now ask how the underlying theory itself is given. For the later treatment of the incompleteness theorems, it is not enough that a theory has formal axioms; rather, its set of axioms must be describable in a precise and effective way. This is exactly where the notions of recursiveness and recursive enumerability enter the picture.

To begin with, we consider sets of natural numbers. Later, by means of a suitable coding, we will also be able to regard syntactic objects such as formulas and proofs as natural numbers.

Definition 2.152. *A set $A \subseteq \mathbb{N}$ is called* recursive *or* decidable *if there exists a procedure that, for every number $n \in \mathbb{N}$, determines after finitely many steps whether $n \in A$ holds or not.*

Equivalently, there exists a computable characteristic function

$$\chi_A : \mathbb{N} \to \{0, 1\},$$

where $\chi_A(n) = 1$ if and only if $n \in A$.

Definition 2.153. *A set $A \subseteq \mathbb{N}$ is called* recursively enumerable *(in computer science also called semi-decidable) if there exists a procedure that outputs all elements of A in a possibly infinite list. Repetitions may occur, and it need not be decidable whether a given n does not belong to A.*

Equivalently, there exists a procedure that, on input n, halts after finitely many steps if and only if $n \in A$.

Intuitively, this means that recursive sets can be *decided*, whereas recursively enumerable sets can at least be *found* or *searched through*, but in general not fully decided. Every recursive set is recursively enumerable, but the converse does not hold in general.

For our purposes, this distinction is of central importance. A formal theory should not consist merely of a bare collection of axioms; rather, its axioms should be describable in an effective way. This is why we will later require that the set of axioms of such a theory be at least recursively enumerable. Then, in principle, it becomes possible to study its theorems systematically and to regard proofs as finite, verifiable objects.

Up to this point, however, we have only spoken about sets of natural numbers. For Gödel's incompleteness theorems, we must now learn how to encode syntactic objects themselves – that is, formulas, terms, and proofs – as natural numbers. It is precisely this encoding that is provided by Gödel numbering, to which we now turn.

Gödel Numbering

The central idea is to regard syntactic objects not merely as formal strings of symbols, but to encode them themselves as natural numbers. This is the decisive step that later makes it possible to formulate statements about formulas and proofs by means of arithmetic.

Since our formal languages consist of only finitely many symbols, and every formula as well as every proof is built from a finite sequence of such symbols, each of these strings can be described uniquely by a natural number. Such an encoding is called a *Gödel numbering.*

Definition 2.154. *A* Gödel numbering *is an effective and injective encoding of the syntactic objects of a formal language – such as symbols, terms, formulas, and proofs – by natural numbers.*

For each syntactic object x, we denote its code by $\ulcorner x \urcorner$.

Intuitively, this means the following: we first assign a natural number to each symbol of the language. A finite sequence of symbols—for example, a term, a formula, or a proof—can then in turn be encoded by a single natural number. There are various ways to choose such an encoding; what matters is not the specific form of the numbering, but that it is unique and effective.

Of particular interest are the following properties:

- From a code, one should be able to recover the underlying syntactic object.
- One should be able to decide or compute effectively whether a number is the code of a symbol, a term, a formula, or a proof.

- Syntactic operations such as "the last part of a formula," "is a proof of," or "follows from" should be traceable at the level of codes.

In this way, syntax becomes arithmetically manageable. What was previously a purely linguistic or symbolic structure is now available in numerical form. This is the decisive step in Gödel's method: once formulas and proofs are encoded as numbers, one can begin to speak about these codes within arithmetic itself.

In the next step, we must therefore show that such encoded syntactic properties and relations can in fact be expressed in arithmetic. That is precisely the content of the next section.

Representability in Arithmetic

Through Gödel numbering, syntactic objects such as formulas and proofs can be viewed as natural numbers. However, this alone does not yet ensure that arithmetic itself can speak about these objects. For precisely this purpose, we need the notion of representability: an arithmetical theory should be able to express certain properties and relations concerning the Gödel numbers of its syntactic objects.

Recall that we write $\overline{n}$ for the *numeral* corresponding to a natural number n, that is, the arithmetical term that represents this number in the theory T. Depending on the signature, this may for instance be the term

$S(S(\ldots S(0)\ldots))$ with the successor symbol applied n times.

Definition 2.155. *Let T be a formalized theory of arithmetic, and let $R \subseteq \mathbb{N}^k$ be a k-ary relation on the natural numbers.*

We say that R is representable in T *if there exists a formula $\varphi_R(x_1, \ldots, x_k)$ in the language of T such that for all $n_1, \ldots, n_k \in \mathbb{N}$ the following holds:*

- *If $R(n_1, \ldots, n_k)$ is true, then T proves the instance*

$$T \vdash \varphi_R(\overline{n_1}, \ldots, \overline{n_k}).$$

- *If $R(n_1, \ldots, n_k)$ is false, then T proves the negation*

$$T \vdash \neg\varphi_R(\overline{n_1}, \ldots, \overline{n_k}).$$

This form of representability therefore means that the theory T can do more than handle individual concrete cases: it captures the relation in question by means of a single uniform formula. Intuitively speaking, the theory can recognize from the codes whether a given syntactic property holds or not.

For our purposes, this is of central importance, because many syntactic relations between formulas and proofs can in fact be described arithmetically. For example, the relation

"x is the Gödel code of a correct proof
of the formula with Gödel code y"

can be expressed by an arithmetical formula. Likewise, one can formulate the relation that a number is the code of a well-formed formula, that one formula follows from others by an admissible rule, or that a proof has a certain structure at a particular place.

The significance of this observation is immediate: once the relation "is a proof of" is arithmetically representable, the concept of provability itself can be formalized within arithmetic. From the proof relation there then arises a proof predicate that allows the theory to speak about its own derivations.

With this, we have reached the crucial technical bridge. Gödel numbering translated syntax into numbers; representability now ensures that these numbers can be meaningfully interpreted and studied within arithmetic. In the next step, this possibility is used to construct formulas that speak about their own provability. For this, we need the *theorem of self-reference*, or equivalently the *fixed-point theorem*.

Self-Reference and the Fixed-Point Theorem

The notions developed so far – Gödel numbering and representability – now make it possible to formulate with precision one of the deepest ideas in logical analysis: a formal theory can make statements about its own statements. This opens the door to self-reference, that is, to formulas that refer indirectly to themselves.

The basic idea is the following. Once formulas and

proofs are encoded by natural numbers and syntactic properties become arithmetically representable, one can construct formulas whose arithmetical content refers to the Gödel code of the formula itself. From this mechanism arises the so-called fixed point: for every suitable property, there is a sentence that asserts exactly this property of its own code.

Theorem 2.156 (Fixed-Point Theorem / Diagonal Lemma). *Let T be a sufficiently strong formalized theory of arithmetic that can represent the syntactic relations needed for Gödel numbering. Then the following holds:*

For every formula $\varphi(x)$ with exactly one free variable, there exists a sentence ψ such that

$$T \vdash \psi \leftrightarrow \varphi(\ulcorner \psi \urcorner).$$

In other words, for every such property $\varphi(x)$, one can construct a sentence ψ that asserts of itself that it has the property φ.

The crucial step is the representability of the diagonal substitution function; it is precisely here that a syntactic operation is turned into an arithmetical statement.

Proof. The idea of the proof is to construct a formula that refers to its own Gödel code. To do so, we first consider the following syntactic operation:

For each natural number n, let $\mathrm{diag}(n)$ be the Gödel code of the formula obtained from the formula with code n by substituting the numeral $\overline{n}$ for its free variable.

Since this operation is effective, it is arithmetically representable. Hence there is a formula

$$\mathrm{Diag}(x, y)$$

with the property that, for every n, the theory T can prove:

$$T \vdash \mathrm{Diag}(\overline{n}, \overline{\mathrm{diag}(n)}).$$

Moreover, $\mathrm{Diag}(x, y)$ is functional, that is: for every x there exists exactly one y such that $\mathrm{Diag}(x, y)$ holds.

Now let $\varphi(x)$ be any formula with exactly one free variable. We define a new formula

$$\theta(x) \;:=\; \exists y\,(\mathrm{Diag}(x, y) \land \varphi(y)).$$

Now let

$$m := \ulcorner\theta\urcorner$$

be the Gödel code of θ, and set

$$\psi \;:=\; \theta(\overline{m}).$$

Then ψ is a sentence, that is, a formula with no free variables.

We now show that $T \vdash \psi \leftrightarrow \varphi(\ulcorner\psi\urcorner)$.

First, by the definition of ψ,

$$\psi \equiv \exists y\,(\mathrm{Diag}(\overline{m}, y) \land \varphi(y)).$$

On the other hand, $\mathrm{diag}(m)$ is exactly the Gödel code of the formula $\theta(\overline{m})$, hence precisely the Gödel code of ψ. Therefore,

$$\mathrm{diag}(m) = \ulcorner\psi\urcorner.$$

By the representability of diag, the theory T in particular proves

$$T \vdash \mathrm{Diag}(\overline{m}, \overline{\ulcorner\psi\urcorner}).$$

Since $\mathrm{Diag}(x, y)$ represents the graph of a function, $\overline{\ulcorner\psi\urcorner}$ is the uniquely determined value corresponding to $\overline{m}$. Hence the existential formula

$$\exists y\,(\mathrm{Diag}(\overline{m}, y) \wedge \varphi(y))$$

is equivalent in T to

$$\varphi(\ulcorner\psi\urcorner).$$

Therefore,

$$T \vdash \psi \leftrightarrow \varphi(\ulcorner\psi\urcorner).$$

This establishes the claimed fixed-point property. □

This theorem is the formal expression of self-reference. It does not say that a formula speaks about itself merely by chance; rather, it shows that such a formula can be constructed systematically and in a controlled way. The construction rests on the fact that the theory can speak about codes of formulas and that the code number of a formula can be inserted back into a formula itself.

Intuitively, the fixed-point theorem can be understood as a form of diagonal substitution: one starts with a formula $\varphi(x)$ that makes a statement about a code x, and then fills the free place in such a way that the formula is ultimately applied to its own code. In this way, one obtains a sentence that captures itself in arithmetical form.

This theorem is especially important because it provides the crucial basis for Gödel-style self-reference. For example, if one chooses as $\varphi(x)$ the statement "the formula coded by x is not provable," one obtains a sentence G such that

$$T \vdash G \leftrightarrow \neg \mathrm{Bew}_T(\ulcorner G \urcorner),$$

that is, a sentence stating that it is itself not provable. This is precisely how the core of Gödel's first incompleteness theorem arises.

The fixed-point theorem also plays a central role in the second incompleteness theorem, since there one constructs a statement concerning the consistency of the theory and at the same time links it, in a suitable way, to its own provability.

The fixed-point theorem is thus the final technical preparation for the actual incompleteness theorems. In the next step, we can formulate the first fundamental limitation result: in sufficiently strong, effectively described, and consistent theories, there are true but unprovable statements.

2.5.11 Gödel's Incompleteness Theorems

Having established the necessary prerequisites, we now turn to the actual limitative results. Gödel's incompleteness theorems reveal a profound gap between what can be expressed within a formal theory and what can be fully proved within that theory.

To recall: Gödel's completeness theorem concerns predicate logic itself. It states that every logically valid statement is also provable. By contrast, the incompleteness theorems do not concern logic as such, but sufficiently strong formalized theories of arithmetic. They show that such theories, provided they are consistent and effectively axiomatized, cannot capture all arithmetical truths in a complete way.

The *first incompleteness theorem* says, informally: In every sufficiently strong, consistent, and effectively axiomatized theory of arithmetic, there are true statements that can neither be proved nor refuted within that theory. Such a theory is therefore incomplete.

The *second incompleteness theorem* strengthens this claim: Such a theory cannot prove its own consistency within the theory itself, provided that it is in fact consistent.

The basic idea behind both theorems rests on self-reference. By means of the fixed-point theorem, one can construct a statement that effectively says: "I am not provable." For the first incompleteness theorem, this yields a statement that – assuming the theory is consistent – is true but unprovable. For the second theorem, a similar form of self-reference is used to show that the theory cannot establish its own consistency.

The following sections present the precise formulations

and proofs of these two fundamental results.

The First G"odel Incompleteness Theorem

We have now prepared all the essential prerequisites: a sufficiently strong theory of arithmetic, an effective description of its axioms, the G"odel numbering of syntactic objects, the representability of arithmetical relations in the theory, and the fixed-point theorem. With these tools, one can show that such a theory cannot capture all arithmetical truths completely.

The basic idea is simple, yet profound: one constructs a sentence that, in effect, states that it is not provable. If the theory is consistent, then this sentence cannot be provable; under G"odel's classical assumption of ω-consistency, it moreover follows that its negation is not provable either. Thus the theory is incomplete.

Theorem 2.157 (First G"odel Incompleteness Theorem). *Let T be a consistent, recursively enumerable, and sufficiently strong theory of arithmetic. Then there exists a sentence G_T in the language of T that is not provable in T.*

If, in addition, T is ω-consistent, then $\neg G_T$ is also not provable in T. In particular, T is then incomplete.

Proof. Let

$$\mathrm{Bew}_T(x, y)$$

be an arithmetical formula expressing that x is the G"odel code of a correct T-proof of the formula whose G"odel code is y. By representability of the proof relation, this formula is available in T. We also define the corresponding provability predicate

$$\mathrm{Prov}_T(y) \;:=\; \exists x\, \mathrm{Bew}_T(x, y).$$

By means of the fixed-point theorem, we now obtain a sentence G_T such that

$$T \vdash G_T \leftrightarrow \neg\mathrm{Prov}_T(\ulcorner G_T \urcorner).$$

Thus, this sentence says exactly what we intended informally: "G_T is not provable."

We first show that G_T is not provable. Suppose, for contradiction, that $T \vdash G_T$. Then there is a concrete T-proof of G_T; let its G"odel code be n. Since the proof relation is arithmetically representable, T can also prove for this concrete code that n is indeed a proof code of G_T, that is,

$$T \vdash \mathrm{Bew}_T(\overline{n}, \ulcorner G_T \urcorner).$$

Hence in particular,

$$T \vdash \mathrm{Prov}_T(\ulcorner G_T \urcorner).$$

On the other hand, the fixed-point equivalence together with $T \vdash G_T$ also yields

$$T \vdash \neg\mathrm{Prov}_T(\ulcorner G_T \urcorner).$$

Therefore T would be inconsistent. Since T is consistent, the assumption $T \vdash G_T$ cannot hold. Consequently,

$$T \nvdash G_T.$$

It remains to consider the negation. Suppose that $T \vdash \neg G_T$. Then from

$$T \vdash G_T \leftrightarrow \neg \mathrm{Prov}_T(\ulcorner G_T \urcorner)$$

we first obtain

$$T \vdash \mathrm{Prov}_T(\ulcorner G_T \urcorner),$$

and hence

$$T \vdash \exists x \, \mathrm{Bew}_T(x, \ulcorner G_T \urcorner).$$

However, for each concrete $n \in \mathbb{N}$, the following holds: if n is not the code of a proof of G_T, then the statement

$$\neg \mathrm{Bew}_T(\overline{n}, \ulcorner G_T \urcorner)$$

is true, and by representability of the proof relation it is provable in T. Since there can be no genuine proof of G_T, each of these statements is provable in T. Thus T proves every individual instance

$$\neg \mathrm{Bew}_T(\overline{n}, \ulcorner G_T \urcorner) \qquad \text{for all } n \in \mathbb{N}.$$

Therefore, if $T \vdash \exists x \, \mathrm{Bew}_T(x, \ulcorner G_T \urcorner)$, we have a formula whose negation is proved for every concrete numeral. This contradicts the ω-consistency of T.

Hence $T \nvdash \neg G_T$ as well. This shows that G_T is undecidable in T; in particular, T is incomplete. □

Remark 2.158. *In more modern presentations, ω-consistency is often replaced by the cleaner assumption* *of mere consistency by using a suitable Rosser sentence instead of Gödel's original sentence. For the line of argument pursued here, however, the classical Gödel formulation is especially transparent, since it makes the role of self-reference, provability, and consistency immediately visible.*

In the next step, we strengthen the result: not only does such a theory contain unprovable arithmetical sentences, but it also cannot prove its own consistency within the theory itself. This leads to the *Second Incompleteness Theorem.*

The Second Incompleteness Theorem

The first incompleteness theorem shows that a sufficiently strong, recursively enumerable, and consistent theory of arithmetic cannot decide all arithmetical truths. The second theorem goes even further: it shows that such a theory cannot prove its own consistency from within the theory itself. It is precisely this point that reveals the limitation of formal systems in a particularly striking way.

To formulate this precisely, we first fix the formal statement of consistency. If $\mathrm{Prov}_T(y)$ is the provability predicate of the theory T, we write

$$\mathrm{Con}(T) \;:=\; \neg\mathrm{Prov}_T(\ulcorner 0 = 1 \urcorner),$$

that is: "There is no T-proof of the statement $0 = 1$."

This formula is the arithmetical formalization of the consistency of T.

Theorem 2.159 (Gödel's Second Incompleteness Theorem). *Let T be a consistent, recursively enumerable, and sufficiently strong theory of arithmetic. Then*

$$T \nvdash \mathrm{Con}(T).$$

In particular, T cannot prove its own consistency.

Proof. We use the Gödel sentence G_T constructed in the previous section, which has the property

$$T \vdash G_T \leftrightarrow \neg \mathrm{Prov}_T(\ulcorner G_T \urcorner).$$

Thus, this sentence says exactly: "G_T is not provable."

The crucial observation is now that the consistency of T, formalized within the theory, is strong enough to justify the non-provability of G_T. Formally, this means that from $\mathrm{Con}(T)$ one can derive within T the statement that there is no proof of G_T. In other words, one can show within T:

$$T \vdash \mathrm{Con}(T) \to \neg \mathrm{Prov}_T(\ulcorner G_T \urcorner).$$

Together with the fixed-point equation for G_T, we therefore obtain

$$T \vdash \mathrm{Con}(T) \to G_T.$$

Now suppose that $T \vdash \mathrm{Con}(T)$. Then modus ponens immediately yields

$$T \vdash G_T.$$

But this contradicts Gödel's first incompleteness theorem, according to which G_T is not provable in a consistent theory T. Therefore, the assumption $T \vdash \mathrm{Con}(T)$ cannot be correct. Consequently,

$$T \nvdash \mathrm{Con}(T).$$

This shows that a sufficiently strong, recursively enumerable, and consistent theory cannot prove its own consistency. □

Remark 2.160. *In more modern presentations, this proof is often formulated in terms of general derivability conditions for the provability predicate. For the exposition pursued here, however, the central idea is sufficient: the first incompleteness theorem yields a sentence that expresses its own unprovability; the second theorem shows that even a proof of the system's own consistency would already be too strong and is therefore not possible within the theory itself.*

Consequences of Gödelian Incompleteness

Gödel's two incompleteness theorems are not merely technical results about formal theories; they mark a fundamental limit on what can be achieved at all within a sufficiently strong axiomatic system. Their significance therefore extends far beyond proof theory in the narrow sense and concerns the foundations of logic, mathematics, and – indirectly – computer science as well.

First, one essential consequence can be stated at the level of logic: for sufficiently strong, recursively enumerable, and consistent theories of arithmetic, completeness in the strong sense is unattainable. There will always be arithmetical statements that are true but cannot be proved within the system. Likewise, the system cannot establish its own consistency internally. The formal theory therefore remains necessarily incomplete and cannot be simultaneously sufficiently strong, effectively describable, and complete.

For mathematics, this result had far-reaching philosophical consequences. It affected in particular Hilbert's program of securing all of mathematics by means of a complete, consistent, and finitely justifiable axiomatic system. Gödel showed that such a goal is unattainable in its original form: as soon as a system can adequately express elementary arithmetic, statements arise that lie, in principle, beyond its power of proof. Thus, Hilbert's ideal of an ultimate and fully closed foundation could no longer be maintained in this form.

This does not mean, however, that formal axiomatics lost its significance. Rather, the perspective on mathematical foundations changed. In place of a single final system, more differentiated programs moved to the forefront, such as the investigation of weaker theories, relative consistency proofs, proof theory, model theory, and recursion theory. Within arithmetic itself, axiomatic systems were also developed that deliberately take Gödelian limits into account and handle them as precisely as possible.

The incompleteness theorems therefore did not bring foundational research to an end; instead, they redirected it into new and productive paths.

For computer science and the theory of computability, the consequences are equally profound. Gödel showed that there are principled limits to the algorithmic capture of mathematical truth. Not every semantically true statement is mechanically decidable or provable. The incompleteness theorems are therefore closely related to questions of computability, decidability, and complexity. They make visible that, even in highly precise formalized contexts, not all problems can be brought fully under algorithmic control.

Conclusion:

- In sufficiently strong formal theories, there exist true but unprovable statements.
- Such a theory cannot prove its own consistency from within.
- The classical ideal of a complete and final axiomatic system cannot be achieved in its original form.
- These results did not bring foundational research to an end, but instead led to new and more differentiated approaches.

Gödel's incompleteness theorems thus mark a point of transition: on the one hand, they reveal the limits of formal provability; on the other hand, they prepare the way for an algorithmic perspective. This is precisely where the next section begins, in which we examine the computability and complexity of logical problems.

2.5.12 Computability and Complexity

After the incompleteness theorems, another fundamental question arises: Which logical problems can in fact be solved algorithmically, and how difficult are they in the most favorable cases? In propositional logic, the central decision problems were still manageable. For first-order predicate logic, however, the picture changes fundamentally: some problems remain semi-decidable, others are undecidable, and only in special cases can one additionally provide a complexity estimate.

We begin by considering three classical decision problems of first-order predicate logic.

Definition 2.161. *The problems* TAUT(FO), SAT(FO), *and* INFER(FO) *are defined as follows:*

- TAUT(FO) *Input: a Σ-formula φ Output: "Yes" if and only if φ is valid in all structures, that is, if it is a tautology of first-order predicate logic.*
- SAT(FO) *Input: a Σ-formula φ Output: "Yes" if and only if φ is satisfiable.*
- INFER(FO) *Input: two Σ-formulas φ, ψ Output: "Yes" if and only if $\varphi \models_{\mathrm{FO}} \psi$ holds.*

For such problems, we distinguish between three fundamental notions of computability.

Definition 2.162. *A problem is called* decidable *if there exists an algorithm that, for every input, terminates after finitely many steps and correctly answers "Yes" or "No".*

It is called semi-decidable *if there exists an algorithm that terminates on every "Yes"-instance, while on "No"-instances it may fail to halt.*

It is called undecidable *if no decision procedure exists for it.*

For first-order predicate logic, the situation is considerably more difficult than for propositional logic: in general, the semantic problems are no longer fully decidable. Nevertheless, the proof-theoretic side remains partially accessible, since with the help of a complete calculus one can still semi-decide valid statements by systematic proof search.

Theorem 2.163. *The problems* TAUT(FO) *and* INFER(FO) *are semi-decidable but undecidable. The problem* SAT(FO) *is not semi-decidable.*

Proof sketch. The semi-decidability of TAUT(FO) follows from the completeness theorem: if a formula is valid, then it has a proof in a complete calculus, and such proofs can be effectively enumerated. The same holds for INFER(FO), because $\varphi \models_{\mathrm{FO}} \psi$ holds exactly when the formula $(\varphi \to \psi)$ is valid.

The undecidability of TAUT(FO) is a classical result due to Church and Turing. Since INFER(FO) already contains validity as a special case, this problem is undecidable as well.

For SAT(FO), the following holds: the unsatisfiability of a formula is semi-decidable, because it is equivalent to the validity of its negation. If SAT(FO) were also semi-decidable, then both a problem and its complement would be semi-decidable, and hence SAT(FO) would be decidable. This contradicts the undecidability of the corresponding validity problems. □

Besides asking *whether* a problem is computable, in the decidable case one is also interested in *how costly* the computation is. This is the role of the problem's complexity.

One particularly important example is model checking for finite structures.

Definition 2.164. *MODELCHECKING Input: a finite structure $\mathcal{A}$ and a formula φ Output: "Yes" if and only if $\mathcal{A} \models_{\mathrm{FO}} \varphi$ holds.*

Theorem 2.165. *The problem MODELCHECKING for finite structures belongs to* PSPACE.

Proof sketch. The formula is evaluated recursively according to its syntactic structure. In essence, one only needs to store the current subformula, an assignment to the free variables, and the current recursion path. This information can be managed using only polynomial space. Hence, the problem belongs to PSPACE. □

For infinite structures, the situation is considerably more complex. If, for a fixed structure $\mathcal{A}$, one considers the problem of deciding whether an input formula is true in $\mathcal{A}$, one speaks of the decision problem of the structure $\mathcal{A}$:

$\mathcal{A}$-Decision Problem

Input: formula φ
Output: Yes iff $\mathcal{A} \models_{\mathrm{FO}} \varphi$.

There is no uniform answer pattern for arbitrary structures: some structures have a decidable theory, while others do not. For sufficiently expressive structures, especially arithmetical ones, this problem is undecidable. In this sense, Church and Turing showed that there is no general algorithm that decides the truth of first-order statements in all sufficiently strong structures.

This yields the following overall picture: in first-order predicate logic, the central semantic problems are generally undecidable. Wherever decidability can be achieved at all, it often comes only at the cost of increased algorithmic effort.

Against this background, two opposing yet equally natural perspectives suggest themselves. On the one hand, one can deliberately restrict the expressive means of logic in order to recover fragments that are more manageable and decidable. An important example of this is description logic, which was developed specifically for the structured representation of knowledge.

On the other hand, one can further increase expressive power by going beyond the capabilities of first-order predicate logic. This leads to second-order predicate logic, which is substantially more powerful, but is subject to even stronger limitations with respect to decidability and computability.

The following sections take up these two directions: first, we consider description logic as a well-defined restriction of predicate logic; afterward, we turn to second-order predicate logic as one of its central extensions.

2.5.13 Exercises and Solutions

In the following, we revisit the central concepts of unification and the resolution rule by means of a few small examples. In each case, decide whether the given statement is true or false. Recall that sets of clauses are understood here as being universally closed.

Exercise 2.166. *1. Consider the first-order signature*

$$\Sigma = \{P/2, S/1, Q/1, h/1, f/1, i/1, e/0\}.$$

and also the following expressions:

$$L_1 = P(v, u), \qquad L_1' = P(e, e),$$

$$C_2 = \{\neg P(e, v)\}, \qquad C_2' = \{\neg S(f(f(v))), Q(w)\},$$

$$\begin{aligned}K_3 = \{&\{\neg P(f(v), e), \neg Q(x)\},\\ &\{P(i(z), w)\},\\ &\{\neg P(e, h(f(y)))\},\\ &\{Q(u)\}\}.\end{aligned}$$

$$\begin{aligned}K_4 = \{&\{\neg Q(f(v)), Q(w)\},\\ &\{P(e, x), Q(i(v))\},\\ &\{P(e, z), \neg S(u)\}\}.\end{aligned}$$

Decide for each of the following statements whether it is true or false.

1. *L_1 and L_1' are unifiable.*
2. *C_2 and C_2' are resolvable.*
3. *We have $K_3 = \mathrm{Res}_\infty(K_3)$.*
4. *We have $K_4 \models_{\mathrm{FO}} \neg Q(z)$.*

Solution 2.167. 1. ***True.*** *The two literals have the same predicate symbols and the same arity. A unifying substitution is*

$$\sigma = \{v \mapsto e,\ u \mapsto e\}.$$

Then we have

$$P(v, u)\sigma = P(e, e) = P(e, e)\sigma.$$

Hence L_1 and L_1' are unifiable. Since the two terms become identical under this substitution, σ is in fact an MGU.

2. ***False.*** *For a resolution step, there must be two complementary literals whose underlying atoms are unifiable. In*

$$C_2 = \{\neg P(e, v)\},$$
$$C_2' = \{\neg S(f(f(v))), Q(w)\}.$$

C_2 contains only a negative P-literal, whereas C_2' contains neither a positive P-literal nor a negative Q-literal. Thus there is no complementary pair of literals on which a resolution step could be performed. Therefore, C_2 and C_2' are not resolvable.

3. ***False.*** *The clause set K_3 is not yet closed under resolution. For example, the clauses*

$$\{\neg P(f(v), e), \neg Q(x)\} \quad \text{and} \quad \{Q(u)\}$$

can be resolved on the literal Q. Using the substitution $\sigma = \{x \mapsto u\}$, we obtain the resolvent

$$\{\neg P(f(v), e)\}.$$

This new resolvent does not belong to K_3. Therefore,

$$K_3 \neq \mathrm{Res}_\infty(K_3).$$

4. ***False.*** *The formula $\neg Q(z)$ is not a logical consequence of K_4. It suffices to give a countermodel in which K_4 is satisfiable and in which $Q(z)$ can still be true.*

Consider, for instance, a structure with the one-element domain $\{a\}$ *in which*

$$\begin{aligned} e^{\mathcal{A}} &= a, \\ P^{\mathcal{A}} &= \{(a,a)\}, \\ Q^{\mathcal{A}} &= \{a\}, \\ S^{\mathcal{A}} &= \emptyset. \end{aligned}$$

and all functions are interpreted as constant functions with value a. *Then all three clauses from* K_4 *are satisfied:*

$$\begin{aligned} &\neg Q(f(v)) \vee Q(w), \\ &P(e,x) \vee Q(i(v)), \\ &P(e,z) \vee \neg S(u). \end{aligned}$$

because in each case at least one disjunct is true. At the same time, $Q(z)$ *is true under the assignment* $z \mapsto a$, *so* $\neg Q(z)$ *is false. Consequently,*

$$K_4 \not\models_{\mathrm{FO}} \neg Q(z).$$

The following exercises revisit the three central layers of the resolution method: first, unification determines whether a resolution step can be applied; next, one checks which new clauses are generated; and finally, suitable countermodels can be used to refute semantic consequences.

In what follows, we review unification, resolution, and the semantic analysis of clause sets by means of a few small examples. In each case, decide whether the statement is true or false.

Exercise 2.168. *Consider the first-order signature*

$$\Sigma = \{S/2, P/1, Q/1, g/1, f/1, i/1, a/0\}.$$

Also consider the following expressions:

$$L_1 = S(f(a), a), \qquad L_1' = S(x, x),$$

$$\sigma_1 = \{y \mapsto a,\ w \mapsto f(g(i(g(i(a)))))\},$$

$$\begin{aligned} C_2 &= \{Q(u), P(y), \neg P(i(i(w)))\}, \\ C_2' &= \{Q(w), \neg Q(y)\}. \end{aligned}$$

$$K_3 = \{\{\neg Q(v), \neg S(w, i(u))\}, \{\neg Q(y)\}\},$$

$$\begin{aligned} K_4 = \{&\{\neg Q(z), \neg P(i(g(f(z))))\}, \\ &\{Q(x)\}, \\ &\{Q(v)\}, \\ &\{Q(f(g(i(y))))\}\}. \end{aligned}$$

For each of the following statements, decide whether it is true or false.

1. σ_1 *is an MGU of* L_1 *and* L_1'.

2. C_2 *and* C_2' *are resolvable.*

3. *We have* $K_3 = \mathrm{Res}_\infty(K_3)$.

4. K_4 *is unsatisfiable.*

Solution 2.169. 1. ***False.*** *The two literals*

$$L_1 = S(f(a), a), \qquad L_1' = S(x, x)$$

are not unifiable. From the first argument position, one would have to obtain

$$x = f(a)$$

and from the second at the same time

$$x = a.$$

Thus $f(a) = a$ would have to hold, which is impossible in the given language. Therefore, no unifier exists at all, and consequently σ_1 cannot be an MGU.

2. ***True.*** *The clauses C_2 and C_2' contain complementary literals, namely $Q(u)$ and $\neg Q(y)$. Since the two clauses are not variable-disjoint, we first rename the variables in C_2', for example by*

$$\nu(w) = w_1, \qquad \nu(y) = y_1.$$

Then we obtain

$$C_2'\nu = \{Q(w_1), \neg Q(y_1)\}.$$

Now $Q(u)$ and $\neg Q(y_1)$ are unifiable, for instance with the MGU

$$\theta = \{u \mapsto y_1\}.$$

The resulting resolvent is

$$\{P(y), \neg P(i(i(w))), Q(w_1)\}.$$

Hence C_2 *and* C_2' *are resolvable.*

3. **True.** *In the clause set*

$$K_3 = \{\{\neg Q(v), \neg S(w, i(u))\}, \{\neg Q(y)\}\}$$

only negative literals occur. However, resolution requires a complementary pair of literals, that is, one positive and one negative literal with the same predicate. No such pair occurs here. Therefore, no further resolution step is possible, and we have

$$K_3 = \mathrm{Res}_\infty(K_3).$$

4. **False.** *The clause set* K_4 *is satisfiable. It suffices to provide a countermodel.*

Consider the one-element structure $\mathcal{A}$ *with domain* $\{a\}$ *in which*

$$\begin{aligned} a^{\mathcal{A}} &= a, \\ f^{\mathcal{A}}(a) &= a, \\ g^{\mathcal{A}}(a) &= a, \\ i^{\mathcal{A}}(a) &= a. \end{aligned}$$

and

$$Q^{\mathcal{A}} = \{a\}, \qquad P^{\mathcal{A}} = \emptyset.$$

Then the clauses in K_4 *are satisfied:*

- *$\neg Q(z) \vee \neg P(i(g(f(z))))$ is true for every z because $\neg P(\dots)$ is always true.*
- *The clauses $\{Q(x)\}$, $\{Q(v)\}$, and $\{Q(f(g(i(y))))\}$ are true because $Q^{\mathcal{A}}$ holds for the unique element.*

Therefore K_4 is satisfiable and hence not unsatisfiable.

These exercises once again illustrate the typical resolution procedure: first, one checks whether there is any unifiable complementary pair of literals at all; next, one determines whether further resolution steps are possible; and finally, one can use a suitable model to show that a set of clauses remains satisfiable.

2.6 Description Logic

In general first-order predicate logic, we have already seen that many of the fundamental problems are either undecidable or can only be handled with considerable algorithmic effort. It is therefore natural to deliberately restrict the expressive power of the logic in order to regain more favorable computational properties. Such deliberately restricted sublanguages are called *fragments* of predicate logic. Depending on the area of application, they often appear in their own notation, but they can still be expressed by formulas of first-order predicate logic.

Among the most important examples of such fragments are *description logics.* They are used to represent knowledge about *concepts*, *individuals*, and *relations* between individuals in a formal way. They are especially valuable wherever knowledge is not only stored but also meant to be systematically analyzed, for example in ontologies, knowledge representation systems, or semantic applications in computer science.

In this subsection, we consider *ALC* (*Attributive Concept Description Language with Complements*) as a fundamental example of such a description logic. ALC is expressive enough to model many typical knowledge relationships, while remaining sufficiently transparent in structure that its semantic and algorithmic properties can still be studied precisely. Typical concepts include "person", "department", or "event"; typical roles include, for example, "works in", "attends", or "manages".

For knowledge bases of this kind, one can ask questions such as the following:

- Is there a person who works in a department that is managed by that same person?
- Does every event attended by a person belong to some department?
- Is there a department in which no person in a leading role is active?

Later on, we will see that formulas of description logic can be translated into statements of first-order predicate logic. ALC thus combines a readable, application-oriented notation with a precise formal semantics.

2.6.1 Concepts, Roles, and Their Interpretation

As in propositional logic and first-order predicate logic, we also distinguish in description logic between syntax and semantics. We first specify which linguistic building blocks are available; we then explain how these building blocks are to be read within an interpretation.

Intuitively, a *concept* describes a set of individuals. For example, the concept *Student* may denote the set of all students. A *role*, by contrast, describes a relation between two individuals, such as *supervises*, when one individual is supervised by another. In addition, we use *individual names*, which allow us to refer to particular distinguished objects.

Definition 2.170. *An* ALC signature *is a tuple*

$$\Gamma = \langle K, R, O \rangle,$$

where $K = \{A_1, \ldots, A_n\}$ *is a finite set of concept symbols,* $R = \{R_1, \ldots, R_m\}$ *is a finite set of role symbols, and* $O = \{a_1, \ldots, a_k\}$ *is a finite set of individual names.*

2.6.2 ALC Concept Descriptions and Semantics

The ALC concept descriptions over a signature are defined inductively.

Definition 2.171. *Let* $\Gamma = \langle K, R, O \rangle$ *be an ALC signature. Then the ALC concept descriptions over* Γ *are defined inductively as follows:*

(ALC0) $\top$ *and* $\bot$ *are ALC concept descriptions.*

(ALC1) Every concept symbol $A \in K$ *is an ALC concept description.*

(ALC2) If C *and* D *are ALC concept descriptions, then* $C \sqcap D$ *and* $C \sqcup D$ *are also ALC concept descriptions.*

(ALC3) If C *is an ALC concept description, then* $\neg C$ *is also an ALC concept description.*

(ALC4) If C *is an ALC concept description and* $R \in R$ *is a role symbol, then* $\forall R.C$ *and* $\exists R.C$ *are also ALC concept descriptions.*

The symbol $\top$ denotes the universal concept, that is, the set of all individuals in the domain under consideration, while $\bot$ denotes the empty concept. Concept descriptions of the form (ALC0) and (ALC1) are called *atomic*; all others are called *complex*. The symbols $\sqcap$, $\sqcup$, and $\neg$ are read as concept conjunction, concept disjunction, and concept negation. The expressions $\forall R.C$

and $\exists R.C$ are called universal and existential restrictions, respectively.

The meaning of these constructions is quite natural. The expression $C \sqcap D$ describes exactly those individuals that belong to both C and D. The expression $C \sqcup D$ denotes those individuals that belong to at least one of the two concepts. $\neg C$ comprises all individuals that are not contained in C. By $\forall R.C$ we mean those individuals whose R-successors all belong to C. In contrast, $\exists R.C$ describes those individuals that are connected via R to at least one individual from C.

An example makes this more concrete. For the signature

$$K = \{Student, Professor, Lecture\},$$
$$R = \{supervises, teaches\}.$$

the expression

$$Student \sqcap \exists\, supervises.Professor$$

describes the set of all individuals who are students and are supervised by at least one professor. The expression

$$\forall\, teaches.Lecture$$

on the other hand denotes all individuals that, via the role *teaches*, are related only to lectures.

We now turn to semantics. An interpretation determines which objects exist in the domain and how the symbols of the signature are assigned to those objects.

Definition 2.172. *Let* $\Gamma = \langle K, R, O\rangle$ *be an* ALC *signature. An* interpretation *for* Γ *is a pair*

$$\mathcal{I} = \langle \Delta^{\mathcal{I}}, \cdot^{\mathcal{I}}\rangle,$$

where $\Delta^{\mathcal{I}}$ *is a nonempty set and* $\cdot^{\mathcal{I}}$ *assigns:*

- *to each concept symbol* $A \in K$ *a set* $A^{\mathcal{I}} \subseteq \Delta^{\mathcal{I}}$,
- *to each role symbol* $R \in R$ *a relation* $R^{\mathcal{I}} \subseteq \Delta^{\mathcal{I}} \times \Delta^{\mathcal{I}}$,
- *to each individual name* $a \in O$ *an element* $a^{\mathcal{I}} \in \Delta^{\mathcal{I}}$.

The set $\Delta^{\mathcal{I}}$ is called the *domain* of $\mathcal{I}$ and contains the objects under discussion. Relative to this domain, the symbols of the signature are interpreted: each concept symbol is understood as a subset of $\Delta^{\mathcal{I}}$, each role symbol as a binary relation on $\Delta^{\mathcal{I}}$, and each individual name denotes a specific element of $\Delta^{\mathcal{I}}$.

Complex concept descriptions are evaluated recursively under an interpretation.

Definition 2.173. *Let* $\mathcal{I} = \langle \Delta^{\mathcal{I}}, \cdot^{\mathcal{I}}\rangle$ *be an interpretation. Then the meaning of a concept description is determined by the following rules:*

$$\top^{\mathcal{I}} = \Delta^{\mathcal{I}}, \qquad \bot^{\mathcal{I}} = \emptyset,$$
$$(C \sqcap D)^{\mathcal{I}} = C^{\mathcal{I}} \cap D^{\mathcal{I}}, \qquad (C \sqcup D)^{\mathcal{I}} = C^{\mathcal{I}} \cup D^{\mathcal{I}},$$
$$(\neg C)^{\mathcal{I}} = \Delta^{\mathcal{I}} \setminus C^{\mathcal{I}},$$
$$(\forall R.C)^{\mathcal{I}} = \{a \in \Delta^{\mathcal{I}} \mid \forall b \in \Delta^{\mathcal{I}} : \ (a, b) \in R^{\mathcal{I}} \Rightarrow b \in C^{\mathcal{I}}\},$$
$$(\exists R.C)^{\mathcal{I}} = \{a \in \Delta^{\mathcal{I}} \mid \exists b \in \Delta^{\mathcal{I}} : \ (a, b) \in R^{\mathcal{I}} \wedge b \in C^{\mathcal{I}}\}.$$

This semantics directly reflects the underlying set-theoretic intuition. $\top$ stands for the entire domain, $\bot$ for the empty set, $\sqcap$ for intersection, $\sqcup$ for union, and $\neg$ for the complement relative to the domain. The universal restriction $\forall R.C$ collects all individuals whose R-successors belong only to C. By contrast, the existential restriction $\exists R.C$ collects those individuals that are connected via R to at least one element of C.

The following small example illustrates these definitions. Consider a university setting with the individuals *anna*, *ben*, *clara*, *course1*, and *course2*. Then we may have:

$$\mathit{Student}^{\mathcal{I}} = \{\mathit{anna}, \mathit{ben}\}, \qquad \mathit{Professor}^{\mathcal{I}} = \{\mathit{clara}\},$$

$$\mathit{supervisedBy}^{\mathcal{I}} = \{(\mathit{anna}, \mathit{clara}), (\mathit{ben}, \mathit{clara})\}.$$

Then, for example, *anna* belongs to

$$\exists\, \mathit{supervisedBy}.\mathit{Professor},$$

because she is supervised by *clara*. Consequently, *anna* is also an element of the concept

$$\mathit{Student} \sqcap \exists\, \mathit{supervisedBy}.\mathit{Professor}.$$

By contrast, *clara* herself belongs to this concept only if she is additionally interpreted as a student.

In the next section, we examine how knowledge about individuals can be expressed in ALC and how general relationships between concepts can be formalized.

Knowledge About Individuals

So far, we have studied the building blocks of description logic in the form of general concept and role descriptions. We now turn to knowledge about concrete objects. In ALC, one can describe not only general notions, but also record statements about individual objects. Such statements concern either the membership of an individual in a concept or a relationship between two individuals.

Statements of the first kind are called *concept assertions.* For example,

$$Student(emma)$$

states that the individual *emma* is a student. Statements of the second kind are called *role assertions.* For instance,

$$attends(emma, seminar1),$$

means that *emma* attends the object *seminar1.* A finite set of such statements describes concrete knowledge about individual objects; such a set is called an *ABox.* The name comes from *assertion*, that is, a statement about individuals.

Definition 2.174. *Let* $\Gamma = \langle K, R, O \rangle$ *be an ALC signature.*

- *For every individual name* $a \in O$ *and every concept description* C *over* Γ*, the expression* $C(a)$ *is called a* concept assertion.

- *For all individual names $a, b \in O$ and every role symbol $R \in R$, the expression $R(a, b)$ is called a* role assertion.

A finite set of concept assertions and role assertions over Γ is called an ABox *over Γ.*

An important point is that an ABox may contain not only atomic concepts. Composite concept descriptions may also occur within a concept assertion. This already allows one to formulate quite refined statements about individual objects.

An example makes this clear. Consider an ALC signature with the concept symbols

$$K = \{Student, Lecturer, Seminar, Library\}$$

and the role symbols

$$R = \{attends, teaches, uses\}.$$

Then an ABox might contain the following statements:

$$\begin{aligned} A = \{&Student(emma), Lecturer(lina), Seminar(seminar1),\\ &attends(emma, seminar1),\\ &teaches(lina, seminar1),\\ &\exists\, attends.Seminar(emma),\\ &\exists\, uses.Library(lina)\} \end{aligned}$$

The last two entries show that concept assertions may also contain complex concept descriptions. The statement

$$\exists\, attends.Seminar(emma)$$

therefore says not only that *emma* attends something, but more specifically that *emma* attends at least one seminar. Correspondingly,

$$\exists\, uses.Library(lina),$$

means that *lina* is connected to at least one library by the role *uses*.

We now turn to the semantics of such assertions. An interpretation satisfies an ABox exactly if all assertions it contains are true under that interpretation.

Definition 2.175. *Let* $\Gamma = \langle K, R, O\rangle$ *be an ALC signature and let* $\mathcal{I} = \langle \Delta^{\mathcal{I}}, \cdot^{\mathcal{I}}\rangle$ *be an interpretation for* Γ*. For every concept assertion* $C(a)$ *and every role assertion* $R(a,b)$ *over* Γ*, we define:*

$$\mathcal{I} \models_{\text{ALC}} C(a) \quad \textit{if and only if} \quad a^{\mathcal{I}} \in C^{\mathcal{I}},$$

$$\mathcal{I} \models_{\text{ALC}} R(a,b) \quad \textit{if and only if} \quad (a^{\mathcal{I}}, b^{\mathcal{I}}) \in R^{\mathcal{I}}.$$

An interpretation $\mathcal{I}$ *satisfies an ABox* A*, written* $\mathcal{I} \models_{\text{ALC}} A$*, if* $\mathcal{I} \models_{\text{ALC}} F$ *holds for every* $F \in A$*.*

In some applications, one additionally requires that different individual names denote different elements of the domain. This extra assumption is called the *Unique Name Assumption* (UNA). In this chapter, we do not adopt it. Different names may therefore, in principle, refer to the same individual.

An example shows how these definitions interact. Consider an interpretation $\mathcal{I}$ with domain

$$\Delta^{\mathcal{I}} = \{emma, lina, seminar, Library\},$$

in which

$$\begin{aligned} Student^{\mathcal{I}} &= \{emma\}, \\ Lecturer^{\mathcal{I}} &= \{lina\}, \\ Seminar^{\mathcal{I}} &= \{seminar1\}, \\ Library^{\mathcal{I}} &= \{library1\}, \\ attends^{\mathcal{I}} &= \{(emma, seminar1)\}, \\ teaches^{\mathcal{I}} &= \{(lina, seminar1)\}, \\ uses^{\mathcal{I}} &= \{(lina, library1)\}. \end{aligned}$$

Then $\mathcal{I}$ satisfies the ABox A. In particular, for example,

$$\mathcal{I} \models_{\mathrm{ALC}} \exists\, attends.Seminar(emma),$$

because *emma* is connected via the role *attends* to an individual interpreted as a seminar. Likewise,

$$\mathcal{I} \models_{\mathrm{ALC}} \exists\, uses.Library(lina)$$

holds, since *lina* stands in the role *uses* to *Library.*

ABoxes therefore collect concrete knowledge about individual objects. In the next step, we turn to the more general knowledge concerning the relationship between concepts and roles, that is, statements that do not merely concern individual entities but entire classes and their relations.

Terminological Knowledge

With the ABox, we have described knowledge about concrete individuals, such as particular persons, events, or objects together with their immediate properties and relations. Often, however, such a purely factual description is not sufficient. In many applications, we also want to capture more general knowledge, namely statements about how concepts are related to one another, which concepts include others, and which concepts turn out to be equivalent. This general, concept-based knowledge is stored in a *TBox.*

Terminological statements therefore connect concepts with one another. For example, they may state that every member of one concept also belongs to another concept, or that two concepts describe exactly the same individuals. Such statements are called *terminological axioms.*

Definition 2.176. *Let* $\Gamma = \langle K, R, O\rangle$ *be an ALC signature. For all concept descriptions C and D over* Γ*, the expressions*

$$C \sqsubseteq D \qquad \textit{and} \qquad C \equiv D$$

are called terminological axioms.

A finite set of terminological axioms over Γ *is called a* TBox *over* Γ.

Axioms of the form $C \sqsubseteq D$ are called *inclusions.* They express that every individual in C also belongs to

D; intuitively, C is therefore a subconcept of D. Axioms of the form $C \equiv D$ are called *equivalences*. They state that both concept descriptions capture exactly the same set of individuals.

In many cases, it is sufficient to build TBoxes using inclusions only. This is not a real restriction, because every equivalence

$$C \equiv D$$

can be rcplaced by the two inclusions

$$C \sqsubseteq D \qquad \text{and} \qquad D \sqsubseteq C$$

.

Semantically, a TBox statement means that the sets denoted by the concept descriptions involved stand in a specific relation to one another.

Definition 2.177. *Let $\Gamma = \langle K, R, O \rangle$ be an ALC signature and let $\mathcal{I} = \langle \Delta^{\mathcal{I}}, \cdot^{\mathcal{I}} \rangle$ be an interpretation for Γ. For concept descriptions C and D over Γ, we define:*

$$\mathcal{I} \models_{\text{ALC}} C \sqsubseteq D \quad \textit{if and only if} \quad C^{\mathcal{I}} \subseteq D^{\mathcal{I}},$$

$$\mathcal{I} \models_{\text{ALC}} C \equiv D \quad \textit{if and only if} \quad C^{\mathcal{I}} = D^{\mathcal{I}}.$$

An interpretation $\mathcal{I}$ satisfies a TBox T, written $\mathcal{I} \models_{\text{ALC}} T$, if $\mathcal{I} \models_{\text{ALC}} F$ holds for every terminological axiom $F \in T$.

Equivalences are especially useful because they allow us to define new concepts in terms of ones that are already known. In this way, knowledge can not only be collected but also organized in a structured manner. A TBox then functions like a formal vocabulary in which concepts are precisely distinguished from one another and systematically connected.

An example from the university domain illustrates this idea. We consider the concepts

$$\mathit{Student},\ \mathit{Instructor},\ \mathit{Assessment},\ \mathit{Active},$$

and the roles

$$\mathit{enrolledIn},\ \mathit{teaches},\ \mathit{grades}.$$

Then a TBox might, for example, contain the following terminological axioms:

$$\mathit{Active} \equiv \mathit{Student} \sqcup \mathit{Instructor},$$

$$\mathit{Student} \sqsubseteq \exists\, \mathit{enrolledIn}.\mathit{Assessment},$$

$$\mathit{Instructor} \sqsubseteq \forall\, \mathit{teaches}.\mathit{Active},$$

$$\mathit{Assessment} \sqsubseteq \exists\, \mathit{grades}.\mathit{Instructor}.$$

The first equivalence states that active individuals are exactly those who are students or instructors. The second inclusion says that every student is enrolled in at least one assessment. The third inclusion describes instructors as individuals who teach only active participants. The fourth inclusion finally states that every assessment is graded by an instructor.

Consider, for example, an interpretation in which *maria* is understood as a student, *prof_schmidt* as an instructor, and *exam1* as an assessment. If, in addition,

$$(\mathit{maria}, \mathit{exam1}) \in \mathit{enrolledIn}^{\mathcal{I}}$$
$$\text{and } (\mathit{exam1}, \mathit{prof\_schmidt}) \in \mathit{grades}^{\mathcal{I}}$$

holds, then this interpretation satisfies the corresponding parts of the TBox. In this way, the TBox does not describe individual facts, but rather general structural relationships between concepts.

ABox and TBox thus complement each other naturally: the ABox contains facts about concrete individuals, while the TBox contains general knowledge about the concepts involved. In the next step, we consider how both can be combined and what kinds of queries can be posed to such knowledge bases.

Knowledge Bases and Queries

ABoxes and TBoxes capture different kinds of knowledge: the ABox describes concrete facts about individual objects, whereas the TBox records general relationships between concepts. In practice, both components are used together. Such a combination is called a *knowledge base*. It therefore contains both factual knowledge about individuals and terminological background knowledge.

Definition 2.178. *Let* $\Gamma = \langle K, R, O\rangle$ *be an ALC signature. An* ALC knowledge base *over* Γ *is a pair*

$$\mathcal{K} = \langle \mathcal{A}, \mathcal{T}\rangle,$$

where $\mathcal{A}$ *is an ABox over* Γ *and* $\mathcal{T}$ *is a TBox over* Γ*.*

An interpretation $\mathcal{I}$ is called a *model* of a knowledge base $\mathcal{K} = \langle \mathcal{A}, \mathcal{T}\rangle$ if it satisfies both the ABox $\mathcal{A}$ and the TBox $\mathcal{T}$. We write

$$\mathcal{I} \models_{\text{ALC}} \mathcal{K}.$$

Various kinds of queries can now be posed to knowledge bases. Such queries are formulated as decision problems. In each case, the question is whether a certain semantic relationship holds in all models of the knowledge base, or whether a concept can occur at all in the given state of knowledge.

Definition 2.179. *Let* $\mathcal{K} = \langle \mathcal{A}, \mathcal{T}\rangle$ *be an ALC knowledge base.*

- *A concept description* C *is called* satisfiable in $\mathcal{K}$ *if there exists a model* $\mathcal{I}$ *of* $\mathcal{K}$ *such that*

  $$C^{\mathcal{I}} \neq \emptyset.$$

- *Two concept descriptions* C *and* D *are called* disjoint with respect to $\mathcal{K}$ *if, for every model* $\mathcal{I}$ *of* $\mathcal{K}$*, we have:*

  $$C^{\mathcal{I}} \cap D^{\mathcal{I}} = \emptyset.$$

- *For a terminological axiom X, we write*

$$\mathcal{K} \models_{\text{ALC}} X,$$

if every model of $\mathcal{K}$ also satisfies X.

Typical queries to an ALC knowledge base include the following.

- **Subsumption** Input: A knowledge base $\mathcal{K} = \langle \mathcal{A}, \mathcal{T} \rangle$ and two concept descriptions C, D. Question: Does $\mathcal{K} \models_{\text{ALC}} C \sqsubseteq D$ hold?

- **Concept Satisfiability** Input: A knowledge base $\mathcal{K} = \langle \mathcal{A}, \mathcal{T} \rangle$ and a concept description C. Question: Is C satisfiable with respect to $\mathcal{K}$?

- **Equivalence** Input: A knowledge base $\mathcal{K} = \langle \mathcal{A}, \mathcal{T} \rangle$ and two concept descriptions C, D. Question: Does $\mathcal{K} \models_{\text{ALC}} C \equiv D$ hold?

- **Disjointness** Input: A knowledge base $\mathcal{K} = \langle \mathcal{A}, \mathcal{T} \rangle$ and two concept descriptions C, D. Question: Are C and D disjoint with respect to $\mathcal{K}$?

An example from the university domain illustrates how the ABox and TBox interact. We consider the concepts

Student, *Instructor*, *Module*, *Exam*

as well as the roles

$$enrolledIn, \quad teaches, \quad grades.$$

A possible TBox is:

$$Student \equiv Person \sqcap \exists\, enrolledIn.Module,$$

$$Instructor \equiv Person \sqcap \exists\, teaches.Module,$$

$$Exam \sqsubseteq \exists\, grades.Instructor.$$

The ABox could then contain the following facts:

$$\begin{aligned}\mathcal{A} = \{&Person(emma),\\ &Person(lina),\\ &Module(logic1),\\ &enrolledIn(emma, logic1),\\ &teaches(lina, logic1),\\ &Exam(exam1),\\ &grades(exam1, lina)\}\end{aligned}$$

This knowledge base then expresses, for example, that *emma* is a student, *lina* is an instructor, and *exam1* is an exam that is graded by an instructor. The TBox provides the general background knowledge, while the ABox contains the concrete instances.

It is especially important that the queries listed above are closely related to one another. Many reasoning problems can be reduced to subsumption.

Proposition 2.180. *Let $\Gamma = \langle K, R, O \rangle$ be an ALC signature. For every ALC knowledge base $\mathcal{K}$ over Γ and all concept descriptions C, D over Γ, the following hold:*

1. *$\mathcal{K} \models_{\text{ALC}} C \equiv D$ if and only if*

$$\mathcal{K} \models_{\text{ALC}} C \sqsubseteq D \quad \text{and} \quad \mathcal{K} \models_{\text{ALC}} D \sqsubseteq C.$$

2. *C and D are disjoint with respect to $\mathcal{K}$ if and only if*

$$\mathcal{K} \models_{\text{ALC}} C \sqcap D \sqsubseteq \bot.$$

3. *C is unsatisfiable in $\mathcal{K}$ if and only if*

$$\mathcal{K} \models_{\text{ALC}} C \sqsubseteq \bot.$$

Proof. The claims follow directly from the semantic definitions.

1. By definition, $C \equiv D$ means that in every model of $\mathcal{K}$, the sets $C^{\mathcal{I}}$ and $D^{\mathcal{I}}$ are equal. This is equivalent to requiring both $C^{\mathcal{I}} \subseteq D^{\mathcal{I}}$ and $D^{\mathcal{I}} \subseteq C^{\mathcal{I}}$, which is precisely the pair of subsumption statements.

2. C and D are disjoint with respect to $\mathcal{K}$ if and only if, for every model $\mathcal{I}$ of $\mathcal{K}$, we have

$$C^{\mathcal{I}} \cap D^{\mathcal{I}} = \emptyset.$$

But since

$$(C \sqcap D)^{\mathcal{I}} = C^{\mathcal{I}} \cap D^{\mathcal{I}} \quad \text{and} \quad \bot^{\mathcal{I}} = \emptyset,$$

this holds if and only if

$$(C \sqcap D)^{\mathcal{I}} \subseteq \bot^{\mathcal{I}}$$

for all models $\mathcal{I}$ of $\mathcal{K}$, that is, exactly when $\mathcal{K} \models_{\mathrm{ALC}} C \sqcap D \sqsubseteq \bot$.

3. C is unsatisfiable in $\mathcal{K}$ if and only if there is no model $\mathcal{I}$ of $\mathcal{K}$ such that $C^{\mathcal{I}} \neq \emptyset$. This is equivalent to saying that for every model $\mathcal{I}$ of $\mathcal{K}$,

$$C^{\mathcal{I}} = \emptyset,$$

that is,

$$C^{\mathcal{I}} \subseteq \bot^{\mathcal{I}}.$$

Hence, this is exactly the case when $\mathcal{K} \models_{\mathrm{ALC}} C \sqsubseteq \bot$.

□

These connections are of great practical importance because they show that many questions can be reduced to one central reasoning task. For this reason, subsumption is often the fundamental computational task both in theory and in implementations.

Finally, there is a remarkable complexity result: the problems of subsumption, concept satisfiability, equivalence, and disjointness are PSPACE-complete for ALC. Thus, although ALC is substantially more manageable than full first-order predicate logic, it is still far from being computationally trivial.

Relations to First-Order Predicate Logic

One important advantage of the description logic ALC is that it can be embedded neatly into first-order predicate logic. This shows that, although ALC has its own application-oriented language, semantically it still remains within the scope of classical first-order logic. For the translation, even two variables, such as x and y, are sufficient. Intuitively, this works because in concept descriptions one only ever moves between the currently considered individual and its direct successors via a role.

We read concept symbols as unary predicates, role symbols as binary predicates, and individual names as constants. To translate concept descriptions into formulas of first-order predicate logic, we use a mapping π that depends on the currently considered individual. Accordingly, we write π_x and π_y, respectively.

Definition 2.181. *Let $\Gamma = \langle K, R, O \rangle$ be an ALC signature. The translation of ALC concept descriptions into formulas of first-order predicate logic is defined recursively by the following rules:*

$$\pi_x(\top) = x = x, \qquad \pi_x(\bot) = x \neq x,$$

$$\pi_x(A) = A(x) \qquad (A \in K),$$

$$\pi_x(C \sqcap D) = \pi_x(C) \wedge \pi_x(D),$$

$$\pi_x(C \sqcup D) = \pi_x(C) \vee \pi_x(D)$$

$$\pi_x(\neg C) = \neg \pi_x(C),$$

$$\pi_x(\forall R.C) = \forall y\,(R(x,y) \to \pi_y(C)),$$
$$\pi_x(\exists R.C) = \exists y\,(R(x,y) \wedge \pi_y(C)).$$

The corresponding rules for π_y are obtained by exchanging the roles of the variables.

The translation can be illustrated immediately by means of examples. For the concept description

$$Person \sqcap \exists\, attends.Module$$

we obtain, for instance,

$$\begin{aligned}&\pi_x(Person \sqcap \exists\, attends.Module)\\ &= Person(x) \wedge \exists y\,(attends(x,y) \wedge Module(y))\end{aligned}$$

Likewise,

$$\forall\, teaches.Seminar$$

is translated into

$$\forall y\,(teaches(x,y) \to Seminar(y)).$$

This makes it clear that there is nothing mysterious about description logic in terms of content: it merely uses a linguistically more compact form for statements that can already be expressed in predicate logic.

TBoxes and ABoxes can likewise be translated systematically. In doing so, we interpret the same signature in predicate-logical terms: concept symbols become unary predicates, role symbols become binary predicates, and individual names become constants.

Definition 2.182. *Let* $\mathcal{T} = \{C_1 \sqsubseteq D_1, \ldots, C_n \sqsubseteq D_n\}$ *be a TBox over* Γ. *We then define*

$$\pi(\mathcal{T}) = \bigwedge_{i=1}^{n} \forall x\,(\pi_x(C_i) \to \pi_x(D_i)).$$

Let $\mathcal{A}$ *be an ABox over* Γ. *Then we set*

$$\pi(\mathcal{A}) = \left(\bigwedge_{C(a)\in\mathcal{A}} \pi_x(C)[x \mapsto a]\right) \wedge \left(\bigwedge_{R(a,b)\in\mathcal{A}} R(a,b)\right).$$

Finally, for an ALC knowledge base $\mathcal{K} = \langle \mathcal{A}, \mathcal{T} \rangle$, *we define*

$$\pi(\mathcal{K}) = \pi(\mathcal{A}) \wedge \pi(\mathcal{T}).$$

A small example from the university context illustrates the idea. Consider the TBox

$$\begin{aligned}\mathcal{T} = \{&\mathit{Student} \sqsubseteq \mathit{Person} \sqcap \exists\, \mathit{attends.Module},\\ &\mathit{Tutor} \equiv \mathit{Person} \sqcap \forall\, \mathit{supervises.Student}\}\end{aligned}$$

and the ABox

$$\begin{aligned}\mathcal{A} = \{&\mathit{Person}(\mathit{eva}),\\ &\mathit{Module}(\mathit{logic1}),\\ &\mathit{attends}(\mathit{eva}, \mathit{logic1}),\\ &\mathit{Tutor}(\mathit{maria})\}.\end{aligned}$$

Then the first TBox inclusion expresses that every student is a person who attends at least one module. The second TBox statement says that a tutor is a person who supervises only students. The ABox supplies concrete instances for this.

The benefit of this translation is that semantic questions in description logic can be reduced to questions in predicate logic. In particular, the following correspondences hold.

Lemma 2.183 (Correspondence Lemma). *Let* $\Gamma = \langle K, R, O \rangle$ *be an* $\mathcal{ALC}$ *signature, let*

$$\mathcal{I} = \langle \Delta^{\mathcal{I}}, \cdot^{\mathcal{I}} \rangle$$

be an $\mathcal{ALC}$ *interpretation for* Γ*, and let* $\mathcal{J}$ *be the first-order structure associated with* $\mathcal{I}$*, having the same domain, the same interpretations of concept symbols as unary predicates, of role symbols as binary predicates, and of individual names as constants.*

Then for every $d \in \Delta^{\mathcal{I}}$ *and every concept description* E *over* Γ*, the following holds:*

$$d \in E^{\mathcal{I}} \iff \mathcal{J} \models \pi_x(E)[x \mapsto d].$$

Proof. We prove the statement by structural induction on the formation of E.

Case 1: $E = \top$**.** By definition,

$$\top^{\mathcal{I}} = \Delta^{\mathcal{I}}.$$

Hence, for every $d \in \Delta^{\mathcal{I}}$,

$$d \in \top^{\mathcal{I}}.$$

On the other hand,

$$\pi_x(\top) = x = x,$$

so $\mathcal{J}$ satisfies the formula $\pi_x(\top)[x \mapsto d]$, since of course $d = d$. Thus the claim holds.

Case 2: $E = \bot$. Then

$$\bot^{\mathcal{I}} = \emptyset.$$

Hence there is no $d \in \Delta^{\mathcal{I}}$ such that

$$d \in \bot^{\mathcal{I}}.$$

On the other hand,

$$\pi_x(\bot) = x \neq x,$$

and this formula is satisfiable under no assignment. Therefore, the required equivalence also holds in this case.

Case 3: $E = A$ with $A \in K$. Then, by definition of the interpretation,

$$d \in A^{\mathcal{I}} \iff d \in A^{\mathcal{J}}.$$

On the other hand,

$$\pi_x(A) = A(x),$$

which is true in $\mathcal{J}$ under the assignment $x \mapsto d$ exactly when $d \in A^{\mathcal{J}}$. This proves the claim.

Case 4: $E = C \sqcap D$. By the induction hypothesis,

$$d \in C^{\mathcal{I}} \iff \mathcal{J} \models \pi_x(C)[x \mapsto d]$$

and

$$d \in D^{\mathcal{I}} \iff \mathcal{J} \models \pi_x(D)[x \mapsto d].$$

Now,

$$(C \sqcap D)^{\mathcal{I}} = C^{\mathcal{I}} \cap D^{\mathcal{I}}.$$

Therefore,

$$\begin{aligned} d \in (C \sqcap D)^{\mathcal{I}} &\iff d \in C^{\mathcal{I}} \text{ and } d \in D^{\mathcal{I}} \\ &\iff \mathcal{J} \models \pi_x(C)[x \mapsto d] \\ &\qquad \text{and } \mathcal{J} \models \pi_x(D)[x \mapsto d] \end{aligned}$$

This is equivalent to

$$\mathcal{J} \models \pi_x(C) \wedge \pi_x(D)[x \mapsto d],$$

that is,

$$\mathcal{J} \models \pi_x(C \sqcap D)[x \mapsto d].$$

Case 5: $E = C \sqcup D$. Again, the induction hypothesis yields:

$$d \in C^{\mathcal{I}} \iff \mathcal{J} \models \pi_x(C)[x \mapsto d],$$
$$d \in D^{\mathcal{I}} \iff \mathcal{J} \models \pi_x(D)[x \mapsto d]$$

Since

$$(C \sqcup D)^{\mathcal{I}} = C^{\mathcal{I}} \cup D^{\mathcal{I}},$$

it follows that

$$\begin{aligned} d \in (C \sqcup D)^{\mathcal{I}} &\iff d \in C^{\mathcal{I}} \\ &\qquad \text{or } d \in D^{\mathcal{I}} \\ &\iff \mathcal{J} \models \pi_x(C)[x \mapsto d] \\ &\qquad \text{or } \mathcal{J} \models \pi_x(D)[x \mapsto d] \end{aligned}$$

This is equivalent to

$$\mathcal{J} \models \pi_x(C) \vee \pi_x(D)[x \mapsto d],$$

that is,

$$\mathcal{J} \models \pi_x(C \sqcup D)[x \mapsto d].$$

Case 6: $E = \neg C$. By the induction hypothesis,

$$d \in C^{\mathcal{I}} \iff \mathcal{J} \models \pi_x(C)[x \mapsto d].$$

Moreover,

$$(\neg C)^{\mathcal{I}} = \Delta^{\mathcal{I}} \setminus C^{\mathcal{I}}.$$

Hence,

$$\begin{aligned} d \in (\neg C)^{\mathcal{I}} &\iff d \notin C^{\mathcal{I}} \\ &\iff \mathcal{J} \not\models \pi_x(C)[x \mapsto d] \\ &\iff \mathcal{J} \models \neg\pi_x(C)[x \mapsto d] \end{aligned}$$

Therefore,

$$\mathcal{J} \models \pi_x(\neg C)[x \mapsto d].$$

Case 7: $E = \forall R.C$. By definition,

$$(\forall R.C)^{\mathcal{I}} = \{a \in \Delta^{\mathcal{I}} \mid \forall b \in \Delta^{\mathcal{I}} : \ (a, b) \in R^{\mathcal{I}} \Rightarrow b \in C^{\mathcal{I}}\}.$$

Thus, for $d \in \Delta^{\mathcal{I}}$,

$$d \in (\forall R.C)^{\mathcal{I}} \iff \forall b \in \Delta^{\mathcal{I}}: \ (d,b) \in R^{\mathcal{I}} \Rightarrow b \in C^{\mathcal{I}}.$$

By the induction hypothesis, for every $b \in \Delta^{\mathcal{I}}$ we have:

$$b \in C^{\mathcal{I}} \iff \mathcal{J} \models \pi_y(C)[y \mapsto b].$$

Therefore, the condition above is equivalent to

$$\forall b \in \Delta^{\mathcal{I}}: \ (d,b) \in R^{\mathcal{I}} \Rightarrow \mathcal{J} \models \pi_y(C)[y \mapsto b].$$

This is precisely the truth condition for

$$\mathcal{J} \models \forall y\, (R(x,y) \to \pi_y(C))\, [x \mapsto d],$$

that is, for

$$\mathcal{J} \models \pi_x(\forall R.C)[x \mapsto d].$$

Case 8: $E = \exists R.C$. Analogously,

$$(\exists R.C)^{\mathcal{I}} = \{a \in \Delta^{\mathcal{I}} \mid \exists b \in \Delta^{\mathcal{I}}: \ (a,b) \in R^{\mathcal{I}} \wedge b \in C^{\mathcal{I}}\}.$$

Thus,

$$d \in (\exists R.C)^{\mathcal{I}} \iff \exists b \in \Delta^{\mathcal{I}}: \ (d,b) \in R^{\mathcal{I}} \wedge b \in C^{\mathcal{I}}.$$

Using the induction hypothesis for C, this is equivalent to

$$\exists b \in \Delta^{\mathcal{I}}: \ (d,b) \in R^{\mathcal{I}} \wedge \mathcal{J} \models \pi_y(C)[y \mapsto b].$$

This is precisely the truth condition for

$$\mathcal{J} \models \exists y\, (R(x,y) \wedge \pi_y(C))\, [x \mapsto d],$$

that is, for

$$\mathcal{J} \models \pi_x(\exists R.C)[x \mapsto d].$$

This completes all cases. Hence, for every concept description E and every $d \in \Delta^{\mathcal{I}}$,

$$d \in E^{\mathcal{I}} \iff \mathcal{J} \models \pi_x(E)[x \mapsto d].$$

□

Theorem 2.184. *Let $\mathcal{K} = \langle \mathcal{A}, \mathcal{T} \rangle$ be an ALC knowledge base over the signature $\Gamma = \langle K, R, O \rangle$, and let C, D be concept descriptions over Γ. By $\pi_x(\cdot)$, $\pi(\mathcal{A})$, $\pi(\mathcal{T})$, and $\pi(\mathcal{K}) = \pi(\mathcal{A}) \wedge \pi(\mathcal{T})$ we denote the translation into predicate logic defined in this section. Then the following hold:*

1. *C is satisfiable in $\mathcal{K}$ if and only if $\pi_x(C) \wedge \pi(\mathcal{K})$ is satisfiable.*

2. *$\mathcal{K} \models_{\text{ALC}} C \sqsubseteq D$ if and only if $\pi_x(C) \wedge \neg\pi_x(D) \wedge \pi(\mathcal{K})$ is unsatisfiable.*

Proof. We use the correspondence introduced in this section between ALC interpretations and predicate-logical structures: concept symbols are interpreted as unary predicates, role symbols as binary predicates, and individual names as constants, and both sides have the same domain.

In addition, we use the correspondence lemma proved just above.

(1) *"⇒":* Suppose that C is satisfiable in $\mathcal{K}$. Then there exists an ALC interpretation I with $I \models_{\text{ALC}} \mathcal{K}$ and some $d \in \Delta^I$ such that $d \in C^I$. Let J be the predicate-logical structure corresponding to I. Since I is a model of $\mathcal{K}$, J satisfies the formula $\pi(\mathcal{K})$. Because $d \in C^I$, the correspondence lemma yields

$$J \models \pi_x(C)[x \mapsto d].$$

Thus J satisfies the conjunction $\pi_x(C) \wedge \pi(\mathcal{K})$, and hence this formula is satisfiable.

"⇐": Conversely, suppose that $\pi_x(C) \wedge \pi(\mathcal{K})$ is satisfiable. Then there exist a predicate-logical structure J and an assignment s such that

$$J \models \pi_x(C) \wedge \pi(\mathcal{K}).$$

In particular, we have

$$J \models \pi(\mathcal{K}) \quad \text{and} \quad J \models \pi_x(C)[x \mapsto s(x)].$$

The structure J can be regarded as an ALC interpretation I by using the same domain and interpreting the predicates and constants as above. From $J \models \pi(\mathcal{K})$ it then follows that $I \models_{\text{ALC}} \mathcal{K}$. Moreover, the correspondence lemma yields from

$$J \models \pi_x(C)[x \mapsto s(x)]$$

the statement

$$s(x) \in C^I.$$

Hence C is satisfiable in $\mathcal{K}$.

(2) *"$\Rightarrow$":* Assume that $\mathcal{K} \models_{\text{ALC}} C \sqsubseteq D$. Then for every model I of $\mathcal{K}$ we have:

$$C^I \subseteq D^I.$$

So there is no model I of $\mathcal{K}$ and no element $d \in \Delta^I$ such that

$$d \in C^I \quad \text{and} \quad d \notin D^I.$$

If the formula $\pi_x(C) \land \neg\pi_x(D) \land \pi(\mathcal{K})$ were satisfiable, there would exist a predicate-logical structure J and an assignment s such that

$$\begin{aligned} &J \models \pi(\mathcal{K}), \\ &J \models \pi_x(C)[x \mapsto s(x)], \\ &J \models \neg\pi_x(D)[x \mapsto s(x)]. \end{aligned}$$

This structure would correspond to an ALC interpretation I with $I \models_{\text{ALC}} \mathcal{K}$ and

$$s(x) \in C^I \quad \text{and} \quad s(x) \notin D^I,$$

contradicting $C^I \subseteq D^I$. Therefore,

$$\pi_x(C) \land \neg\pi_x(D) \land \pi(\mathcal{K})$$

is unsatisfiable.

"$\Leftarrow$": Now assume that $\pi_x(C) \land \neg\pi_x(D) \land \pi(\mathcal{K})$ is unsatisfiable. We show that this implies $\mathcal{K} \models_{\text{ALC}} C \sqsubseteq D$. To this end, let I be an arbitrary model of $\mathcal{K}$ and let

$d \in C^I$ be arbitrary. Let J be the predicate-logical structure corresponding to I. Since $I \models_{\text{ALC}} \mathcal{K}$, we also have

$$J \models \pi(\mathcal{K}).$$

From $d \in C^I$, the correspondence lemma yields

$$J \models \pi_x(C)[x \mapsto d].$$

If in addition $d \notin D^I$, then we would also have

$$J \models \neg\pi_x(D)[x \mapsto d]$$

holding. But then $\pi_x(C) \wedge \neg\pi_x(D) \wedge \pi(\mathcal{K})$ would be satisfiable, contradicting the assumption. Therefore, we must have $d \in D^I$ in every case. Since I and d were arbitrary, it follows that

$$C^I \subseteq D^I$$

for all models I of $\mathcal{K}$, and hence $\mathcal{K} \models_{\text{ALC}} C \sqsubseteq D$. □

Thus, ALC is not only an autonomous language of knowledge representation, but also a well-defined fragment of first-order predicate logic. It is precisely this embedding that explains why many algorithmic properties of description logic can later be studied using the methods of classical logic.

Further Description Logics

ALC is a particularly intuitive and, at the same time, highly typical example of a description logic. It already contains the most important logical building blocks for describing concepts and relations. For precisely this reason, it serves well as a starting point for understanding the broader landscape of description logics: depending on the application, the language is deliberately extended or restricted in order to gain either greater expressive power or better algorithmic properties.

One simple but often useful restriction is the logic AL. It essentially corresponds to ALC without general negation on concepts. Such reduced languages are easier to handle in some situations because they offer fewer possibilities for indirect self-reference and for the formation of complex opposing concepts. In practice, this kind of restriction is especially interesting when one wants a structure with inference rules that are as transparent as possible.

On the other hand, there are numerous extensions that increase the expressive power of ALC. An important class of such extensions concerns *number restrictions*. These make it possible to specify how many successors an individual may or must have via a role. For example, one can formulate statements such as:

$$\geq 2\, \mathit{supervises} \qquad \text{or} \qquad \leq 3\, \mathit{takes}.$$

In the first case, the expression describes all individuals

that are connected to at least two other individuals via the role *supervises*. In the second case, it captures those individuals that stand in the relation *takes* to at most three objects.

So-called *qualified number restrictions* are even more expressive. Here one counts not only the number of successors, but also restricts the concept to which those successors must belong. Thus,

$$\geq 2\, takes.Module$$

describes all students who take at least two modules. Similarly,

$$\leq 1\, supervises.Professor$$

denotes all persons who are supervised by at most one professor. Constructions of this kind are very useful in applications because they can express structural constraints directly.

Another extension involves *nominals*. These are singleton concepts of the form

$$\{a\},$$

that is, concepts containing exactly one object denoted by the individual name a. Nominals make it possible to incorporate specific known objects directly into concept descriptions. For instance, one may use the concept

$$\exists\, supervises.\{maria\}$$

to describe all individuals that are related, via the role *supervises*, to exactly the object denoted by *maria*. In knowledge representation, this is especially useful when certain key objects are meant to be marked explicitly.

There are also important generalizations on the side of roles. One frequently used construction is the *inverse role*. For a role R, one writes

$$R^-$$

for the reversed relation. Semantically, this means:

$$(R^-)^{\mathcal{I}} = \{(b, a) \mid (a, b) \in R^{\mathcal{I}}\}.$$

If, for example, *supervises* is the role relating supervisors to supervisees, then the inverse role *supervises*$^-$ describes the opposite perspective, namely "is supervised by". Such inversions are particularly useful when one wants to analyze the same knowledge base from different points of view.

Besides concept and role constructors, the relationships between roles are also important. For this purpose, one often uses an *RBox*. It may contain statements such as

$$R \sqsubseteq S$$

which means that every R-relation is also an S-relation. In a university setting, for example, one might stipulate that every *supervises* relation is at the same time an *advises* relation. Another important property is the *transitivity* of a role. If a role R is transitive, then from $(a, b) \in R^{\mathcal{I}}$ and $(b, c) \in R^{\mathcal{I}}$ it always follows that $(a, c) \in R^{\mathcal{I}}$. This is often useful for relations such as *isPartOf* or *isPrerequisiteFor*.

The names of many description logics are built from such extensions. For instance, an additional letter such as N indicates number restrictions, Q qualified number restrictions, O nominals, I inverse roles, H role hierarchies, and S transitivity. In this way, an entire family of logics arises, each tailored to different requirements. Depending on which components are added or omitted, the balance between expressive power and computational tractability changes considerably.

It is precisely this diversity that makes description logics so important in practice. They are used in many variants in ontologies and on the Semantic Web. A well-known standard system in this context is *OWL* (*Web Ontology Language*), which is based on description-logical ideas and is used in tools such as *Protégé*. Description logics thus form a bridge between logical precision and application-oriented knowledge representation.

Exercises and Solutions

In the following, we practice the syntax and semantics of ALC as well as its translation into predicate logic by means of a small example. For each of the statements below, decide whether it is true or false.

Exercise 2.185. *Consider the ALC signature* $\Gamma = (K, R, O)$ *with*

$$K = \{E, D, B\}, \qquad R = \{T\}, \qquad O = \{d, e, h, c\},$$

together with the interpretation $\mathcal{I}$ *for* Γ *given by*

$$\Delta^{\mathcal{I}} = \{d, e, h, c\},$$

$$E^{\mathcal{I}} = \{e, h\}, \qquad D^{\mathcal{I}} = \{d\}, \qquad B^{\mathcal{I}} = \{e, c\},$$
$$T^{\mathcal{I}} = \{(e, d), (d, d), (e, h), (e, c)\}.$$

Moreover, let

$$\varphi_4 = \forall y\, (T(x, y) \rightarrow E(y)) \vee \neg D(x).$$

Determine for each of the following statements whether it is true or false.

1. $\exists T.B \sqcap \neg D$ *is an ALC concept description over* Γ.
2. $(\neg\exists T.(E \sqcap B))^{\mathcal{I}} = \{d, e, h, c\}$.
3. $\mathcal{I} \models_{\mathrm{ALC}} E \sqsubseteq D$.
4. $\pi_x(\forall T.E \sqcup \neg D) = \varphi_4$.

Solution 2.186. *1.* ***True.*** *The expressions* $\exists T.B$ *and* $\neg D$ *are both ALC concept descriptions:* $\exists T.B$ *is an existential restriction, and* $\neg D$ *is the negation* *of a concept symbol. Since ALC is closed under concept conjunction, it follows that*

$$\exists T.B \sqcap \neg D$$

is also a well-formed ALC concept description over Γ.

2. **True.** *We first determine the inner expression:*

$$E^{\mathcal{I}} \cap B^{\mathcal{I}} = \{e, h\} \cap \{e, c\} = \{e\}.$$

Next, we check whether there is a pair in $T^{\mathcal{I}}$ *whose second component lies in this set. The* T*-pairs are*

$$(e, d),\ (d, d),\ (e, h),\ (e, c).$$

None of these pairs ends in e. *Therefore,*

$$(\exists T.(E \sqcap B))^{\mathcal{I}} = \emptyset.$$

Hence,

$$(\neg\exists T.(E \sqcap B))^{\mathcal{I}} = \Delta^{\mathcal{I}} \setminus \emptyset = \{d, e, h, c\}.$$

3. **False.** *For* $E \sqsubseteq D$ *to hold, we would need:*

$$E^{\mathcal{I}} \subseteq D^{\mathcal{I}}.$$

But here we have

$$E^{\mathcal{I}} = \{e, h\}, \qquad D^{\mathcal{I}} = \{d\}.$$

In particular, $e \in E^{\mathcal{I}}$, but $e \notin D^{\mathcal{I}}$. Therefore,

$$E^{\mathcal{I}} \not\subseteq D^{\mathcal{I}},$$

and thus

$$\mathcal{I} \not\models_{\text{ALC}} E \sqsubseteq D.$$

4. ***True.*** *We first translate the two parts separately:*

$$\pi_x(\forall T.E) = \forall y\,(T(x,y) \to E(y)),$$

$$\pi_x(\neg D) = \neg D(x).$$

Since the translation preserves disjunction as $\vee$, it follows that

$$\pi_x(\forall T.E \sqcup \neg D) = \forall y\,(T(x,y) \to E(y)) \vee \neg D(x).$$

This is exactly φ_4.

These exercises illustrate how closely syntax, semantics, and translation are connected in ALC: first, one checks whether an expression is well formed; then one computes its meaning in an interpretation; and finally, if needed, one translates it into a first-order formula.

2.7 Second-Order Predicate Logic

In this final section of the chapter, we extend first-order predicate logic to a significantly more powerful language: second-order predicate logic. The crucial difference is that we can quantify not only over individual objects in the domain, but also over sets, relations, and predicates in general. As a result, the logic gains considerable expressive power, but at the same time many metatheoretical properties become more difficult than in the first-order setting.

In computer science, this logic plays an important role above all in descriptive complexity theory. There one investigates which classes of computable problems can be characterized by logical formulas. Before turning to that topic, we first introduce the syntax and semantics of second-order predicate logic and explain the difference from first-order logic by means of a few examples.

2.7.1 Syntax and Semantics

As in first-order predicate logic, we again work with signatures, that is, with predicate, function, and constant symbols whose meaning is fixed only through an interpretation. The actual difference from the first-order setting does not lie in these symbols, but in the language itself: in addition to variables for individuals, we now also have variables for relations, together with quantification over these variables.

In the first-order setting, variables range over individual elements of the domain. In the second-order setting, we additionally introduce variables that range over relations. Such second-order variables are assigned a fixed arity. A unary second-order variable therefore stands for a set of objects, a binary second-order variable for a binary relation, and so on.

Intuitively, one may think of it as follows: in the first-order setting, one asks, for example, whether there is a particular object with a given property. In the second-order setting, one can additionally ask whether there is a set with a certain property or whether a relation itself satisfies certain conditions.

Definition 2.187. *Let Σ be a signature. The set of second-order predicate logic formulas over Σ is defined inductively. Let $x, y, z, \dots$ be first-order variables and $X, Y, Z, \dots$ be second-order variables, where each second-order variable has a fixed arity.*

The second-order Σ-formulas are formed as follows:

- *If P is a predicate symbol from Σ of arity n and $t_1, \dots, t_n$ are Σ-terms, then*

$$P(t_1, \dots, t_n)$$

  is a second-order formula.

- *If X is a second-order variable of arity n and $t_1, \dots, t_n$ are Σ-terms, then*

$$X(t_1, \dots, t_n)$$

  is a second-order formula.

- *If φ and ψ are second-order formulas, then*

$$(\varphi \wedge \psi), \qquad (\varphi \vee \psi)$$

are also second-order formulas.

- *If φ is a second-order formula, then*

$$\neg\varphi$$

  is also a second-order formula.

- *If x is a first-order variable and φ is a second-order formula, then*

$$\forall x\,\varphi \qquad \text{and} \qquad \exists x\,\varphi$$

  are also second-order formulas.

- *If X is a second-order variable and φ is a second-order formula, then*

$$\forall X\,\varphi \qquad \text{and} \qquad \exists X\,\varphi$$

  are also second-order formulas.

Every formula of first-order predicate logic is therefore automatically also a formula of second-order logic. Second-order logic thus extends first-order logic without altering it.

The semantics are extended accordingly. In addition to a structure, we now also require an assignment for the second-order variables. Such an assignment maps each first-order variable to an element of the domain and each second-order variable to a relation of the appropriate arity over the domain.

Definition 2.188. *Let Σ be a signature and let A be a Σ-structure with underlying set $|A|$. A* second-order assignment *is a function*

$$\gamma : (V \cup V_2) \to |A| \cup \bigcup_{n \in \mathbb{N}_0} \mathcal{P}(|A|^n),$$

that assigns to each first-order variable an element of $|A|$, and to each second-order variable X of arity n an n-ary relation on $|A|$.

If γ is an assignment and x is a first-order variable, we write $\gamma[x \mapsto a]$ for the assignment that differs from γ only in mapping x to a. If X is a second-order variable and R is a suitable relation, we write $\gamma[X \mapsto R]$ for the assignment modified accordingly.

A pair $\langle A, \gamma \rangle$ consisting of a structure and an assignment is called an *interpretation* for second-order formulas.

The evaluation of formulas proceeds recursively. For atomic formulas, we again check whether the corresponding relation holds in the structure; for quantifiers over second-order variables, we now quantify over all appropriate subsets or relations of the domain.

Definition 2.189. *Let Σ be a signature and $\mathcal{I} = \langle A, \gamma \rangle$ an interpretation. Then each formula φ is assigned its truth value $[\![\varphi]\!]_{\mathcal{I}} \in \{\text{true}, \text{false}\}$ as follows:*

$$[\![P(t_1, \dots, t_n)]\!]_{\mathcal{I}} = \text{true}$$
$$\iff ([\![t_1]\!]_{\mathcal{I}}, \dots, [\![t_n]\!]_{\mathcal{I}}) \in P^A$$

$$\llbracket X(t_1, \ldots, t_n) \rrbracket_{\mathcal{I}} = \text{true}$$
$$\iff (\llbracket t_1 \rrbracket_{\mathcal{I}}, \ldots, \llbracket t_n \rrbracket_{\mathcal{I}}) \in \gamma(X)$$

$$\llbracket \varphi \wedge \psi \rrbracket_{\mathcal{I}} = \text{true}$$
$$\iff \llbracket \varphi \rrbracket_{\mathcal{I}} = \text{true } \textit{and } \llbracket \psi \rrbracket_{\mathcal{I}} = \text{true}$$

$$\llbracket \varphi \vee \psi \rrbracket_{\mathcal{I}} = \text{true}$$
$$\iff \llbracket \varphi \rrbracket_{\mathcal{I}} = \text{true } \textit{or } \llbracket \psi \rrbracket_{\mathcal{I}} = \text{true}$$

$$\llbracket \neg \varphi \rrbracket_{\mathcal{I}} = \text{true}$$
$$\iff \llbracket \varphi \rrbracket_{\mathcal{I}} = \text{false}$$

$$\llbracket \exists x\, \varphi \rrbracket_{\mathcal{I}} = \text{true}$$
$$\iff \exists a \in |A| : \llbracket \varphi \rrbracket_{\langle A, \gamma[x \mapsto a] \rangle} = \text{true}$$

$$\llbracket \forall x\, \varphi \rrbracket_{\mathcal{I}} = \text{true}$$
$$\iff \forall a \in |A| : \llbracket \varphi \rrbracket_{\langle A, \gamma[x \mapsto a] \rangle} = \text{true}$$

$$\llbracket \exists X\, \varphi \rrbracket_{\mathcal{I}} = \text{true}$$
$$\iff \exists R \subseteq |A|^{\mathrm{Ary}(X)} : \llbracket \varphi \rrbracket_{\langle A, \gamma[X \mapsto R] \rangle} = \text{true}$$

$$\llbracket \forall X\, \varphi \rrbracket_{\mathcal{I}} = \text{true}$$
$$\iff \forall R \subseteq |A|^{\mathrm{Ary}(X)} : \llbracket \varphi \rrbracket_{\langle A, \gamma[X \mapsto R] \rangle} = \text{true}$$

Second-order logic therefore quantifies not only over individual elements, but also over arbitrary relations of the appropriate arity on the domain. This makes the language considerably more expressive than first-order logic. At the same time, this also explains why many metatheoretical properties become more difficult: quantification over sets and relations allows us to speak in a much stronger and more indirect way.

A simple example illustrates this additional expressive power. Consider a signature with a unary predicate P and a formula

$$\exists X\,(\exists x\, X(x) \land \forall x\,(X(x) \to P(x))).$$

This formula says that there exists a nonempty subset of the P-elements. In first-order logic, this cannot be expressed in the same form without additional machinery, because there one cannot quantify over sets.

Another example comes from graph theory. Let E be a binary relation symbol. Then

$$\exists X\,\forall x\,\forall y\,(E(x,y) \to (X(x) \leftrightarrow \neg X(y)))$$

describes the property that the graph is bipartite: there is a partition of the vertices into a set X and its complement such that every edge connects the two sides. Statements of this kind show clearly how naturally second-order quantification arises in structural problems.

In the next section, we will see that this greater expressive power comes at a price. Second-order predicate logic is substantially more powerful than first-order logic, but it is also algorithmically much harder to handle.

2.7.2 Properties and Expressive Power

Second-order predicate logic extends first-order logic at only one point, but precisely this point changes many things fundamentally: we may now quantify not only over elements of the domain, but also over sets and relations. As a result, a large part of the familiar syntactic machinery remains intact, yet the expressive power increases significantly.

To begin with, many of the formal properties known from first-order predicate logic carry over unchanged to second-order logic. This applies in particular to the usual transformation rules for connectives and quantifiers, the substitution lemma, the coincidence lemma, and the local dependence on assignments. In this section, we therefore focus on those properties that are new to second-order logic or that behave differently here than they do in the first-order setting.

The familiar dualities and commutation rules also hold for second-order quantifiers, as long as no variables are captured.

Proposition 2.190. *The following equivalences hold:*

1. $\exists X\, \varphi \equiv \neg \forall X\, \neg \varphi$,

2. $\varphi \equiv \forall X\, \varphi$, *if* $X \notin FV(\varphi)$,

3. $\forall X\, \varphi \equiv \neg \exists X\, \neg \varphi$,

4. $\varphi \equiv \exists X\, \varphi$, *if* $X \notin FV(\varphi)$,

5. $\forall X \forall Y\, \varphi \equiv \forall Y \forall X\, \varphi$,

6. $\forall X(\varphi \wedge \psi) \equiv (\forall X\, \varphi) \wedge (\forall X\, \psi)$,

7. $\exists X \exists Y \, \varphi \equiv \exists Y \exists X \, \varphi$,

8. $\exists X (\varphi \vee \psi) \equiv (\exists X \, \varphi) \vee (\exists X \, \psi)$.

Proof. We prove each equivalence directly using the semantics of second-order quantifiers.

(1) Let $I = \langle A, \gamma \rangle$ be an arbitrary interpretation. Then:

$$I \models \exists X.\, \varphi$$

holds if and only if there exists a relation $S \subseteq A^{\mathrm{Ary}(X)}$ such that

$$\langle A, \gamma[X \mapsto S] \rangle \models \varphi.$$

This is the case if and only if it is *not* true that for all relations $S \subseteq A^{\mathrm{Ary}(X)}$,

$$\langle A, \gamma[X \mapsto S] \rangle \models \neg\varphi,$$

that is, if and only if

$$I \models \neg\forall X.\, \neg\varphi.$$

(2) Let $X \notin FV(\varphi)$. Then the truth value of φ does not depend on the assignment of X. Hence, for every relation $S \subseteq A^{\mathrm{Ary}(X)}$, we have:

$$\langle A, \gamma \rangle \models \varphi \iff \langle A, \gamma[X \mapsto S] \rangle \models \varphi.$$

Therefore:

$$\langle A, \gamma \rangle \models \varphi \iff \langle A, \gamma \rangle \models \forall X.\, \varphi.$$

(3) Let $I = \langle A, \gamma \rangle$. Then:

$$I \models \forall X.\, \varphi$$

holds if and only if for all relations $S \subseteq A^{\text{Ary}(X)}$,

$$\langle A, \gamma[X \mapsto S] \rangle \models \varphi.$$

This is the case if and only if there is no relation $S \subseteq A^{\text{Ary}(X)}$ such that

$$\langle A, \gamma[X \mapsto S] \rangle \models \neg\varphi,$$

that is, if and only if

$$I \models \neg\exists X.\, \neg\varphi.$$

(4) Again let $X \notin FV(\varphi)$. Then for every relation $S \subseteq A^{\text{Ary}(X)}$,

$$\langle A, \gamma \rangle \models \varphi \quad \Longleftrightarrow \quad \langle A, \gamma[X \mapsto S] \rangle \models \varphi.$$

In particular, such a relation S exists, so that

$$\langle A, \gamma \rangle \models \exists X.\, \varphi$$

holds if and only if

$$\langle A, \gamma \rangle \models \varphi.$$

(5) Let $I = \langle A, \gamma \rangle$. Then:

$$I \models \forall X \forall Y.\, \varphi$$

holds if and only if for all relations $S \subseteq A^{\text{Ary}(X)}$ and all relations $T \subseteq A^{\text{Ary}(Y)}$,

$$\langle A, \gamma[X \mapsto S, Y \mapsto T]\rangle \models \varphi.$$

Since the choice of S and T is independent, this is equivalent to:

$$I \models \forall Y \forall X.\, \varphi.$$

(6) Let $I = \langle A, \gamma\rangle$. Then:

$$I \models \forall X(\varphi \wedge \psi)$$

holds if and only if for all relations $S \subseteq A^{\text{Ary}(X)}$,

$$\langle A, \gamma[X \mapsto S]\rangle \models \varphi \wedge \psi.$$

This is the case if and only if for all S, both

$$\langle A, \gamma[X \mapsto S]\rangle \models \varphi \quad \text{and} \quad \langle A, \gamma[X \mapsto S]\rangle \models \psi$$

hold, that is, if and only if

$$I \models (\forall X.\, \varphi) \wedge (\forall X.\, \psi).$$

(7) Let $I = \langle A, \gamma\rangle$. Then:

$$I \models \exists X \exists Y.\, \varphi$$

holds if and only if there exist relations $S \subseteq A^{\text{Ary}(X)}$ and $T \subseteq A^{\text{Ary}(Y)}$ such that

$$\langle A, \gamma[X \mapsto S, Y \mapsto T]\rangle \models \varphi.$$

Since the existence of S and T is independent, this is equivalent to:

$$I \models \exists Y \exists X.\, \varphi.$$

(8) Let $I = \langle A, \gamma \rangle$. Then:

$$I \models \exists X(\varphi \vee \psi)$$

holds if and only if there exists a relation $S \subseteq A^{\mathrm{Ary}(X)}$ such that

$$\langle A, \gamma[X \mapsto S] \rangle \models \varphi \vee \psi.$$

This is the case if and only if either

$$\langle A, \gamma[X \mapsto S] \rangle \models \varphi \quad \text{or} \quad \langle A, \gamma[X \mapsto S] \rangle \models \psi$$

holds, that is, if and only if

$$I \models (\exists X.\, \varphi) \vee (\exists X.\, \psi).$$

□

Even more interesting is the fact that, in second-order logic, relations and functions themselves can be treated as objects. This is one of the main reasons for its greater expressive power.

Proposition 2.191. *Let X be an n-ary second-order variable, and let Y be an $(n+1)$-ary second-order variable. Then, for every formula φ in which Y does not occur, the following holds:*

$$\forall x\, \exists X\, \varphi \equiv \exists Y\, \forall x\, \varphi',$$

where φ' is obtained from φ by replacing every atomic formula of the form $X(t_1, \ldots, t_n)$ with $Y(x, t_1, \ldots, t_n)$.

Proof. We prove the semantic equivalence. Let $I = \langle A, \gamma \rangle$ be an arbitrary interpretation throughout (that is, a structure together with a second-order assignment). We use the semantics of second-order quantifiers as given in the definitions above.

"$\Rightarrow$": Assume that $I \models \forall x\, \exists X.\, \varphi$. By the semantics, this means that for every $a \in A$ there exists a relation

$$S_a \subseteq A^n$$

(more precisely, S_a has the arity of X) such that

$$\langle A, \gamma[X \mapsto S_a] \rangle \models \varphi \quad \text{under the assignment } x \mapsto a.$$

(In short: for every a there is some S_a such that $I[X \mapsto S_a, x \mapsto a] \models \varphi$.)

We now construct a single $(n+1)$-ary relation $R \subseteq A \times A^n$ by

$$R \;:=\; \{\, (a, \bar{t}) \in A \times A^n \mid \bar{t} \in S_a \,\}.$$

Set $\gamma' := \gamma[Y \mapsto R]$. We claim that for every $a \in A$,

$$\langle A, \gamma'[x \mapsto a] \rangle \models \varphi'.$$

Indeed, in φ' all occurrences of $X(u_1, \ldots, u_n)$ have been replaced by $Y(x, u_1, \ldots, u_n)$. For a fixed a and a tuple $\bar{t}$, the formula $Y(a, \bar{t})$ holds in $\langle A, \gamma' \rangle$ if and only if $(a, \bar{t}) \in R$, that is, if and only if $\bar{t} \in S_a$. Hence, under the assignment $x \mapsto a$, the truth of every atomic statement in φ of the form $X(\dots)$ is, in φ', equivalent to the truth of the corresponding statement under the assignment $X \mapsto S_a$. For all other atomic subformulas (those not containing X), nothing changes, and by assumption Y does not occur in φ. Therefore, from $I[X \mapsto S_a, x \mapsto a] \models \varphi$ it follows immediately that $\langle A, \gamma'[x \mapsto a]\rangle \models \varphi'$. Since this holds for every $a \in A$, we obtain $\langle A, \gamma'\rangle \models \forall x.\varphi'$, and thus $I \models \exists Y\, \forall x.\, \varphi'$ (with the newly constructed relation R serving as the witness).

"$\Leftarrow$": Assume that $I \models \exists Y\, \forall x.\, \varphi'$. Then there exists an $(n+1)$-ary relation $R \subseteq A \times A^n$ such that

$$\langle A, \gamma[Y \mapsto R]\rangle \models \forall x.\, \varphi'.$$

For each $a \in A$, define the n-ary relation

$$S_a := \{\, \bar{t} \in A^n \mid (a, \bar{t}) \in R \,\}.$$

Since $\langle A, \gamma[Y \mapsto R]\rangle \models \forall x.\varphi'$, for every $a \in A$ we have:

$$\langle A, \gamma[Y \mapsto R, x \mapsto a]\rangle \models \varphi'.$$

By the construction of φ' and S_a, it follows, analogously to the argument above, that the truth of φ' under the assignment $Y \mapsto R$ and $x \mapsto a$ is exactly the same as the truth of φ under the assignment $X \mapsto S_a$ and $x \mapsto a$. Thus, for every $a \in A$,

$$\langle A, \gamma[X \mapsto S_a, x \mapsto a]\rangle \models \varphi.$$

By the definition of the semantics, it follows that $\langle A, \gamma \rangle \models \forall x \, \exists X . \, \varphi$, that is, $I \models \forall x \, \exists X . \, \varphi$.

Since both directions have been established, the desired equivalence follows:

$$\forall x \, \exists X . \, \varphi \;\equiv\; \exists Y \, \forall x . \, \varphi' .$$

□

In general, formulas of second-order predicate logic can be transformed into a normal form in which all second-order quantifiers appear at the front. We say that a second-order formula is in *second-order prenex form* if it has the shape

$$Q_1 X_1 \, Q_2 X_2 \, \cdots \, Q_n X_n . \, \psi$$

where each Q_i is either $\forall$ or $\exists$, and ψ contains no further second-order quantifiers. First-order quantifiers may still occur in ψ.

Using the equivalences from Proposition 2.180 together with the commutation rule from Proposition 2.181, every second-order formula can be transformed into an equivalent formula of this form.

Second-order quantification also makes it possible to express properties that cannot be stated directly in first-order logic. A classical example is the assertion that a relation characterizes equality:

$$\mathrm{Eq}(=) \;:=\; \forall x \forall y \Big(x = y \leftrightarrow \forall X \, (X(x) \rightarrow X(y)) \Big).$$

In this way, equality can be described internally in second-order logic in a very robust manner. Likewise, functions can be encoded by relations that are total and functional. In other words, function variables can be simulated by suitable relations, provided their graphs are viewed as sets of tuples.

Proposition 2.192. *For every formula φ of second-order predicate logic, there exists an equivalent formula of second-order predicate logic in second-order prenex normal form.*

Proof. We show how to construct, from an arbitrary formula φ, an equivalent formula in the required prenex normal form. In doing so, we freely use the equivalences from Propositions 2.180 and 2.181, as well as the usual rules for renaming bound variables (in order to avoid name clashes). The proof idea is constructive: we move all second-order quantifiers outward step by step until all second-order quantifiers stand at the beginning; first-order quantifiers may remain inside the formula. Since every formula contains only finitely many quantifiers, the procedure terminates.

1. Reduction to a suitable form:
First, we bring φ into a form in which negations occur only in front of atomic formulas (negation normal form) and in which all bound variables are pairwise distinct (by systematically renaming bound variables). Both can be achieved by means of the well-known equivalences (double negation, De Morgan, renaming of bound variables);

these steps do not alter the structure with respect to second-order quantifiers, but they simplify the subsequent procedure.

2. Procedure for pulling out second-order quantifiers:
We repeat the following step as long as the formula still contains a second-order quantifier inside a logical operator or inside a first-order quantifier.

Consider a subformula of the form

$$\tau = \Phi(Q_2X.\psi)$$

where $Q_2 \in \{\forall, \exists\}$ is a *second-order* quantifier and $\Phi(\cdot)$ is a context formula representing either a Boolean connective (conjunction/disjunction/negation) or a first-order quantifier ($\forall x$ or $\exists x$). We move $Q_2X.\psi$ to a position where it is "outermost" by means of the following semantically justified transformations:

1. *If Φ is a Boolean connective:*
   Use the equivalences from Proposition 2.180, for example

   $$\forall X.(\alpha \wedge \beta) \equiv (\forall X.\alpha) \wedge (\forall X.\beta)$$
   $$\exists X.(\alpha \vee \beta) \equiv (\exists X.\alpha) \vee (\exists X.\beta)$$

   together with commutativity/associativity in order to pull Q_2X outward step by step across the Boolean connectives. By repeated application of these rules, every second-order quantifier can be pulled out across conjunctions/disjunctions without changing the semantics. (When crossing a negation, one uses the duality $\exists X.\varphi \equiv \neg\forall X.\neg\varphi$ and $\forall X.\varphi \equiv \neg\exists X.\neg\varphi$.)

2. *If Φ is a first-order quantifier, for example $\forall x$ or $\exists x$:*
   Here we use Proposition 2.181 and the duality rules from Proposition 2.180 in order to move the second-order quantifier to the desired position. The following cases arise, each permitting semantically justified transformations:

   - If the subformula has the form $\forall x.\, Q_2 X.\, \psi$ and $Q_2 = \exists$, we apply Proposition 2.181:
     $$\forall x\, \exists X.\, \psi \;\equiv\; \exists Y\, \forall x.\, \psi',$$
     where ψ' is obtained by replacing each atom $X(t_1, \ldots, t_n)$ with $Y(x, t_1, \ldots, t_n)$ (possibly after renaming bound variables). Thus the second-order quantifier $\exists X$ has been moved in front of the first-order quantifier $\forall x$ (albeit as $\exists Y$ with arity increased from n to $n+1$).
   - If the subformula is $\exists x.\, Q_2 X.\, \psi$ and $Q_2 = \exists$, then the existential quantifiers may be interchanged:
     $$\exists x\, \exists X.\, \psi \;\equiv\; \exists X\, \exists x.\, \psi,$$
     since both x and X are existentially quantified independently; thus X has been moved one level outward.

- If the subformula is $\exists x. \forall X. \psi$, we first write $\forall X.\psi \equiv \neg\exists X.\neg\psi$ (Proposition 2.180), so that the previously treated case becomes applicable; after the transformation, the duality is converted back.
- Analogous transformations apply when the outer first-order quantifier is $\forall x$ and the inner second-order quantifier is $\forall X$ or $\exists X$: one uses the combination of duality (Prop. 2.180) and Prop. 2.181, possibly after renaming bound variables, to move the second-order quantifier to the desired outer position.

It is important that each of these applications preserves semantics and that bound variables are renamed whenever necessary in order to avoid collisions.

3. Termination and result:
Each step moves at least one second-order quantifier one position further outward (across a connective or quantifier boundary). Since φ contains only finitely many second-order quantifiers, the procedure terminates after finitely many applications. In the end, one obtains a formula equivalent to φ in which no second-order quantifier occurs inside a logical operator or a first-order quantifier, i.e. all second-order quantifiers stand at the beginning. Hence the resulting formula has the desired form

$$Q_1 X_1 \, Q_2 X_2 \, \dots \, Q_m X_m . \, \psi(X_1, \dots, X_m),$$

where ψ contains no second-order quantifiers anymore. (First-order quantifiers may remain in ψ; the definition of "second-order prenex normal form" requires only that all second-order quantifiers stand at the beginning.)

4. Variable freshness / naming convention: Throughout the procedure, we always ensure that bound variables are renamed in such a way that no collisions arise (neither between first-order nor between second-order variables). This is always possible, since only finitely many names occur in φ.

Thus, we have constructively shown how to obtain, from every second-order formula, an equivalent formula in second-order prenex normal form. □

Particularly important is the ability to capture global properties of structures. For example, one can express that a structure is finite by stating that every injective self-map must also be surjective. Conversely, one can characterize infinitude by requiring an injective but non-surjective mapping. Such formulas are not possible in this form at the first-order level.

An elegant example comes from graph theory. Consider a binary relation E as the edge relation of a graph. In second-order logic, one can then express that the graph is bipartite: there exists a set of vertices X such that every edge connects a vertex in X with a vertex outside X. Thus, second-order logic allows us to make statements not only about individual vertices or edges, but also about subsets of the graph as a whole.

This greater expressive power comes at a price. First, compactness is lost.

Theorem 2.193. *Second-order predicate logic is not compact.*

Proof. For each $n \in \mathbb{N}$, let $\chi_{\geq n}$ be the formula stating that the domain contains at least n distinct elements:

$$\chi_{\geq n} := \exists x_1 \cdots \exists x_n \bigwedge_{1 \leq i < j \leq n} x_i \neq x_j.$$

Now consider, in addition, a second-order formula Fin expressing that the domain is finite. Such a formula is available in second-order logic because finiteness can be characterized via the behavior of all injective functions.

The set

$$\Gamma := \{\mathrm{Fin}, \chi_{\geq 1}, \chi_{\geq 2}, \chi_{\geq 3}, \dots\}$$

is unsatisfiable, because Fin requires a finite domain, whereas the formulas $\chi_{\geq n}$ together force the domain to have arbitrarily large finite cardinality. However, every finite subset of Γ is satisfiable: if it contains only $\chi_{\geq 1}, \dots, \chi_{\geq k}$, then a domain with exactly k elements suffices. Therefore, the logic is not compact. □

The negative result on the algorithmic side is even stronger. In second-order logic, the central semantic decision problems are not merely difficult; in general, they are not even semidecidable.

Theorem 2.194. *The problems* TAUT(SO), SAT(SO), *and* INFER(SO) *are undecidable and not semidecidable.*

Proof sketch. Second-order logic is expressive enough to speak about sets, relations, and thus large parts of arithmetic and combinatorial structure. As a result, well-known undecidable problems can be reduced to semantic questions in second-order logic. Unlike the first-order case, there is therefore no complete algorithmic search procedure that can reliably semidecide tautology, satisfiability, or inference. □

This makes the fundamental tension of second-order predicate logic especially clear: it is far more expressive than first-order logic, but precisely for that reason it loses key desirable properties such as compactness and algorithmic tractability. A resolution calculus in the sense familiar from first-order logic does not exist here.

In practice, this high expressive power is nevertheless extremely useful, especially whenever one wants to formulate structural properties of entire objects or relations. For general automated theorem proving, however, second-order logic is too powerful. This is why applications often focus on suitable fragments that offer a better balance between expressive power and computability.

2.7.3 Logic and Descriptive Complexity

Descriptive complexity studies which complexity classes can be characterized by logical formulas. Its perspective is the reverse of that of classical complexity theory: instead of asking how a problem is computed by a machine, it asks which kind of logic can describe problems of exactly the same difficulty. This is especially interesting for decision problems over finite structures, such as graphs, relations, or finite databases.

$$\begin{aligned}
\text{P} &= \text{the class of problems that are}\\
&\quad\ \text{decidable deterministically in polynomial time,}\\
\text{NP} &= \text{the class of problems whose}\\
&\quad\ \text{yes-instances can be verified}\\
&\quad\ \text{in polynomial time,}\\
\text{coNP} &= \text{the class of problems whose}\\
&\quad\ \text{no-instances can be verified}\\
&\quad\ \text{in polynomial time.}
\end{aligned}$$

These classes are central reference points in computer science. The class P stands for problems that are algorithmically well manageable, whereas NP and coNP may already be considerably more difficult.

For the class NP, there is a particularly elegant logical characterization. Here, second-order logic plays the decisive role: one first quantifies over suitable relations and then formulates only a first-order statement about those relations.

Definition 2.195. *A formula of the form*

$$\exists X_1 \, \exists X_2 \cdots \exists X_n \, \psi,$$

where ψ *contains no second-order quantifiers, is called an* existential second-order formula, *or briefly an* ∃SO-formula.

Such formulas are especially natural when, for example, one first selects a suitable set of vertices or edges in a graph and then checks whether this choice satisfies a certain property. A typical pattern is therefore: "Guess a structure, and then verify the property."

Theorem 2.196 (Fagin's Theorem). *Over finite structures, exactly the decision problems in* NP *can be characterized by an ∃SO-formula.*

The significance of this theorem is remarkable: what complexity theory understands as nondeterministic polynomial time corresponds, in logical terms, exactly to existential quantification over relations. This establishes a bridge between algorithmic difficulty and logical expressive power.

A classical example is the question of whether a finite graph has a Hamiltonian cycle. Logically, one can first guess a relation for the selected edges or for the ordering of the vertices and then verify whether exactly this choice describes a closed tour through all vertices. The problem therefore belongs to NP and can accordingly be formulated as an ∃SO-property.

By the duality of yes-instances and no-instances, one obtains the counterpart for coNP.

Definition 2.197. *A formula of the form*

$$\forall X_1 \forall X_2 \cdots \forall X_n \psi,$$

where ψ *contains no second-order quantifiers, is called a* universal second-order formula, *or briefly a* $\forall$SO-formula.

Corollary 2.198. *Over finite structures, exactly the decision problems in* coNP *can be characterized by a* $\forall SO$ *formula.*

For the class P, the situation is considerably more subtle. Precisely because P stands for problems that are efficiently solvable in practice, a logical characterization would be especially desirable. To this day, however, no generally accepted logical characterization of P is known. There are even conjectures that such a characterization, in the expected form, may not exist at all; in this context, Gurevich's conjecture is often mentioned.

Descriptive complexity thus illustrates particularly well how closely logic and computer science are connected: logical quantification over relations can express exactly the kind of nondeterministic guessing and checking that is central to complexity theory. At the same time, it makes visible that expressive power and computability often stand in a certain tension.

2.7.4 Exercises and Solutions

In the following, we practice the syntax and semantics of second-order predicate logic using a few small examples. The main goal is to classify formulas correctly, understand their meaning in a concrete structure, and recognize the additional expressive power of second-order logic. In each case, decide whether the statement is true or false.

Exercise 2.199. *Consider the signature*

$$\Sigma = \{P/1, E/2, c/0\}$$

and the Σ*-structure* $\mathcal{A}$ *given by*

$$|A| = \{a, b, c\}, \qquad c^{\mathcal{A}} = a,$$

$$P^{\mathcal{A}} = \{a, b\}, \qquad E^{\mathcal{A}} = \{(a, b), (b, c), (c, a)\}.$$

In addition, let the following formulas be given:

$$\psi_1 = \exists X\, (\exists x\, X(x) \wedge \forall x\, (X(x) \to P(x))),$$

$$\psi_2 = \forall X \Big((\forall x\, (X(x) \to P(x))) \to \exists x\, X(x) \Big),$$

$$\psi_3 = \exists X\, \forall x\, \forall y\, (E(x, y) \to (X(x) \leftrightarrow \neg X(y))),$$

$$\psi_4 = \forall X\, (X(c) \to P(c)).$$

For each of the following statements, decide whether it is true or false.

1. ψ_1 *is an* $\exists SO$ *formula.*

2. $\mathcal{A} \models \psi_2$.

3. $\mathcal{A} \models \psi_3$.

4. ψ_4 *is logically equivalent to* $P(c)$.

Solution 2.200. 1. ***True.*** *The formula* ψ_1 *begins with exactly one second-order quantifier, namely* $\exists X$*, and no further second-order quantifiers occur in the remaining part of the formula. Hence it has the form of an existential second-order formula. The quantifiers* $\exists x$ *and* $\forall x$ *are only first-order quantifiers and are irrelevant for this classification.*

2. ***False.*** *The formula*

$$\psi_2 = \forall X\Big((\forall x\,(X(x) \to P(x))) \to \exists x\,X(x)\Big)$$

asserts that every subset of $P^{\mathcal{A}}$ *is nonempty. But this is impossible, since we may also assign the empty set to* X*. For* $X = \emptyset$*, the formula*

$$\forall x\,(X(x) \to P(x))$$

is true, but

$$\exists x\,X(x)$$

is false. Therefore, the implication is false for this assignment, and so overall we have

$$\mathcal{A} \not\models \psi_2.$$

3. **False.** *The formula*

$$\psi_3 = \exists X\, \forall x\, \forall y\, (E(x,y) \to (X(x) \leftrightarrow \neg X(y)))$$

expresses that the vertices of the graph described by E can be partitioned into two classes such that every edge runs between the two classes. This is exactly the property of being bipartite.

However, the graph described by $E^{\mathcal{A}}$ is a triangle:

$$(a,b),\ (b,c),\ (c,a).$$

A triangle is not bipartite. Indeed, if $a \in X$, then $(a,b) \in E^{\mathcal{A}}$ implies that $b \notin X$; from $(b,c) \in E^{\mathcal{A}}$ it follows that $c \in X$; and finally, $(c,a) \in E^{\mathcal{A}}$ requires that $a \notin X$. This contradicts the assumption $a \in X$. Hence

$$\mathcal{A} \not\models \psi_3.$$

4. **True.** *The formula*

$$\psi_4 = \forall X\, (X(c) \to P(c))$$

is logically equivalent to $P(c)$.

If $P(c)$ holds, then $X(c) \to P(c)$ is true for every assignment of X, and therefore so is $\forall X\, (X(c) \to P(c))$. Conversely, if $P(c)$ does not hold, choose $X = \{c^{\mathcal{A}}\}$. Then $X(c)$ is true, but $P(c)$ is false, so $X(c) \to P(c)$ is false, and consequently $\forall X\, (X(c) \to P(c))$ is false as well. Therefore,

$$\forall X\, (X(c) \to P(c)) \equiv P(c).$$

These exercises illustrate the characteristic style of reasoning in second-order predicate logic: first one identifies the form of an expression, then checks its semantics in a concrete structure, and finally uses the additional quantification over relations to express properties that cannot be formulated in first-order logic, or not in the same way.

Chapter 3

Modal Logic

Propositional logic investigates how statements with the truth values "true" and "false" can be combined. Modal logic adds a further layer: not only *what* is said, but also *under which conditions* or *in what sense* it holds becomes part of the analysis. We may thus distinguish, for example, between what is necessary, what is merely possible, what is permitted or forbidden, and what someone knows, believes, or considers possible.

Even simple everyday sentences show why such a refinement is useful:

- It might still rain this evening.
- A password must not be shared.
- After pressing the switch, the lamp must light up.
- Anna knows that the train leaves at 8 o'clock.

In all of these examples, the issue is not merely the truth value of a statement, but an additional layer of meaning. Modal logic makes precisely this additional content amenable to exact formal description.

In this chapter, we consider three central areas of application of modal logic. First, we study general modal logic, in which modal notions such as possibility and necessity are introduced formally. Second, we turn to temporal logics, which allow statements about temporal developments, processes, and future states to be expressed. Third, we examine epistemic logics, that is, logics of knowledge, belief, and the distribution of information among several agents.

General modal logic provides the foundation for all further variants. It is not tied to a single area of application, but instead supplies a formal framework for different kinds of modal statements. Typical examples are statements of the form "it is necessary that φ" or "it is possible that φ". This abstract basic form is introduced in Section 3.1. There we also encounter the semantic framework conditions, in particular the so-called Kripke frames and their interpretation.

Temporal logics, by contrast, describe how statements change over time. Here the central questions are, for example, whether something holds in every future state, whether something may occur at some point, or whether a system permanently possesses a certain property. Such logics are especially useful in computer science, for instance in the specification and verification of programs and technical systems. In Section 3.2 we introduce linear temporal logic (LTL), while Section 3.3 treats computation tree logic (CTL), in which future developments may branch. A comparison of both perspectives then follows in Section 3.3.4.

Finally, epistemic logics investigate states of knowledge and belief. Here one no longer asks only whether a statement is true, but also who knows what, who possesses which information, and how states of information and communication influence one another. Examples include statements such as "Agent a knows that the resource is free" or "Alice considers it possible that Bob has received the message". Section 3.4 discusses various epistemic approaches. It begins with the foundations of the epistemic interpretation of modal logic, then turns to multi-agent systems in Section 3.4.2. Finally, Section 3.4.3 introduces Public Announcement Logic (PAL), which makes it possible to formalize public announcements of information and their consequences for knowledge.

The roots of modal logic go back a long way. Questions of necessity, possibility, and contingency were already discussed in ancient and medieval philosophy. The form of formal treatment commonly used today, however, was developed mainly in the 20th century and received a decisive impulse from possible-worlds semantics. As a result, modal logic became a flexible tool that plays an important role in philosophy, linguistics, computer science, artificial intelligence, and theoretical computer science alike.

This chapter presents modal logic not merely as another formal subfield, but as a natural extension of logic itself: as soon as statements are evaluated not only in terms of whether they are true or false, but also in terms of whether they are possible, necessary, permitted, known, or temporally constrained, a much richer formal language emerges.

3.1 General Modal Logic

General modal logic extends propositional logic by adding expressions that make it possible to capture *possibility* and *necessity* in a formal way. Whereas propositional logic asks only whether a formula is true or false, modal logic additionally allows us to ask *in what sense* a statement holds. For example, one can distinguish between a statement that must hold in all conceivable cases and a statement that may hold only in some cases.

Typical readings include, for example:

- "It is necessary that the alarm is triggered."
- "It is possible that the train is delayed."
- "It is necessary that no error state is ever reached from a safe initial state."

The two modal operators used in this chapter are the *box* symbol $\Box$ for necessity and the *diamond* symbol $\Diamond$ for possibility. Accordingly, $\Box\varphi$ can be read as "φ is necessarily true" and $\Diamond\varphi$ as "φ is possible".

3.1.1 Syntax and Semantics

The language of modal logic builds directly on propositional logic. In addition to the familiar connectives, it introduces two new operators, namely $\Box$ and $\Diamond$.

Definition 3.1 (Syntax of Modal Logic). *Let* Σ *be a propositional signature. The set of modal logic formulas over* Σ*, denoted by* $\mathrm{ML}(\Sigma)$*, is defined inductively by the following rules:*

$$\begin{array}{ll}
\text{(ML0)} & A \in \Sigma \;\Rightarrow\; A \in \mathrm{ML}(\Sigma), \\
\text{(ML1)} & \varphi, \psi \in \mathrm{ML}(\Sigma) \;\Rightarrow\; (\varphi \lor \psi) \in \mathrm{ML}(\Sigma), \\
\text{(ML2)} & \varphi, \psi \in \mathrm{ML}(\Sigma) \;\Rightarrow\; (\varphi \land \psi) \in \mathrm{ML}(\Sigma), \\
\text{(ML3)} & \varphi \in \mathrm{ML}(\Sigma) \;\Rightarrow\; \neg\varphi \in \mathrm{ML}(\Sigma), \\
\text{(ML4)} & \varphi \in \mathrm{ML}(\Sigma) \;\Rightarrow\; \Box\varphi \in \mathrm{ML}(\Sigma), \\
\text{(ML5)} & \varphi \in \mathrm{ML}(\Sigma) \;\Rightarrow\; \Diamond\varphi \in \mathrm{ML}(\Sigma).
\end{array}$$

Rules (ML0) through (ML3) are already familiar from propositional logic. A formula of the form A is called *atomic.* All formulas that are not atomic are called *complex.* A *literal* is either an atomic formula or the negation of an atomic formula.

The connectives $\land$, $\lor$, and $\neg$ are read exactly as in propositional logic. What is new are the modal operators. Their usual informal reading is:

$\Box\varphi$ "φ holds in all relevant cases"

$\Diamond\varphi$ "φ holds in at least one relevant case"

The precise meaning of "relevant" is determined only by the semantics.

As in propositional logic, we use the abbreviations

$$\begin{aligned} &\varphi \rightarrow \psi := \neg\varphi \vee \psi \\ \text{and} \quad &\varphi \leftrightarrow \psi := (\varphi \rightarrow \psi) \wedge (\psi \rightarrow \varphi) \end{aligned}$$

For parentheses, we follow the familiar conventions from propositional logic. In addition, $\Box$, $\Diamond$, and $\neg$ bind more strongly than $\wedge$ and $\vee$. In particular,

$$\Box A \vee B$$

is to be read as

$$(\Box A) \vee B$$

and not as $\Box(A \vee B)$.

Example 3.2. *The following formulas are well-formed, and their intuitive reading is:*

$\Diamond(\Box A \wedge B)$

It is possible that A holds necessarily and that B holds at the same time.

$\Box\Diamond B$

Necessarily, B is possible.

$\Diamond A \rightarrow \Box B$

If A is possible, then B holds necessarily.

$\Box A \rightarrow \Diamond B$

If A holds necessarily, then B is possible.

The semantics of modal logic is no longer truth-functional in the narrow sense. Simple propositional interpretations are therefore not sufficient. Instead, we work with multiple states connected by an accessibility relation. Each state carries its own propositional valuation. As a result, the same statement can be evaluated differently in different states.

Definition 3.3 (Kripke Structure). *Let Σ be a propositional signature. A* Kripke structure *over Σ is a triple*

$$\mathcal{K} = \langle W, R, V \rangle,$$

where

- *W is a nonempty set of states,*
- *$R \subseteq W \times W$ is a relation on W, which we call the* accessibility relation,
- *$V : W \to \mathcal{P}(\Sigma)$ is a function that assigns to each state exactly those propositional variables that are true at that state.*

For a state $w \in W$, we write

$$R(w) = \{\, w' \in W \mid (w, w') \in R \,\}$$

for the set of states accessible from w.

Thus, the function V does not directly describe the truth values of individual formulas, but initially only which atomic propositions hold at a state. Propositional variables that are not contained in $V(w)$ are accordingly taken to be false at w.

Remark 3.4. *The term "accessibility relation" is to be understood semantically. The relation R does not merely state which nodes in a graph are connected, but which states are taken into account, from the perspective of a given state, when evaluating modal operators.*

Since formulas are evaluated not merely on a structure, but always relative to a distinguished state, we work with *pointed* Kripke structures.

Definition 3.5 (Pointed Kripke Structure). *A* pointed Kripke structure *for Σ is a pair*

$$\langle \mathcal{K}, w \rangle,$$

where $\mathcal{K} = \langle W, R, V \rangle$ is a Kripke structure over Σ and $w \in W$ is a state of $\mathcal{K}$.

We now define the evaluation of formulas.

Definition 3.6 (Evaluation). *Let Σ be a propositional signature and let $\langle \mathcal{K}, w \rangle$ be a pointed Kripke structure for Σ, where $\mathcal{K} = \langle W, R, V \rangle$. The evaluation function*

$$[\![\cdot]\!]_{\langle \mathcal{K}, w \rangle}$$

assigns to each formula in $\mathrm{ML}(\Sigma)$ *a truth value in* $\{\mathsf{true}, \mathsf{false}\}$*, as follows:*

$$[\![A]\!]_{\langle \mathcal{K}, w \rangle} = \mathsf{true} \iff A \in V(w)$$

$$[\![\neg\varphi]\!]_{\langle \mathcal{K}, w \rangle} = \mathsf{true} \iff [\![\varphi]\!]_{\langle \mathcal{K}, w \rangle} = \mathsf{false}$$

$$\llbracket \varphi \wedge \psi \rrbracket_{\langle \mathcal{K}, w \rangle} = \mathsf{true} \iff \llbracket \varphi \rrbracket_{\langle \mathcal{K}, w \rangle} = \mathsf{true} \textit{ and } \llbracket \psi \rrbracket_{\langle \mathcal{K}, w \rangle} = \mathsf{true}$$

$$\llbracket \varphi \vee \psi \rrbracket_{\langle \mathcal{K}, w \rangle} = \mathsf{true} \iff \llbracket \varphi \rrbracket_{\langle \mathcal{K}, w \rangle} = \mathsf{true} \textit{ or } \llbracket \psi \rrbracket_{\langle \mathcal{K}, w \rangle} = \mathsf{true}$$

$$\llbracket \Box \varphi \rrbracket_{\langle \mathcal{K}, w \rangle} = \mathsf{true} \iff \forall w' \in R(w) : \llbracket \varphi \rrbracket_{\langle \mathcal{K}, w' \rangle} = \mathsf{true}$$

$$\llbracket \Diamond \varphi \rrbracket_{\langle \mathcal{K}, w \rangle} = \mathsf{true} \iff \exists w' \in R(w) : \llbracket \varphi \rrbracket_{\langle \mathcal{K}, w' \rangle} = \mathsf{true}$$

The two modal operators thereby acquire a very intuitive meaning: $\Box\varphi$ requires that φ holds in all states accessible from w, whereas $\Diamond\varphi$ is already satisfied if there exists at least one such state in which φ holds. These two operators are dual to each other; informally, $\Diamond\varphi$ corresponds to the statement $\neg\Box\neg\varphi$.

Definition 3.7 (Satisfaction and Model). *Let $\varphi \in$ ML(Σ). Then a pointed Kripke structure $\langle \mathcal{K}, w \rangle$ is called a* model *of φ if*

$$\llbracket \varphi \rrbracket_{\langle \mathcal{K}, w \rangle} = \mathsf{true}$$

holds. In this case, we also write

$$\langle \mathcal{K}, w \rangle \models_{\mathrm{ML}(\Sigma)} \varphi,$$

or, more briefly,

$$\mathcal{K}, w \models_{\mathrm{ML}(\Sigma)} \varphi.$$

Example 3.8. *Let $\Sigma = \{A, B\}$ and $\mathcal{K} = \langle W, R, V \rangle$ be given by*

$$W = \{s_0, s_1, s_2\},$$

$$R = \{(s_0, s_1), (s_0, s_2), (s_1, s_1), (s_2, s_2)\},$$

$$V(s_0) = \{A\}, \qquad V(s_1) = \{A, B\}, \qquad V(s_2) = \{A\}.$$

Then, in the pointed structure $\langle \mathcal{K}, s_0 \rangle$*, we have:*

$$\langle \mathcal{K}, s_0 \rangle \models_{\mathrm{ML}(\Sigma)} \Diamond B,$$

because the state s_1 *is accessible from* s_0*, and* B *is true there. By contrast,*

$$\langle \mathcal{K}, s_0 \rangle \not\models_{\mathrm{ML}(\Sigma)} \Box B,$$

since $s_2 \in R(s_0)$*, but* $B \notin V(s_2)$*.*

For formulas without modal operators, modal-logical evaluation coincides with that of propositional logic. The real added value of Kripke semantics therefore becomes apparent only when $\Box$ or $\Diamond$ occur.

In the following sections, we investigate further properties of modal formulas, in particular logical equivalence, embeddings into predicate logic, and methods for proving satisfiability and entailment.

3.1.2 Concepts and Properties

So far, we have evaluated modal formulas only at individual states. In many situations, however, it is more natural to take the entire Kripke structure as the object of reference. This allows us to ask whether a formula holds in *all* states, whether it is satisfiable *somewhere*, and when two formulas have the same modal-logical meaning.

Definition 3.9 (Global Satisfaction). *Let Σ be a propositional signature and let $\varphi \in \mathrm{ML}(\Sigma)$. For a Kripke structure $\mathcal{K} = \langle W, R, V \rangle$, we define:*

$$\mathcal{K} \models_{\mathrm{ML}(\Sigma)} \varphi \Leftrightarrow \forall w \in W : \mathcal{K}, w \models_{\mathrm{ML}(\Sigma)} \varphi$$

Furthermore, we write

$$\models_{\mathrm{ML}(\Sigma)} \varphi,$$

if φ holds in every Kripke structure and at every state.

Definition 3.10 (Satisfiability and Validity). *Let Σ be a propositional signature and let $\varphi \in \mathrm{ML}(\Sigma)$. Then:*

- *φ is called Σ-satisfiable if there exists at least one model of φ,*
- *φ is called Σ-contradictory if φ has no model,*
- *φ is called Σ-valid if $\models_{\mathrm{ML}(\Sigma)} \varphi$ holds.*

A valid formula is also called a tautology.

Remark 3.11. *As in propositional logic, every tautology of the form $A \vee \neg A$ is also valid in modal logic. The difference is that modal formulas additionally depend on the accessibility structure.*

Example 3.12. *Let*

$$\mathcal{K} = \langle W, R, V \rangle \quad \textit{with} \quad W = \{u, v\}, \qquad R = \{(u, v)\},$$

$$V(u) = \emptyset, \qquad V(v) = \{A\}.$$

Then, in the pointed model $\langle \mathcal{K}, u\rangle$, *we have:*

$$\langle \mathcal{K}, u\rangle \models_{\mathrm{ML}(\Sigma)} \Diamond A,$$

because the state v *is accessible from* u, *and* A *is true there.*

By contrast,

$$\langle \mathcal{K}, v\rangle \not\models_{\mathrm{ML}(\Sigma)} \Diamond A,$$

because v *has no successor. Thus,* $\Diamond A$ *is satisfiable, but not valid.*

Moreover, in $\langle \mathcal{K}, u\rangle$ *we have:*

$$\langle \mathcal{K}, u\rangle \models_{\mathrm{ML}(\Sigma)} \Box A \qquad \textit{and} \qquad \langle \mathcal{K}, u\rangle \not\models_{\mathrm{ML}(\Sigma)} A.$$

Therefore, $\Box A \to A$ *is not valid in general.*

Remark 3.13. *A common misconception is to assume that* $\Box\varphi$ *or* $\Diamond\varphi$ *automatically also checks the current state itself. This is the case only if* $R(w, w)$ *holds. In general, both operators refer exclusively to the states in* $R(w)$.

Definition 3.14 (Equivalence). *Let* $\varphi, \psi \in \mathrm{ML}(\Sigma)$. *Then* φ *and* ψ *are called* equivalent, *written*

$$\varphi \equiv_{\mathrm{ML}(\Sigma)} \psi,$$

if for every pointed Kripke structure $\langle \mathcal{K}, w\rangle$ *we have:*

$$\langle \mathcal{K}, w\rangle \models_{\mathrm{ML}(\Sigma)} \varphi \Leftrightarrow \langle \mathcal{K}, w\rangle \models_{\mathrm{ML}(\Sigma)} \psi$$

Proposition 3.15 (Central Modal Logical Equivalences). *Let* Σ *be a propositional signature. For all formulas* $\varphi, \psi \in \mathrm{ML}(\Sigma)$, *the following hold:*

1. $\neg\Box\varphi \equiv_{\mathrm{ML}(\Sigma)} \Diamond\neg\varphi$,
2. $\neg\Diamond\varphi \equiv_{\mathrm{ML}(\Sigma)} \Box\neg\varphi$,
3. $\Box(\varphi \wedge \psi) \equiv_{\mathrm{ML}(\Sigma)} \Box\varphi \wedge \Box\psi$,
4. $\Diamond(\varphi \vee \psi) \equiv_{\mathrm{ML}(\Sigma)} \Diamond\varphi \vee \Diamond\psi$,
5. $\Diamond(\varphi \to \psi) \equiv_{\mathrm{ML}(\Sigma)} \Box\varphi \to \Diamond\psi$.

Proof. Let $\mathcal{K} = \langle W, R, V\rangle$ be an arbitrary Kripke structure and $w \in W$ an arbitrary world. Then:

(1)

$$\begin{aligned}\mathcal{K}, w \models \neg\Box\varphi &\iff \mathcal{K}, w \not\models \Box\varphi \\ &\iff \exists v \in R(w) : \mathcal{K}, v \not\models \varphi \\ &\iff \exists v \in R(w) : \mathcal{K}, v \models \neg\varphi \\ &\iff \mathcal{K}, w \models \Diamond\neg\varphi.\end{aligned}$$

(2)

$$\neg\Diamond\varphi \equiv_{\mathrm{ML}(\Sigma)} \neg\neg\Box\neg\varphi \equiv_{\mathrm{ML}(\Sigma)} \Box\neg\varphi.$$

More precisely, (1) immediately yields

$$\neg\Box\varphi \equiv_{\mathrm{ML}(\Sigma)} \Diamond\neg\varphi.$$

Applying this equivalence to $\neg\varphi$, we obtain

$$\neg\Box\neg\varphi \equiv_{\mathrm{ML}(\Sigma)} \Diamond\neg\neg\varphi,$$

and thus, by double-negation elimination,

$$\neg\Diamond\varphi \equiv_{\mathrm{ML}(\Sigma)} \Box\neg\varphi.$$

(3)

$$\begin{aligned}
\mathcal{K}, w \models \Box(\varphi \wedge \psi) &\iff \forall v \in R(w) : \mathcal{K}, v \models \varphi \wedge \psi \\
&\iff \forall v \in R(w) : (\mathcal{K}, v \models \varphi \\
&\qquad \wedge \mathcal{K}, v \models \psi) \\
&\iff (\forall v \in R(w) : \mathcal{K}, v \models \varphi) \\
&\qquad \wedge (\forall v \in R(w) : \mathcal{K}, v \models \psi) \\
&\iff \mathcal{K}, w \models \Box\varphi \wedge \mathcal{K}, w \models \Box\psi \\
&\iff \mathcal{K}, w \models \Box\varphi \wedge \Box\psi
\end{aligned}$$

(4)

$$\begin{aligned}
\mathcal{K}, w \models \Diamond(\varphi \vee \psi) &\iff \exists v \in R(w) : \mathcal{K}, v \models \varphi \vee \psi \\
&\iff \exists v \in R(w) : (\mathcal{K}, v \models \varphi \\
&\qquad \vee \mathcal{K}, v \models \psi) \\
&\iff (\exists v \in R(w) : \mathcal{K}, v \models \varphi) \\
&\qquad \vee (\exists v \in R(w) : \mathcal{K}, v \models \psi) \\
&\iff \mathcal{K}, w \models \Diamond\varphi \vee \mathcal{K}, w \models \Diamond\psi \\
&\iff \mathcal{K}, w \models \Diamond\varphi \vee \Diamond\psi
\end{aligned}$$

(5)

$$\begin{aligned}
\Diamond(\varphi \to \psi) &\equiv_{\mathrm{ML}(\Sigma)} \Diamond(\neg\varphi \vee \psi) \\
&\equiv_{\mathrm{ML}(\Sigma)} \Diamond\neg\varphi \vee \Diamond\psi \\
&\equiv_{\mathrm{ML}(\Sigma)} \neg\Box\varphi \vee \Diamond\psi \\
&\equiv_{\mathrm{ML}(\Sigma)} \Box\varphi \to \Diamond\psi.
\end{aligned}$$

This proves all the claims. □

Embedding into Predicate Logic

Modal formulas can be translated into formulas of first-order predicate logic. In this translation, worlds are treated as objects, the accessibility relation is represented by a binary predicate, and propositional variables are interpreted as unary predicates.

Let $\Sigma = \{A, B, C, \dots\}$ be a propositional signature, and let

$$\Sigma_{\mathrm{FO}} = \{P_A/1, P_B/1, P_C/1, \dots, R/2\}$$

be the corresponding first-order signature.

Definition 3.16 (Standard Translation). *The standard translation* ST_i *is defined inductively by*

$$\mathrm{ST}_i(A) = P_A(x_i),$$

$$\mathrm{ST}_i(\varphi \wedge \psi) = \mathrm{ST}_i(\varphi) \wedge \mathrm{ST}_i(\psi)$$
$$\mathrm{ST}_i(\varphi \vee \psi) = \mathrm{ST}_i(\varphi) \vee \mathrm{ST}_i(\psi)$$
$$\mathrm{ST}_i(\neg\varphi) = \neg\mathrm{ST}_i(\varphi),$$
$$\mathrm{ST}_i(\Box\varphi) = \forall x_{i+1}\big(R(x_i, x_{i+1}) \to \mathrm{ST}_{i+1}(\varphi)\big),$$
$$\mathrm{ST}_i(\Diamond\varphi) = \exists x_{i+1}\big(R(x_i, x_{i+1}) \wedge \mathrm{ST}_{i+1}(\varphi)\big).$$

Given a pointed Kripke structure $\langle \mathcal{K}, w\rangle$ with $\mathcal{K} = \langle W, R, V\rangle$, we define the corresponding first-order structure

$$\mathcal{A}_{\mathcal{K}} = \langle W, P_A^{\mathcal{A}_{\mathcal{K}}}, P_B^{\mathcal{A}_{\mathcal{K}}}, \dots, R^{\mathcal{A}_{\mathcal{K}}}\rangle$$

by

$$\begin{aligned} P_A^{\mathcal{A}_\mathcal{K}} &= \{ u \in W \mid A \in V(u) \} \\ P_B^{\mathcal{A}_\mathcal{K}} &= \{ u \in W \mid B \in V(u) \} \\ R^{\mathcal{A}_\mathcal{K}} &= R \end{aligned}$$

Moreover, let β_w be the assignment such that $\beta_w(x_0) = w$.

Theorem 3.17 (Compatibility of Modal and First-Order Semantics). *For every formula $\varphi \in \mathrm{ML}(\Sigma)$ and every pointed Kripke structure $\langle \mathcal{K}, w \rangle$, the following holds:*

$$\begin{aligned} &\langle \mathcal{K}, w \rangle \models_{\mathrm{ML}(\Sigma)} \varphi \\ &\quad \iff \langle \mathcal{A}_\mathcal{K}, \beta_w \rangle \models_{\mathrm{FO}(\Sigma_{\mathrm{FO}})} \mathrm{ST}_0(\varphi) \end{aligned}$$

Proof. We prove the statement by structural induction on the formation of φ.

Base case: $\varphi = A$ for a propositional variable A. Then

$$\mathrm{ST}_0(A) = P_A(x_0).$$

We have

$$\begin{aligned} \langle \mathcal{K}, w \rangle \models_{\mathrm{ML}(\Sigma)} A &\iff A \in V(w) \\ &\iff w \in P_A^{\mathcal{A}_\mathcal{K}} \\ &\iff \langle \mathcal{A}_\mathcal{K}, \beta_w \rangle \models_{\mathrm{FO}(\Sigma_{\mathrm{FO}})} P_A(x_0) \end{aligned}$$

Inductive step for the connectives: For $\varphi = \psi \wedge \chi$, we have

$$\mathrm{ST}_0(\psi \wedge \chi) = \mathrm{ST}_0(\psi) \wedge \mathrm{ST}_0(\chi).$$

By the induction hypothesis, it follows that

$$\langle \mathcal{K}, w \rangle \models \psi \wedge \chi \iff \langle \mathcal{K}, w \rangle \models \psi \text{ and } \langle \mathcal{K}, w \rangle \models \chi$$

$$\iff \langle \mathcal{A}_{\mathcal{K}}, \beta_w \rangle \models \mathrm{ST}_0(\psi) \text{ and } \langle \mathcal{A}_{\mathcal{K}}, \beta_w \rangle \models \mathrm{ST}_0(\chi)$$

$$\iff \langle \mathcal{A}_{\mathcal{K}}, \beta_w \rangle \models \mathrm{ST}_0(\psi) \wedge \mathrm{ST}_0(\chi).$$

The cases for $\neg$ and $\vee$ are analogous.

Inductive step for $\Diamond$: For $\varphi = \Diamond\psi$, we have

$$\mathrm{ST}_0(\Diamond\psi) = \exists x_1 \big(R(x_0, x_1) \wedge \mathrm{ST}_1(\psi)\big).$$

Now,

$$\langle \mathcal{K}, w \rangle \models \Diamond\psi \iff \exists v \in R(w) : \ \langle \mathcal{K}, v \rangle \models \psi.$$

By the induction hypothesis, this is equivalent to

$$\exists v \in W : \ \langle \mathcal{A}_{\mathcal{K}}, \beta_w[x_1 \mapsto v] \rangle \models \mathrm{ST}_1(\psi) \quad \text{and} \quad R(w, v).$$

This is precisely the satisfaction of

$$\exists x_1 \big(R(x_0, x_1) \wedge \mathrm{ST}_1(\psi)\big)$$

under β_w.

The case $\Box\psi$ is dual and is handled by universal quantification over all successors.

This completes the proof. $\square$

Definition 3.18 (Negation Normal Form). *A modal logic formula φ is said to be in* negation normal form *(NNF) if negations occur only directly in front of atomic formulas.*

Proposition 3.19. *For every formula $\varphi \in \mathrm{ML}(\Sigma)$, there exists an equivalent formula in negation normal form.*

Proof. We describe a recursive transformation. First, if necessary, we eliminate the abbreviations $\rightarrow$ and $\leftrightarrow$ by setting

$$\psi \rightarrow \chi := \neg\psi \vee \chi, \qquad \psi \leftrightarrow \chi := (\psi \rightarrow \chi) \wedge (\chi \rightarrow \psi).$$

After that, it suffices to push negations inward.

To this end, we define two transformations nnf_+ and nnf_-, where

$$\mathsf{nnf}_+(\theta) \equiv \theta \qquad \text{and} \qquad \mathsf{nnf}_-(\theta) \equiv \neg\theta$$

is intended to hold. For atomic formulas and logical connectives, we set:

$$\mathsf{nnf}_+(A) = A, \qquad \mathsf{nnf}_-(A) = \neg A,$$

$$\begin{aligned}
\mathsf{nnf}_+(\alpha \wedge \beta) &= \mathsf{nnf}_+(\alpha) \wedge \mathsf{nnf}_+(\beta)\\
\mathsf{nnf}_+(\alpha \vee \beta) &= \mathsf{nnf}_+(\alpha) \vee \mathsf{nnf}_+(\beta)\\
\mathsf{nnf}_-(\alpha \wedge \beta) &= \mathsf{nnf}_-(\alpha) \vee \mathsf{nnf}_-(\beta)\\
\mathsf{nnf}_-(\alpha \vee \beta) &= \mathsf{nnf}_-(\alpha) \wedge \mathsf{nnf}_-(\beta)
\end{aligned}$$

$$\mathsf{nnf}_+(\neg\alpha) = \mathsf{nnf}_-(\alpha), \qquad \mathsf{nnf}_-(\neg\alpha) = \mathsf{nnf}_+(\alpha).$$

For the modal operators, we use the dualities from the previous proposition:

$$\mathsf{nnf}_+(\Box\alpha) = \Box\, \mathsf{nnf}_+(\alpha), \qquad \mathsf{nnf}_+(\Diamond\alpha) = \Diamond\, \mathsf{nnf}_+(\alpha),$$

$$\mathsf{nnf}_-(\Box\alpha) = \Diamond\, \mathsf{nnf}_-(\alpha), \qquad \mathsf{nnf}_-(\Diamond\alpha) = \Box\, \mathsf{nnf}_-(\alpha).$$

Termination is clear, since each step refers only to proper subformulas. Moreover, in the end negations occur only in front of atomic formulas.

Semantic correctness is proved by structural induction on the formation of the formula. The base cases are immediate. For the induction steps, we use De Morgan's laws together with the modal dualities

$$\neg\Box\varphi \equiv_{\mathrm{ML}(\Sigma)} \Diamond\neg\varphi, \qquad \neg\Diamond\varphi \equiv_{\mathrm{ML}(\Sigma)} \Box\neg\varphi.$$

Hence every formula can be transformed into an equivalent formula in negation normal form. □

3.1.3 Logical Consequence and the Tableau Calculus

In this section, we study how modal formulas can be analyzed algorithmically. In particular, we want to determine whether a formula is satisfiable, valid, or derivable from another formula. To do so, we use a tableau calculus that systematically generates a tree of constraints from a given formula.

Decision Problems

We begin with the three fundamental decision problems of modal logic:

$$\mathrm{Sat(ML)} \qquad \mathrm{Taut(ML)} \qquad \mathrm{Infer(ML)}.$$

These are defined as follows:

$$\begin{aligned}
\mathrm{Sat(ML)} &: \text{Input } \varphi; \\
&\quad \text{Output Yes iff } \varphi \text{ is satisfiable} \\
\mathrm{Taut(ML)} &: \text{Input } \varphi; \\
&\quad \text{Output Yes iff } \varphi \text{ is valid} \\
\mathrm{Infer(ML)} &: \text{Input } \varphi, \psi; \\
&\quad \text{Output Yes iff } \varphi \models_{\mathrm{ML}(\Sigma)} \psi
\end{aligned}$$

Definition 3.20 (Logical Consequence). *Let Σ be a propositional signature, and let $\varphi, \psi \in \mathrm{ML}(\Sigma)$. We write*

$$\varphi \models_{\mathrm{ML}(\Sigma)} \psi$$

if and only if for every pointed Kripke structure $\langle \mathcal{K}, w \rangle$ over Σ, the following holds:

$$\langle \mathcal{K}, w \rangle \models_{\mathrm{ML}(\Sigma)} \varphi \quad \Longrightarrow \quad \langle \mathcal{K}, w \rangle \models_{\mathrm{ML}(\Sigma)} \psi.$$

Proposition 3.21. *Let Σ be a propositional signature, and let $\varphi, \psi \in \mathrm{ML}(\Sigma)$. Then:*

1. *$\varphi \models_{\mathrm{ML}(\Sigma)} \psi$ if and only if $\varphi \wedge \neg\psi$ is unsatisfiable.*

2. *$\varphi \models_{\mathrm{ML}(\Sigma)} \psi$ if and only if $\varphi \rightarrow \psi$ is valid.*

Proof. We first prove (1).

$\Rightarrow$: Assume that $\varphi \models_{\mathrm{ML}(\Sigma)} \psi$. Then there can be no model that satisfies φ while falsifying ψ. For if $\langle \mathcal{K}, w \rangle$ were a model of $\varphi \wedge \neg\psi$, then it would in particular be a model of φ, and hence by assumption also a model of ψ. But this contradicts $\neg\psi$. Therefore, $\varphi \wedge \neg\psi$ is unsatisfiable.

$\Leftarrow$: Assume that $\varphi \wedge \neg\psi$ is unsatisfiable. Let $\langle \mathcal{K}, w \rangle$ be an arbitrary model of φ. If $\langle \mathcal{K}, w \rangle$ were not a model of ψ, then $\neg\psi$ would hold there, and hence $\langle \mathcal{K}, w \rangle$ would be a model of $\varphi \wedge \neg\psi$, contradicting unsatisfiability. Therefore every model of φ must also satisfy ψ, and thus $\varphi \models_{\mathrm{ML}(\Sigma)} \psi$.

For (2), we use the definition of implication together with part (1):

$$\begin{aligned}
\varphi \models_{\mathrm{ML}(\Sigma)} \psi &\iff \varphi \wedge \neg\psi \text{ is unsatisfiable} \\
&\iff \neg(\varphi \wedge \neg\psi) \text{ is valid} \\
&\iff \neg\varphi \vee \neg\neg\psi \text{ is valid} \\
&\iff \neg\varphi \vee \psi \text{ is valid} \\
&\iff \varphi \to \psi \text{ is valid.}
\end{aligned}$$

□

Remark 3.22. *The reduction of consequence to unsatisfiability is especially important in practice: to show that ψ follows from φ, it is enough to test the formula $\varphi \wedge \neg\psi$ for contradiction.*

Basic Idea of the Tableau Calculus

The tableau calculus decomposes a formula according to its syntactic structure. The idea is to construct, step by step, a tree-like structure of requirements from a given formula. A node of the form $\langle i, \chi \rangle$ is read as saying that the formula χ is to hold at state i. A node of the form (i, j) introduces a successor state.

This becomes especially intuitive for $\Diamond$-formulas: from $\Diamond\chi$ we know that there must be some reachable state in which χ holds. The tableau calculus therefore creates a fresh state and places χ there. By contrast, for $\Box\chi$, the formula χ must hold at all successor states that have already been introduced. These two ideas underlie the entire calculus.

In what follows, we work with formulas in negation normal form. Therefore, every formula is first transformed into an equivalent NNF.

Definition 3.23 (Initial Tableau). *Let φ be a formula in negation normal form. The initial tableau for φ is defined as*

$$\text{StartTableau}(\varphi) = \langle \{\langle 1, \varphi \rangle\}, \emptyset, \langle 1, \varphi \rangle, C \rangle,$$

where the unique node is initially unmarked, that is,

$$C(\langle 1, \varphi \rangle) = \times.$$

Definition 3.24 (Tableau). *A tableau is a tree*

$$T = \langle V, E, r, C \rangle,$$

where

- *V is the set of nodes,*

- *E is the set of edges,*
- *r is the root, and*
- *$C : V \to \{\checkmark, \times\}$ is a marking function.*

A node of the form $\langle i, \chi \rangle$ is called a formula node, *and a node of the form (i, j) is called an* edge node. *A node is called* marked *if it is labeled with $\checkmark$.*

A tableau is called saturated *if every node is marked.*

A leaf is called closed *if, on the path to the root, both $\langle i, L \rangle$ and $\langle i, \neg L \rangle$ occur for some literal L. A tableau is called* closed *if all of its leaves are closed. Otherwise, it is called* open.

Tableau Rules

These rules are applied to formulas in negation normal form.

- **Literal Rule:** Choose an unmarked node $\langle i, L \rangle$, where L is a literal, and mark it.
- **Conjunction Rule:** Choose an unmarked node $\langle i, \varphi \wedge \psi \rangle$. Mark it and, below every leaf that depends on this node, first add $\langle i, \varphi \rangle$ and then below it $\langle i, \psi \rangle$.
- **Disjunction Rule:** Choose an unmarked node $\langle i, \varphi \vee \psi \rangle$. Mark it and branch at every affected leaf into two branches: one branch contains $\langle i, \varphi \rangle$, the other $\langle i, \psi \rangle$.

- **Diamond Rule:** Choose an unmarked node $\langle i, \Diamond\varphi \rangle$. Mark it and, at every affected leaf, add a fresh state j together with an edge node (i, j) and a formula node $\langle j, \varphi \rangle$. In addition, all already marked box formulas $\langle i, \Box\psi \rangle$ must be propagated to this new state.

- **Box Rule:** Choose an unmarked node $\langle i, \Box\varphi \rangle$. Mark it and propagate φ to all successors of i that already occur on the path.

The rules are not tied to a fixed order. Therefore, different tableaux can arise from the same initial formula.

Example 3.25. *Consider the formula*

$$\Diamond A \wedge \Box \neg A.$$

It is unsatisfiable. A tableau begins with $\langle 1, \Diamond A \wedge \Box \neg A \rangle$*. After applying the conjunction rule, the two requirements*

$$\langle 1, \Diamond A \rangle \qquad \textit{and} \qquad \langle 1, \Box \neg A \rangle$$

appear. The diamond rule creates a fresh successor 2 *with* $\langle 2, A \rangle$*. At the same time, the box rule requires that* $\langle 2, \neg A \rangle$ *also holds. Thus, on the same branch we obtain the contradiction*

$$\langle 2, A \rangle \quad \textit{and} \quad \langle 2, \neg A \rangle,$$

so the branch closes.

Example 3.26. *The formula*

$$\Diamond A \wedge \Box A$$

is satisfiable, by contrast. A simple model consists of two states w_0, w_1 with $R(w_0) = \{w_1\}$ and $A \in V(w_1)$. Then both $\Diamond A$ and $\Box A$ hold at w_0: there is an accessible state where A is true, and all accessible states satisfy A as well.

Obtaining a Model from an Open Tableau

A saturated open tableau already contains enough information to read off a Kripke model from it.

Definition 3.27 (Induced Kripke Structure). *Let $T = \langle V, E, r, C \rangle$ be a saturated tableau for a formula φ, and let v be an open leaf of T. Write V_v for the set of all nodes on the path from v to the root r. We then define the Kripke structure induced by v*

$$\mathcal{K}_v^T = \langle W, R, V \rangle$$

by

$$W = \{\, i \mid \langle i, \chi \rangle \in V_v \,\}, \qquad R = \{\, (i, j) \mid (i, j) \in V_v \,\},$$

$$V(i) = \{\, A \in \Sigma \mid \langle i, A \rangle \in V_v \,\}.$$

Example 3.28. *In an open tableau for*

$$\Box(A \vee \neg C) \wedge \Diamond(\neg A \vee \Diamond B)$$

an open leaf may, for example, have the form $\langle 3, B \rangle$*. Then the induced structure may look as follows:*

$$W = \{1, 2, 3\}, \qquad R = \{(1, 2), (2, 3)\},$$

$$V(1) = \emptyset, \qquad V(2) = \{A\}, \qquad V(3) = \{B\}.$$

Thus, the open branch already describes a small model in which the original formula is satisfied.

Theorem 3.29 (Correctness of the Tableau Calculus). *Let* φ *be a modal formula and let* T *be a saturated tableau for* φ*. Then the following hold:*

1. φ *is satisfiable if and only if* T *is open.*

2. φ *is unsatisfiable if and only if* T *is closed.*

3. *If* v *is an open leaf of* T*, then*

$$\langle \mathcal{K}_v^T, 1 \rangle \models_{\mathrm{ML}(\Sigma)} \varphi.$$

Proof. The tableau rules are designed in such a way that they mirror the semantics of the connectives and modal operators step by step.

For (1) and (2): An open tableau contains a branch without contradiction. By saturation, all requirements on this branch have been fully unfolded. A closed tableau branch, by contrast, contains a contradiction in the form of a literal together with its negation, and therefore cannot support a model.

More precisely, if φ is satisfiable, then there exists a model $\langle \mathcal{K}, w \rangle$. The tableau rules are semantically sound, so every satisfied branch can be extended in such a way that it remains open. Conversely, an open saturated branch provides exactly the information needed to define a Kripke structure satisfying the formula.

For (3): Let v be an open leaf. Then the induced structure $\mathcal{K}_v^T$ is constructed so that every requirement appearing on the branch is satisfied at the corresponding state. This is immediate for literals, since an open branch contains no contradiction. For conjunctions and disjunctions, it follows from the way the rules decompose the formula into subrequirements. For $\Diamond$-formulas, a new successor is introduced, and for $\Box$-formulas, the already existing successors are equipped with the corresponding requirements.

Hence the structure induced by v satisfies the root label $\langle 1, \varphi \rangle$, and therefore

$$\langle \mathcal{K}_v^T, 1 \rangle \models_{\mathrm{ML}(\Sigma)} \varphi.$$

□

Corollary 3.30 (Tree-Model Property and Finite Model Property). *Every satisfiable formula of modal logic has a tree-shaped model. Moreover, every satisfiable modal formula has a finite model.*

Proof. We prove the two statements separately.

Tree-Model Property: Let φ be satisfiable. Then, by the soundness theorem, a saturated tableau for φ is open. An open branch of this tableau induces a Kripke structure whose underlying graph is a tree, since the tableau itself has a tree-like form. By Theorem 3.29(3), this structure satisfies the formula φ. Hence, φ has a tree-shaped model.

Finite Model Property: Let φ be satisfiable. Then, once again, there exists an open saturated tableau for φ. Since only subformulas of φ can occur, the set of possible labels is finite. In particular, under a fair tableau construction, genuinely new information can arise only finitely many times before a previously encountered configuration repeats. In that case, the construction can be terminated or recurring situations can be identified. This yields a finite open tableau.

By definition, a finite open tableau yields a finite Kripke structure. By Theorem 3.29(3), this structure satisfies φ. Therefore, every satisfiable formula has a finite model. □

Remark 3.31. *With a bit more effort, one can even show that the problems* Sat(ML), Taut(ML), *and* Infer(ML) *are PSPACE-complete. For the conceptual development of this chapter, however, the main point is this: the tableau calculus provides not only a decision procedure, but, in the open case, also a concrete model.*

3.1.4 Exercises and Solutions

In the following exercises, you will work both with the semantics of modal logic and with the standard translation into predicate logic. Pay careful attention to whether a statement is meant locally at a state or globally in an entire Kripke structure.

Exercise 3.32. *Consider the propositional signature*

$$\Sigma = \{B, C, E, F\}.$$

Let the following Kripke structures be given:

$$\begin{aligned}
\mathcal{K}_1 &= \langle W_1, R_1, V_1 \rangle \\
W_1 &= \{w_1, w_2, w_3, w_4, w_5\} \\
R_1 &= \{(w_1, w_3), (w_1, w_5), (w_2, w_4), \\
&\qquad (w_3, w_4), (w_5, w_4), (w_5, w_5)\}
\end{aligned}$$

$$\begin{aligned}
&V_1(w_1) = \{B, C\}, \quad V_1(w_2) = \{B\}, \quad V_1(w_3) = \{E, F\}, \\
&V_1(w_4) = \{C\}, \qquad V_1(w_5) = \{E\}.
\end{aligned}$$

Furthermore, let

$$\begin{aligned}
\mathcal{K}_2 &= \langle W_2, R_2, V_2 \rangle \\
W_2 &= \{w_1, w_2\} \\
R_2 &= \{(w_1, w_1)\} \\
V_2(w_1) &= \{F\} \\
V_2(w_2) &= \emptyset
\end{aligned}$$

and

$$\varphi_2 = \Box E \wedge \neg B \vee \Box F.$$

In addition, we consider

$$\begin{aligned}
\mathcal{K}_3 &= \langle W_3, R_3, V_3 \rangle \\
W_3 &= \{w_1, w_2, w_3\} \\
R_3 &= \{(w_1, w_1), (w_2, w_1), \\
&\qquad (w_3, w_2), (w_3, w_3)\} \\
V_3(w_1) &= \{B, E, F\} \\
V_3(w_2) &= \{B, C, E\} \\
V_3(w_3) &= \{B, C, F\}
\end{aligned}$$

as well as

$$\varphi_3 = \Diamond\neg\neg C, \qquad \varphi_4 = \neg\Box\neg\neg(\neg\Box B \wedge B).$$

Determine for each of the following statements whether it is true or false:

1. *We have* $R_1(w_4) = w_1$.

2. *We have* $\mathcal{K}_2 \models_{\mathrm{ML}(\Sigma)} \varphi_2$.

3. *We have* $[\![\varphi_3]\!]_{\langle \mathcal{K}_3, w_3 \rangle} = \mathsf{true}$.

4. *We have*

$$\begin{aligned}
\mathrm{ST}_i(\varphi_4) = \neg\forall x_2(R(x_1, x_2) \rightarrow \\
\neg\neg(\neg\forall x_3(R(x_2, x_3) \rightarrow \\
P_B(x_3)) \wedge P_B(x_2))).
\end{aligned}$$

Solution 3.33. 1. ***False.***

First, $R_1(w_4)$ is not a world, but the set of all worlds accessible from w_4. Since no pair in R_1 has w_4 as its first component, we have

$$R_1(w_4) = \emptyset.$$

Thus, the statement $R_1(w_4) = w_1$ is already false both type-theoretically and substantively.

2. ***True.***

   First, we read the formula using the usual precedence rules as

   $$\varphi_2 = (\Box E \wedge \neg B) \vee \Box F.$$

   To show that $\mathcal{K}_2 \models_{\mathrm{ML}(\Sigma)} \varphi_2$, we must check both states.

   For w_1, we have $R_2(w_1) = \{w_1\}$. Hence,

   $$\mathcal{K}_2, w_1 \models \Box F,$$

   since $F \in V_2(w_1)$. Therefore, the right disjunct is already true, and hence

   $$\mathcal{K}_2, w_1 \models \varphi_2.$$

   For w_2, we have $R_2(w_2) = \emptyset$. Therefore, both $\Box E$ and $\Box F$ are vacuously true, and in particular

   $$\mathcal{K}_2, w_2 \models (\Box E \wedge \neg B) \vee \Box F.$$

   Hence the formula holds in all states of $\mathcal{K}_2$, so

   $$\mathcal{K}_2 \models_{\mathrm{ML}(\Sigma)} \varphi_2.$$

3. ***True.***

   We evaluate the formula in $\langle \mathcal{K}_3, w_3 \rangle$*:*

   $$[\![\Diamond\neg\neg C]\!]_{\langle \mathcal{K}_3, w_3 \rangle} = \mathsf{true}$$

   if and only if there exists a successor $u \in R_3(w_3)$ *in which* $\neg\neg C$ *holds.*

   Now,

   $$R_3(w_3) = \{w_2, w_3\}.$$

   In w_2*, we have* $C \in V_3(w_2)$*, so*

   $$[\![C]\!]_{\langle \mathcal{K}_3, w_2 \rangle} = \mathsf{true},$$

   and therefore also

   $$[\![\neg\neg C]\!]_{\langle \mathcal{K}_3, w_2 \rangle} = \mathsf{true}.$$

   Thus, we have already found an accessible state in which the inner formula is true. Consequently,

   $$[\![\varphi_3]\!]_{\langle \mathcal{K}_3, w_3 \rangle} = \mathsf{true}.$$

4. ***True.***

   First, we compute the standard translation of

   $$\varphi_4 = \neg\Box\neg\neg(\neg\Box B \land B).$$

For $i = 1$, this yields:

$$\begin{aligned}\mathrm{ST}_1(\varphi_4) &= \neg\mathrm{ST}_1(\Box\neg\neg(\neg\Box B \wedge B)) \\ &= \neg\forall x_2\Big(R(x_1, x_2) \to \\ &\quad \mathrm{ST}_2(\neg\neg(\neg\Box B \wedge B))\Big) \\ &= \neg\forall x_2\Big(R(x_1, x_2) \to \\ &\quad \neg\neg(\mathrm{ST}_2(\neg\Box B) \wedge \mathrm{ST}_2(B))\Big) \\ &= \neg\forall x_2\Big(R(x_1, x_2) \to \\ &\quad \neg\neg(\neg\mathrm{ST}_2(\Box B) \wedge P_B(x_2))\Big) \\ &= \neg\forall x_2\Big(R(x_1, x_2) \to \\ &\quad \neg\neg(\neg\forall x_3(R(x_2, x_3) \to \\ &\qquad P_B(x_3)) \wedge P_B(x_2))\Big).\end{aligned}$$

This is exactly the formula stated above. Hence, the statement is correct.

In summary:

(1) false, *(2) true,* *(3) true,* *(4) true.*

Exercise 3.34. *Consider the propositional signature*

$$\Sigma = \{A, E, H\}.$$

Let the saturated tableau T be given by the following relevant branch description: The root is labeled with the *formula*

$$\langle 1,\ (A \vee \Box\neg E) \wedge \Diamond E\rangle.$$

After applying the $\wedge$*-rule, the two nodes*

$$\langle 1,\ A \vee \Box\neg E\rangle \qquad \text{and} \qquad \langle 1,\ \Diamond E\rangle$$

are obtained. On one branch, the disjunction is resolved so that the branch continues via

$$\langle 1,\ \Box\neg E\rangle.$$

From the diamond node, a fresh successor 2 *is introduced, and on the corresponding branch the nodes*

$$(1,2), \qquad \langle 2,\neg E\rangle, \qquad \langle 2, A\rangle$$

then occur. All leaves of the tableau are closed.

Also consider the following formulas and Kripke structures:

$$\varphi_1 = E \wedge \Diamond\Diamond(\neg H \vee \neg H),$$

$$\mathcal{K}_3 = \langle W_3, R_3, V_3\rangle$$

$$\text{with } W_3 = \{w_1, w_2\}$$

$$R_3 = \{(w_1, w_2)\}$$

$$V_3(w_1) = \emptyset, \qquad V_3(w_2) = \emptyset,$$

and

$$\varphi_4 = (A \vee \Box\neg E) \wedge \Diamond E.$$

For each of the following statements, decide whether it is true or false.

1. φ_1 is in negation normal form.

2. The tableau T is open.

3. We have $\mathcal{K}_v^T = \mathcal{K}_3$ for the leaf node $v = \langle 2, \neg E\rangle$.

4. We have $\varphi_4 \models_{\mathrm{ML}(\Sigma)} A$.

Solution 3.35. *1.* ***True.***

The formula

$$\varphi_1 = E \wedge \Diamond\Diamond(\neg H \vee \neg H)$$

is already in negation normal form. Negations occur only directly in front of atomic formulas, namely in the literal $\neg H$. The remaining connectives are permitted in NNF.

2. ***False.***

The tableau T is closed, *because by assumption all leaves are closed. A tableau is open if and only if it has at least one open leaf. That is not the case here.*

3. ***False.***

From the described branch, we obtain for the node $v = \langle 2, \neg E\rangle$ an induced Kripke structure in which the atom A also occurs at state 2. In particular,

$$V_v^T(2) = \{A\}.$$

In the given structure $\mathcal{K}_3$, however, we have

$$V_3(w_2) = \emptyset.$$

Hence $\mathcal{K}_v^T$ cannot coincide with $\mathcal{K}_3$. The structures already differ in the atomic valuation of w_2.

4. ***True.***

We show that $\varphi_4 \wedge \neg A$ is unsatisfiable. It follows immediately that

$$\varphi_4 \models_{\mathrm{ML}(\Sigma)} A.$$

Suppose there were a model $\langle \mathcal{K}, w \rangle$ such that

$$\mathcal{K}, w \models \varphi_4 \wedge \neg A.$$

Then at w we have in particular

$$(A \vee \Box\neg E) \wedge \Diamond E \quad \textit{and} \quad \neg A.$$

From $A \vee \Box\neg E$ together with $\neg A$, it follows that

$$\Box\neg E.$$

On the other hand, $\Diamond E$ requires that there is an accessible state in which E is true. This contradicts $\Box\neg E$, since $\Box\neg E$ says that all accessible states satisfy $\neg E$.

Therefore $\varphi_4 \wedge \neg A$ is unsatisfiable, and consequently

$$\varphi_4 \models_{\mathrm{ML}(\Sigma)} A.$$

In summary, we have:

(1) true, (2) false, (3) false, (4) true.

3.2 Linear Temporal Logic

Linear temporal logic (LTL) extends propositional logic by making it possible to speak not only about a single state, but also about the temporal evolution of a system. Whereas propositional logic describes what holds at a given moment, LTL makes it possible to express what happens next, what will occur at some future point, or what is required to hold throughout the entire future.

The basic idea is straightforward: we do not regard time as an unordered collection of moments, but as a sequence of points in time that occur one after another. At each point, the state of a system may change. It is precisely such developments that one wants to describe and verify in many applications. In computer science, this concerns, for example, programs, operating systems, communication protocols, or control systems. One may require, for instance, that after an error state a normal operating state is eventually reached again, that a requested action is in fact carried out later, or that a safety condition is preserved permanently.

A typical example is an automated backup system. One might require that after a backup run has started, a success message eventually appears. Likewise, one could stipulate that after an alarm, normal operation must not continue indefinitely, but that an inspection phase has to occur first. Properties of this kind can be formulated precisely in LTL and then examined along possible system executions.

In what follows, we first introduce the syntax and meaning of LTL formulas. We then discuss important equivalences and normal forms. After that, we turn to the connection with Kripke structures and finally show how LTL can be compared with first-order predicate logic.

3.2.1 Syntax and Semantics

Like propositional logic, LTL is a formal language in which formulas are built from propositional variables, logical connectives, and further operators. The difference is that LTL additionally contains operators that refer to temporal progression. These operators make it possible to formulate statements not only about the present moment, but also about the future.

Definition 3.36 (Syntax of LTL). *Let Σ be a propositional signature. The LTL formulas over Σ are defined inductively as follows:*

1. *Every propositional variable in Σ is an LTL formula.*
2. *$\top$ and $\bot$ are LTL formulas.*
3. *If φ and ψ are LTL formulas, then $\neg\varphi$, $\varphi \wedge \psi$, and $\varphi \vee \psi$ are also LTL formulas.*
4. *If φ is an LTL formula, then $X\varphi$, $G\varphi$, and $F\varphi$ are also LTL formulas.*
5. *If φ and ψ are LTL formulas, then $\varphi\, U\, \psi$ is also an LTL formula.*

The basic logical connectives behave exactly as they do in propositional logic. What is new are the temporal operators. Their meanings are as follows:

- $X\varphi$: *φ holds at the next point in time.*
- $G\varphi$: *From now on, φ always holds.*
- $F\varphi$: *At some point in the future, φ holds.*
- $\varphi U \psi$: *φ holds until ψ occurs.*

The operator X stands for *next*, G for *globally*, F for *finally*, and U for *until*. What matters most is that these operators do not describe merely a property at a single moment, but rather a property of a temporal evolution.

As useful abbreviations, we use the following, just as in propositional logic:

$$\varphi \to \psi := \neg\varphi \vee \psi$$

and

$$\varphi \leftrightarrow \psi := (\varphi \to \psi) \wedge (\psi \to \varphi)$$

In addition, we define the operator

$$\varphi W \psi \ := \ \varphi U \psi \vee G\varphi.$$

It is called *weak until* and expresses that φ is required to hold until ψ occurs, without requiring that ψ must ever occur at all.

For LTL formulas, the usual parenthesis rules and binding priorities apply. As in propositional logic, negation binds most strongly. The unary operators X, G, and F also have high binding strength. The operator U binds more strongly than $\wedge$ and $\vee$, but less strongly than negation.

Example 3.37. *Some typical LTL formulas are:*

- $\mathsf{Error} \wedge \neg\mathsf{Maintenance}$*: At the present time, an error is present, but no maintenance is taking place.*
- $X\,\mathsf{Approval}$*: The approval is present at the next point in time.*
- $G\,\neg\mathsf{Crash}$*: A crash never occurs.*
- $F\,\mathsf{Success}$*: Success occurs at some point in the future.*
- $\mathsf{Request}\,U\,\mathsf{Response}$*: A request remains pending until a response arrives.*
- $G(\mathsf{Login} \to F\,\mathsf{Confirmation})$*: Every login eventually leads to a confirmation.*

Frequently occurring general combinations are:

- $\varphi \to F\psi$ "If φ holds, then at some point ψ will also hold."
- $F\varphi \to F\psi$ "If φ holds at some point, then ψ will also hold at some point."
- $FG\varphi$ "From some point in the future onward, φ holds permanently."

- $GF\varphi$ "At every point in time, φ will eventually hold again."

The formula $GF\varphi$ is also interpreted as saying that "φ holds infinitely often."

To interpret LTL precisely, we do not consider isolated states, but entire temporal evolutions. Such an evolution specifies, for each point in time, which propositional variables hold there.

Definition 3.38 (LTL Interpretation). *Let Σ be a propositional signature. An LTL interpretation over Σ is an infinite sequence*

$$\pi = t_0 t_1 t_2 t_3 \ldots$$

of sets from $P(\Sigma)$.

The set t_i describes exactly the propositional variables that hold at time i. Thus, t_0 may be understood as the current situation, t_1 as the situation at the next step, t_2 as the situation after that, and so on. Time progresses in a discrete and linear way: at every point in time, there is exactly one next moment.

For an interpretation $\pi = t_0 t_1 t_2 t_3 \ldots$, we write π_i for the "tail of the sequence starting at time i," that is,

$$\pi_i = t_i t_{i+1} t_{i+2} \ldots .$$

This allows formulas to be evaluated at any point in time, not only at the beginning of a sequence.

Definition 3.39 (Satisfaction of LTL Formulas). *Let Σ be a propositional signature. The satisfaction relation $\models_{\text{LTL}}$ between LTL interpretations and LTL formulas is defined inductively as follows:*

$$
\begin{array}{rl}
\pi \models_{\text{LTL}} A & \textit{if } A \in t_0, \\
\pi \models_{\text{LTL}} \top & \textit{always}, \\
\pi \models_{\text{LTL}} \bot & \textit{never}, \\
\pi \models_{\text{LTL}} \neg\varphi & \textit{if } \pi \not\models_{\text{LTL}} \varphi, \\
\pi \models_{\text{LTL}} \varphi \wedge \psi & \textit{if } \pi \models_{\text{LTL}} \varphi \textit{ and } \pi \models_{\text{LTL}} \psi, \\
\pi \models_{\text{LTL}} \varphi \vee \psi & \textit{if } \pi \models_{\text{LTL}} \varphi \textit{ or } \pi \models_{\text{LTL}} \psi, \\
\pi \models_{\text{LTL}} X\varphi & \textit{if } \pi_1 \models_{\text{LTL}} \varphi, \\
\pi \models_{\text{LTL}} G\varphi & \textit{if } \pi_i \models_{\text{LTL}} \varphi \textit{ for all } i \in \mathbb{N}_0, \\
\pi \models_{\text{LTL}} F\varphi & \textit{if there exists } i \in \mathbb{N}_0 \textit{ such that } \pi_i \models_{\text{LTL}} \varphi, \\
\pi \models_{\text{LTL}} \varphi U \psi & \textit{if there exists } j \in \mathbb{N}_0 \textit{ such that } \pi_j \models_{\text{LTL}} \psi \\
 & \quad \textit{and } \pi_i \models_{\text{LTL}} \varphi \textit{ for all } 0 \leq i < j.
\end{array}
$$

An LTL interpretation is called a *model* of a formula φ if it satisfies φ. We denote the set of all models of φ by $\text{Mod}_{\text{LTL}}(\varphi)$. The presentation above can also be understood in terms of word notation: an LTL interpretation is an infinite sequence of sets of propositional atoms and can therefore be regarded as an infinite word over $P(\Sigma)$.

This perspective is primarily useful from a technical point of view, but it is of secondary importance for the conceptual understanding. What remains decisive is the interpretation as a temporal evolution in which t_i describes the propositional variables that hold at time i.

3.2.2 Equivalences and Normal Forms

In this section, we examine equality and transformations of LTL formulas. As in propositional logic, it is often not important *how* a property is formulated, but only *which* temporal behaviors it describes. Two formulas are therefore considered equivalent if they are satisfied in exactly the same LTL interpretations.

Definition 3.40 (Equivalence of LTL Formulas). *Let Σ be a propositional signature. Two LTL formulas φ and ψ over Σ are called* equivalent, *written*

$$\varphi \equiv_{\mathrm{LTL}} \psi,$$

if for every LTL interpretation π over Σ we have:

$$\pi \models_{\mathrm{LTL}} \varphi \quad \textit{if and only if} \quad \pi \models_{\mathrm{LTL}} \psi.$$

Many transformation rules from propositional logic remain valid in LTL as well. In particular, the coincidence lemma and the substitution lemma also hold here. To begin with, the familiar basic logical laws are sufficient.

Proposition 3.41 (Classical Equivalences). *For all LTL formulas φ and ψ, the following hold:*

$$\neg(\varphi \wedge \psi) \equiv_{\mathrm{LTL}} \neg\varphi \vee \neg\psi, \qquad \neg(\varphi \vee \psi) \equiv_{\mathrm{LTL}} \neg\varphi \wedge \neg\psi,$$

$$\neg\neg\varphi \equiv_{\mathrm{LTL}} \varphi, \qquad \top \equiv_{\mathrm{LTL}} \varphi \vee \neg\varphi, \qquad \bot \equiv_{\mathrm{LTL}} \varphi \wedge \neg\varphi.$$

Proof. All statements follow immediately from the classical semantics of $\neg, \wedge, \vee$ and from the fact that every

LTL interpretation either satisfies a formula or does not satisfy it.

Let π be an arbitrary LTL interpretation.

For the first equivalence, we have:

$$\begin{aligned}\pi \models_{\mathrm{LTL}} \neg(\varphi \wedge \psi) &\iff \pi \not\models_{\mathrm{LTL}} (\varphi \wedge \psi)\\ &\iff \pi \not\models_{\mathrm{LTL}} \varphi \quad \text{or} \quad \pi \not\models_{\mathrm{LTL}} \psi\end{aligned}$$

and hence

$$\pi \models_{\mathrm{LTL}} \neg(\varphi \wedge \psi) \iff \pi \models_{\mathrm{LTL}} \neg\varphi \vee \neg\psi.$$

The second De Morgan law follows analogously.

For double negation, we obtain:

$$\pi \models_{\mathrm{LTL}} \neg\neg\varphi \iff \pi \not\models_{\mathrm{LTL}} \neg\varphi \iff \pi \models_{\mathrm{LTL}} \varphi.$$

For $\top$ and $\bot$, the claim is equally direct: $\varphi \vee \neg\varphi$ is true in every interpretation, whereas $\varphi \wedge \neg\varphi$ cannot be true in any interpretation. □

In the next step, we consider formulas containing temporal operators. Some of these can be transformed just like the connectives of propositional logic. It is especially important that some operators can be expressed in terms of others, while other combinations are precisely not valid in general.

Proposition 3.42 (Important transformations for temporal operators)**.** *For all LTL formulas φ, ψ, χ, the following hold:*

$$G(\varphi \wedge \psi) \equiv_{\mathrm{LTL}} G\varphi \wedge G\psi, \qquad F(\varphi \vee \psi) \equiv_{\mathrm{LTL}} F\varphi \vee F\psi,$$

$$X(\varphi \wedge \psi) \equiv_{\mathrm{LTL}} X\varphi \wedge X\psi, \qquad X(\varphi \vee \psi) \equiv_{\mathrm{LTL}} X\varphi \vee X\psi,$$

$$(\varphi \wedge \psi)\, U\, \chi \equiv_{\mathrm{LTL}} (\varphi\, U\, \chi) \wedge (\psi\, U\, \chi)$$

$$\varphi\, U\, (\psi \vee \chi) \equiv_{\mathrm{LTL}} (\varphi\, U\, \psi) \vee (\varphi\, U\, \chi)$$

Proof. We prove the statements one by one.

For $G(\varphi \wedge \psi)$, we have:

$$\pi \models_{\mathrm{LTL}} G(\varphi \wedge \psi) \iff \text{for all } i \in \mathbb{N}_0 : \ \pi_i \models_{\mathrm{LTL}} \varphi \wedge \psi.$$

This is the case exactly when, for every i, both $\pi_i \models_{\mathrm{LTL}} \varphi$ and $\pi_i \models_{\mathrm{LTL}} \psi$ hold. Hence:

$$\pi \models_{\mathrm{LTL}} G(\varphi \wedge \psi) \iff \pi \models_{\mathrm{LTL}} G\varphi \wedge G\psi.$$

The statement for $F(\varphi \vee \psi)$ follows analogously from the existential semantics of F.

For X, the calculation is immediate:

$$\begin{aligned} \pi \models_{\mathrm{LTL}} X(\varphi \wedge \psi) &\iff \pi_1 \models_{\mathrm{LTL}} \varphi \wedge \psi \\ &\iff \pi_1 \models_{\mathrm{LTL}} \varphi \quad \text{and} \quad \pi_1 \models_{\mathrm{LTL}} \psi \end{aligned}$$

thus

$$\pi \models_{\mathrm{LTL}} X(\varphi \wedge \psi) \iff \pi \models_{\mathrm{LTL}} X\varphi \wedge X\psi.$$

Distributivity over $\vee$ follows in the same way.

For the until statement $(\varphi \wedge \psi)U\chi$, let j be a witness for the fact that $\pi \models_{\mathrm{LTL}} (\varphi \wedge \psi)U\chi$ holds. Then $\pi_j \models_{\mathrm{LTL}} \chi$, and for all $i < j$ we have $\pi_i \models_{\mathrm{LTL}} \varphi \wedge \psi$, hence in particular $\pi_i \models_{\mathrm{LTL}} \varphi$ and $\pi_i \models_{\mathrm{LTL}} \psi$. Therefore, the same witness j works for both $\varphi U\chi$ and $\psi U\chi$.

Conversely, if $\pi \models_{\mathrm{LTL}} (\varphi U \chi) \wedge (\psi U \chi)$, then there are witnesses j_1, j_2 for the two until formulas. Let $j = \min\{j_1, j_2\}$. Then $\pi_j \models_{\mathrm{LTL}} \chi$, and for all $i < j$, both φ and ψ satisfy the required condition. Hence $(\varphi \wedge \psi) U \chi$ also holds.

The distributive law for U over $\vee$ on the right-hand side is proved in the same way. □

Not all seemingly natural transformations are correct. This is especially important for F and G.

Example 3.43 (Two Counterexamples). *The following equivalences do* not *hold in general:*

$$F(\varphi \wedge \psi) \equiv_{\mathrm{LTL}} F\varphi \wedge F\psi, \qquad G(\varphi \vee \psi) \equiv_{\mathrm{LTL}} G\varphi \vee G\psi.$$

A counterexample to the first formula can be obtained, for instance, using the propositions

active *and* approved.

Consider a temporal evolution in which active *holds only at the first moment and* approved *only later. Then both* Factive *and* Fapproved *are true, but* active ∧ approved *is never true at the same time. Hence* F(active ∧ approved) *is false.*

For the second formula, it suffices to consider a run in which waiting *and* working *alternate. Then at every point at least one of the two statements is true, so* G(waiting ∨ working) *holds. However, neither* Gwaiting *nor* Gworking *is satisfied.*

Some operators behave idempotently or exhibit absorption. This means that applying these operators repeatedly does not produce any new semantic content.

Proposition 3.44 (Idempotence and Absorption). *For all LTL formulas φ and ψ, the following holds:*

$$GG\varphi \equiv_{\mathrm{LTL}} G\varphi, \qquad FF\varphi \equiv_{\mathrm{LTL}} F\varphi,$$

$$\varphi U(\varphi U\psi) \equiv_{\mathrm{LTL}} \varphi U\psi, \qquad (\varphi U\psi)U\psi \equiv_{\mathrm{LTL}} \varphi U\psi,$$

$$GFG\varphi \equiv_{\mathrm{LTL}} FG\varphi, \qquad FGF\varphi \equiv_{\mathrm{LTL}} GF\varphi.$$

Proof. The first two equivalences are immediate:

$$\begin{aligned} \pi \models_{\mathrm{LTL}} GG\varphi &\iff \forall i \in \mathbb{N}_0 : \pi_i \models_{\mathrm{LTL}} G\varphi \\ &\iff \forall i, j \in \mathbb{N}_0 : \pi_{i+j} \models_{\mathrm{LTL}} \varphi \end{aligned}$$

that is, precisely when $\pi \models_{\mathrm{LTL}} G\varphi$. The argument for $FF\varphi$ is entirely analogous.

For $\varphi U(\varphi U\psi)$, it suffices to observe the following: if $\pi \models_{\mathrm{LTL}} \varphi U\psi$, then the formula $\varphi U\psi$ already holds at the current position. Hence $\pi \models_{\mathrm{LTL}} \varphi U(\varphi U\psi)$ follows automatically. Conversely, a witness for the outer until-formula together with a witness for the inner until-formula immediately yields a witness for $\varphi U\psi$.

For $(\varphi U\psi)U\psi$, the idea is similar: once ψ holds at some point in time, it is enough for all earlier positions that $\varphi U\psi$ already holds there. Conversely, a witness for $(\varphi U\psi)U\psi$ already induces a witness for $\varphi U\psi$. Thus the two formulas describe the same set of temporal behaviours.

The two absorption laws for $GFG\varphi$ and $FGF\varphi$ follow directly from the meanings of G and F: in both cases, no additional semantic content is added when these operators are nested repeatedly. $\square$

Even more useful than isolated equivalences are the so-called duality laws. They describe how negation interacts with the temporal operators.

Proposition 3.45 (Temporal Dualities and Unfolding Laws). *For all LTL formulas φ and ψ, the following hold:*

$$\begin{aligned}\neg X\varphi &\equiv_{\mathrm{LTL}} X\neg\varphi\\ \neg G\varphi &\equiv_{\mathrm{LTL}} F\neg\varphi\\ \neg F\varphi &\equiv_{\mathrm{LTL}} G\neg\varphi\end{aligned}$$

$$\begin{aligned}G\varphi &\equiv_{\mathrm{LTL}} \varphi \wedge XG\varphi\\ F\varphi &\equiv_{\mathrm{LTL}} \varphi \vee XF\varphi\\ F\varphi &\equiv_{\mathrm{LTL}} \top U\varphi\\ \varphi U\psi &\equiv_{\mathrm{LTL}} \psi \vee (\varphi \wedge X(\varphi U\psi)).\end{aligned}$$

Moreover, for the negation of Until we have:

$$\neg(\varphi U\psi) \equiv_{\mathrm{LTL}} ((\varphi \wedge \neg\psi)U(\neg\varphi \wedge \neg\psi)) \vee G(\varphi \wedge \neg\psi).$$

Proof. We first prove the first three equivalences.

For X, this follows directly:

$$\begin{aligned}\pi \models_{\mathrm{LTL}} \neg X\varphi &\iff \pi \not\models_{\mathrm{LTL}} X\varphi\\ &\iff \pi_1 \not\models_{\mathrm{LTL}} \varphi\\ &\iff \pi_1 \models_{\mathrm{LTL}} \neg\varphi\\ &\iff \pi \models_{\mathrm{LTL}} X\neg\varphi\end{aligned}$$

For G and F, we use the definitions:

$$\pi \models_{\text{LTL}} \neg G\varphi \iff \neg(\forall i \in \mathbb{N}_0 : \pi_i \models_{\text{LTL}} \varphi)$$

and this is equivalent to

$$\begin{aligned}\exists i \in \mathbb{N}_0 : \pi_i \not\models_{\text{LTL}} \varphi &\iff \exists i \in \mathbb{N}_0 : \pi_i \models_{\text{LTL}} \neg\varphi \\ &\iff \pi \models_{\text{LTL}} F\neg\varphi.\end{aligned}$$

The third equation follows analogously.

The two unfolding laws

$$G\varphi \equiv_{\text{LTL}} \varphi \wedge XG\varphi \qquad \text{and} \qquad F\varphi \equiv_{\text{LTL}} \varphi \vee XF\varphi$$

are immediate once one observes that $G\varphi$ requires φ to hold now and at every future point, whereas $F\varphi$ requires only that φ hold at some point in the future. The present occurrence of φ is therefore either already covered or irrelevant to the truth of the formula.

The equivalence

$$F\varphi \equiv_{\text{LTL}} \top U\varphi$$

follows directly from the semantics of U: since $\top$ is true at every point in time, $\top U\varphi$ says exactly that φ eventually holds.

For the unfolding of Until, let us consider a path π.

"$\Rightarrow$": Assume that $\pi \models_{\text{LTL}} \varphi U\psi$. Then there is a least point in time j such that

$$\pi_j \models_{\text{LTL}} \psi$$

and for all $i < j$ we have $\pi_i \models_{\text{LTL}} \varphi$. If $j = 0$, then already $\pi \models_{\text{LTL}} \psi$. If $j > 0$, then $\pi \models_{\text{LTL}} \varphi \wedge X(\varphi U \psi)$: φ holds at the current point, and from the next point onward the remaining suffix of the path again satisfies $\varphi U \psi$. Hence in both cases,

$$\pi \models_{\text{LTL}} \psi \vee (\varphi \wedge X(\varphi U \psi)).$$

"$\Leftarrow$": If $\pi \models_{\text{LTL}} \psi$, then $\varphi U \psi$ is satisfied immediately. If $\pi \models_{\text{LTL}} \varphi \wedge X(\varphi U \psi)$, then φ holds now and the remaining suffix of the path again satisfies $\varphi U \psi$. Repeating this argument, one eventually reaches a point at which ψ holds. Therefore $\varphi U \psi$ holds as well.

Finally, consider the negation of Until. The statement

$$\neg(\varphi U \psi)$$

means that there is no point in time at which ψ holds while φ has held at every earlier point. This can happen in two ways: either ψ never holds and φ remains true forever, in which case the path satisfies $G(\varphi \wedge \neg\psi)$; or there is a first point at which φ fails before ψ appears, in which case we have

$$(\varphi \wedge \neg\psi) U (\neg\varphi \wedge \neg\psi).$$

Together, these two cases yield exactly the right-hand side of the claimed equivalence. □

With these transformations in hand, we can now introduce two normal forms. The first pushes negations so that they appear only directly in front of propositional variables; the second allows only the temporal operators X and U.

Definition 3.46 (Normal Forms). *Let φ be an LTL formula.*

- *φ is said to be in* negation normal form *(NNF) if negation occurs only directly in front of propositional variables.*
- *φ is said to be in XU* normal form *if only the temporal operators X and U occur in φ.*

Proposition 3.47. *For every LTL formula, there exists an equivalent formula in negation normal form.*

Proof. We define a transformation $\mathrm{nnf}(\cdot)$ recursively according to the structure of a formula. In doing so, negations are pushed inward step by step by means of the classical laws and the temporal dualities.

For atomic formulas and constants, we set

$$\mathrm{nnf}(A) = A, \qquad \mathrm{nnf}(\top) = \top, \qquad \mathrm{nnf}(\bot) = \bot.$$

For Boolean connectives and temporal operators, we define recursively:

$$\begin{aligned}\mathrm{nnf}(\varphi \wedge \psi) &= \mathrm{nnf}(\varphi) \wedge \mathrm{nnf}(\psi)\\ \mathrm{nnf}(\varphi \vee \psi) &= \mathrm{nnf}(\varphi) \vee \mathrm{nnf}(\psi)\end{aligned}$$

$$\begin{aligned}
\mathrm{nnf}(X\varphi) &= X\,\mathrm{nnf}(\varphi)\\
\mathrm{nnf}(G\varphi) &= G\,\mathrm{nnf}(\varphi)\\
\mathrm{nnf}(F\varphi) &= F\,\mathrm{nnf}(\varphi)\\
\mathrm{nnf}(\varphi U\psi) &= \mathrm{nnf}(\varphi)U\,\mathrm{nnf}(\psi).
\end{aligned}$$

For negations, we use the transformations from the preceding propositions:

$$\mathrm{nnf}(\neg A) = \neg A, \qquad \mathrm{nnf}(\neg\neg\varphi) = \mathrm{nnf}(\varphi),$$

$$\begin{aligned}
\mathrm{nnf}(\neg(\varphi \wedge \psi)) &= \mathrm{nnf}(\neg\varphi) \vee \mathrm{nnf}(\neg\psi)\\
\mathrm{nnf}(\neg(\varphi \vee \psi)) &= \mathrm{nnf}(\neg\varphi) \wedge \mathrm{nnf}(\neg\psi)
\end{aligned}$$

$$\begin{aligned}
\mathrm{nnf}(\neg X\varphi) &= X\,\mathrm{nnf}(\neg\varphi)\\
\mathrm{nnf}(\neg G\varphi) &= F\,\mathrm{nnf}(\neg\varphi)\\
\mathrm{nnf}(\neg F\varphi) &= G\,\mathrm{nnf}(\neg\varphi)
\end{aligned}$$

and for Until we use the equivalence from the previous proposition:

$$\begin{aligned}
\mathrm{nnf}(\neg(\varphi U\psi)) = \mathrm{nnf}(\,&((\varphi \wedge \neg\psi)U(\neg\varphi \wedge \neg\psi))\\
&\vee G(\varphi \wedge \neg\psi))
\end{aligned}$$

Correctness follows by structural induction on the formula. At each step, an equivalent transformation is used, so the semantics is preserved. At the same time, negations are pushed farther and farther inward until they eventually occur only in front of propositional variables. Therefore, $\mathrm{nnf}(\varphi)$ is an equivalent formula in negation normal form. □

Proposition 3.48. *For every LTL formula, there exists an equivalent formula in XU normal form.*

Proof. Here, too, we define a recursive transformation. The goal is to eliminate the operators F, G, and W. The basic idea is simple: F can be replaced by U, and G can be described using F together with negation.

We define:

$$\tau(A) = A, \qquad \tau(\top) = \top, \qquad \tau(\bot) = \bot,$$

$$\tau(\neg\varphi) = \neg\tau(\varphi)$$

$$\tau(\varphi \wedge \psi) = \tau(\varphi) \wedge \tau(\psi)$$

$$\tau(\varphi \vee \psi) = \tau(\varphi) \vee \tau(\psi)$$

$$\tau(X\varphi) = X\tau(\varphi), \qquad \tau(\varphi U \psi) = \tau(\varphi) U \tau(\psi),$$

$$\tau(F\varphi) = \top U \tau(\varphi), \qquad \tau(G\varphi) = \neg(\top U \neg\tau(\varphi)).$$

After this definition, only the temporal operators X and U occur in $\tau(\varphi)$. In fact, F and G are completely replaced, while all other constructions remain unchanged. That $\tau(\varphi)$ is equivalent to φ again follows by structural induction and from the equivalences already proved,

$$F\varphi \equiv_{\text{LTL}} \top U \varphi \qquad \text{and} \qquad G\varphi \equiv_{\text{LTL}} \neg F \neg\varphi.$$

Therefore, for every LTL formula there exists an equivalent formula in XU normal form. □

These two normal forms are particularly useful when one wants to standardize formulas, simplify them, or prepare them for later proofs. For the further development of the theory, it is therefore often helpful to first transform LTL formulas into one of these forms.

3.2.3 Relation to Kripke Structures

So far, we have defined LTL directly on infinite temporal evolutions. In applications, however, such an evolution usually does not arise in isolation, but rather as a possible execution of a system. Kripke structures provide exactly the right model for this setting: a state represents a concrete system configuration, and the edges indicate which successor states are reachable from it.

The basic idea is as follows. If we start in a state s and then follow an outgoing edge to a successor state, we obtain one possible evolution of the system. If we continue this process indefinitely, we get a path through the Kripke structure. For each state along such a path, we can record which propositional variables are true there. In this way, a path induces an LTL behavior.

Definition 3.49 (Paths). *Let Σ be a propositional signature, and let $K = \langle W, R, V \rangle$ be a Kripke structure for Σ. For a state $s \in W$, the set of paths in K that begin in s is defined by*

$$\begin{aligned} \text{Paths}(K, s) = \{&s_0 s_1 s_2 \ldots \mid \\ &s_0 = s \wedge \\ &(s_i, s_{i+1}) \in R \ \forall i\} \end{aligned}$$

The set of all paths of K is

$$\text{Paths}(K) = \bigcup_{s \in W} \text{Paths}(K, s).$$

A path therefore describes, first of all, only the sequence of visited states. For LTL, however, the states themselves are not what matters, but rather which propositional variables hold in them. We therefore translate a path into a sequence of sets of propositional variables.

Definition 3.50 (Trace). *Let $K = \langle W, R, V\rangle$ be a Kripke structure, and let*

$$\rho = s_0 s_1 s_2 \ldots$$

be a sequence of states of K. Then

$$\mathrm{Trace}(K, \rho) = V(s_0)V(s_1)V(s_2)\ldots$$

is called the trace associated with ρ.

A trace is thus the actual temporal observation of a path: instead of the states themselves, we record only which propositions hold at the individual time points.

Definition 3.51 (Traces). *Let $K = \langle W, R, V\rangle$ be a Kripke structure for Σ, and let $s \in W$ be a state. The set of traces in K that begin in s is*

$$\mathrm{Traces}(K, s) = \{\mathrm{Trace}(K, \rho) \mid \rho \in \mathrm{Paths}(K, s)\}.$$

The set of all traces of K is

$$\mathrm{Traces}(K) = \bigcup_{s \in W} \mathrm{Traces}(K, s).$$

Example 3.52. *Consider a Kripke structure over the signature $\Sigma = \{\mathsf{Request}, \mathsf{Response}, \mathsf{Error}\}$. State u_1 represents an ongoing login process, u_2 a waiting phase, u_3 the* *successful completion, u_4 an error state, and u_5 a state in which both* Response *and* Error *hold.*

A possible path is

$$\rho = u_1 u_2 u_3 u_3 u_3 \ldots$$

with trace

$$\begin{aligned} \mathrm{Trace}(K, \rho) = &\{\mathsf{Request}\}\{\mathsf{Request}\} \\ &\{\mathsf{Response}\}\{\mathsf{Response}\}\{\mathsf{Response}\}\ldots \end{aligned}$$

This trace describes an execution in which a request initially remains pending and is then answered successfully.

By contrast, a sequence such as

$$u_1 u_3 u_3 u_3 \ldots$$

can certainly be written down as a string, but it is a path only if there is in fact an edge from u_1 to u_3. Without such an edge, the sequence does not belong to $\mathrm{Paths}(K, u_1)$*, and its trace therefore does not belong to* $\mathrm{Traces}(K, u_1)$*.*

For LTL, we shall from now on consider only Kripke structures in which every state has at least one successor. Then every path can be extended indefinitely.

Definition 3.53 (Non-Terminating Kripke Structure)**.** *A Kripke structure $K = \langle W, R, V \rangle$ is called* non-terminating *if every state $s \in W$ has at least one outgoing edge.*

In non-terminating Kripke structures, all paths are infinite. Accordingly, all traces are infinite as well and can be viewed directly as LTL interpretations.

We can now say when an LTL formula holds at a state of a Kripke structure. The idea is particularly strict: a state satisfies a formula precisely when *all* possible evolutions starting from that state satisfy the formula. Thus, LTL describes properties that are robust under all possible future developments of a system.

Definition 3.54 (Satisfaction in a Kripke Structure). *Let Σ be a propositional signature, let $K = \langle W, R, V \rangle$ be a non-terminating Kripke structure for Σ, and let $s \in W$ be a state. Then, for an LTL formula φ, we have:*

$$s \models_K^{\mathrm{LTL}} \varphi \iff$$
$$\forall \pi \in \mathrm{Traces}(K, s) : \pi \models_{\mathrm{LTL}} \varphi$$

Alternatively, we also write

$$\langle K, s \rangle \models^{\mathrm{LTL}} \varphi.$$

This definition makes the role of LTL in system modelling very clear: a formula is not tested on a single path, but on all possible executions that start from a given state.

Example 3.55. *Consider a Kripke structure over the signature*

$$\Sigma = \{\mathsf{Ready}, \mathsf{Sending}, \mathsf{Confirmed}\}.$$

Let the states be understood as follows: in an initial state, Ready *holds; afterwards, perhaps* Sending*; and later, possibly* Confirmed.

We consider the formulas

$$\varphi = FG\,\mathsf{Confirmed}\wedge$$
$$\psi = (\mathsf{Ready} \vee \mathsf{Sending})\,W\,\mathsf{Confirmed}$$

Then a state s may well satisfy ψ, even though not every possible evolution starting from s ever leads to Confirmed. *For φ, the situation is stricter: every possible evolution from s must eventually enter a phase in which* Confirmed *holds permanently.*

A state t from which there exists an evolution that shows only Ready *forever and never* Confirmed, *by contrast, does not satisfy FG* Confirmed. *A single counter-evolution is therefore enough to falsify the formula at that state.*

One further point is especially important and may seem surprising at first glance: negation is generally not dual at the level of states. For individual LTL interpretations, it is of course true that either a formula or its negation is satisfied. For states in Kripke structures, however, the situation is different, because quantification ranges over *all* possible traces.

Example 3.56. *There may be a state s such that neither*

$$s \models_K^{\mathrm{LTL}} X\,\mathsf{Confirmed}$$

nor

$$s \models_K^{\mathrm{LTL}} \neg X\,\mathsf{Confirmed}$$

holds. This happens exactly when, starting from s, some evolutions show Confirmed *at the next step while others do not. In that case,* XConfirmed *is not true on all traces, and its negation is not true on all traces either.*

This establishes the connection between LTL and Kripke structures: states describe system situations, paths describe possible developments, traces translate these developments into temporal evolutions, and LTL formulas describe properties that are meant to hold on all such evolutions.

3.2.4 Distinguishing States in Kripke Structures

In the previous sections, we considered LTL formulas as properties of temporal evolutions. We now turn to the question of how such formulas apply to states in Kripke structures. The central idea is this: From the perspective of LTL, two states are indistinguishable exactly when they satisfy the same temporal properties.

Definition 3.57 (LTL Equivalence of States). *Let* Σ *be a propositional signature, and let* $K = \langle W, R, V \rangle$ *be a non-terminating Kripke structure for* Σ*. Two states* $s, w \in W$ *are called* LTL-equivalent with respect to K, *written*

$$s \equiv_K^{\mathrm{LTL}} w,$$

if for every LTL formula φ *over* Σ*, we have:*

$$\langle K, s \rangle \models^{\mathrm{LTL}} \varphi \quad \textit{if and only if} \quad \langle K, w \rangle \models^{\mathrm{LTL}} \varphi.$$

It is important to distinguish this notion from the formal equivalence of formulas introduced earlier: there, two formulas are compared with one another. Here, by contrast, we compare states of a Kripke structure.

Intuitively, $s \equiv_K^{\text{LTL}} w$ means that, from the standpoint of LTL, the future looks exactly the same when starting from s as when starting from w. Thus, if the same formulas hold in both states, then LTL cannot distinguish between them.

An even stronger, but very natural, notion of comparison is trace equivalence. Two states are trace-equivalent if exactly the same observation sequences are possible from each of them.

Definition 3.58 (Trace Equivalence). *Let Σ be a propositional signature, and let $K = \langle W, R, V \rangle$ be a Kripke structure for Σ. Two states $s, w \in W$ are called* trace-equivalent with respect to K, *written*

$$s \equiv_K^{\text{Trace}} w,$$

if

$$\text{Traces}(K, s) = \text{Traces}(K, w)$$

holds.

Trace equivalence is a very direct form of indistinguishability: if two states generate exactly the same traces, then they are also indistinguishable for any logic that operates solely on those traces.

Example 3.59. *Consider a Kripke structure over*

$$\Sigma = \{\mathsf{ready}, \mathsf{waiting}, \mathsf{done}\}.$$

Suppose there are two distinct states s and w that both admit the same possible evolution: initially, ready *holds; afterward, the system may enter a waiting loop; and later,* done *becomes true. The two states are syntactically different, but their observable future behavior is the same.*

In such a case,

$$\text{Traces}(K, s) = \text{Traces}(K, w),$$

so s and w are trace-equivalent. Consequently, they are also LTL-equivalent.

The following proposition shows that trace equivalence always implies LTL equivalence.

Proposition 3.60. *Let Σ be a propositional signature, and let $K = \langle W, R, V\rangle$ be a non-terminating Kripke structure for Σ. If two states $s, w \in W$ are trace-equivalent with respect to K, then they are also LTL-equivalent with respect to K.*

Proof. Let $s, w \in W$ with $s \equiv_K^{\text{Trace}} w$. Then

$$\text{Traces}(K, s) = \text{Traces}(K, w).$$

Now consider an arbitrary LTL formula φ. If $\langle K, s\rangle \models^{\text{LTL}} \varphi$, then all traces in Traces(K, s) satisfy the formula φ. Since the sets of traces are equal, the same is true for all traces in Traces(K, w), and therefore $\langle K, w\rangle \models^{\text{LTL}} \varphi$.

The converse direction is identical. Hence

$$\langle K, s\rangle \models^{\text{LTL}} \varphi \quad \text{if and only if} \quad \langle K, w\rangle \models^{\text{LTL}} \varphi$$

for all formulas φ. Thus, s and w are LTL-equivalent. □

To analyze differences between states more finely, we now consider finite initial segments of traces. For two traces, we are first interested only in whether they look the same in the first $k + 1$ steps.

Definition 3.61 (Agreement on an Initial Segment). *Let*

$$\pi_1 = t_0 t_1 t_2 \ldots \qquad \textit{and} \qquad \pi_2 = r_0 r_1 r_2 \ldots$$

be two LTL traces, and let $k \in \mathbb{N}_0$. We write

$$\pi_1 \equiv_k \pi_2,$$

if

$$t_i = r_i \quad \textit{for all } i \in \{0, \ldots, k\}.$$

For a set of atoms $V \subseteq \Sigma$, we define a formula χ_V that holds exactly when precisely the atoms from V are true at the current moment:

$$\chi_V := \bigwedge_{A \in V} A \wedge \bigwedge_{A \in \Sigma \setminus V} \neg A.$$

If $\pi = t_0 t_1 t_2 \ldots$ is a trace, we define

$$\chi_\pi^k := \chi_{t_0} \wedge X\chi_{t_1} \wedge XX\chi_{t_2} \wedge \cdots \wedge X^k \chi_{t_k}.$$

This formula describes exactly the initial segment of the trace π up to that point.

Example 3.62. *For* $\Sigma = \{\mathsf{req}, \mathsf{ack}, \mathsf{done}\}$ *and the trace*

$$\pi = \{\mathsf{req}\}\{\mathsf{req}, \mathsf{ack}\}\{\mathsf{done}\}\dots$$

we obtain:

$$\chi_\pi^0 = \mathsf{req} \wedge \neg\mathsf{ack} \wedge \neg\mathsf{done},$$

$$\chi_\pi^1 = (\mathsf{req} \wedge \neg\mathsf{ack} \wedge \neg\mathsf{done}) \wedge X(\mathsf{req} \wedge \mathsf{ack} \wedge \neg\mathsf{done}).$$

Thus, the formula χ_π^1 *describes exactly the first two time points of this trace.*

The following observation is central: if two states are LTL-equivalent, then every path starting from one state can be approximated by finite initial segments of paths starting from the other state.

Proposition 3.63. *Let* Σ *be a propositional signature, and let* $K = \langle W, R, V\rangle$ *be a non-terminating Kripke structure for* Σ. *If* $s, w \in W$ *are LTL-equivalent, then the following holds: for every path* $\rho_s \in \mathrm{Paths}(K, s)$ *and every* $k \in \mathbb{N}_0$, *there exists a path* $\rho_w^k \in \mathrm{Paths}(K, w)$ *such that*

$$\mathrm{Trace}(K, \rho_s) \equiv_k \mathrm{Trace}(K, \rho_w^k).$$

Proof. Let $\rho_s \in \mathrm{Paths}(K, s)$ be arbitrary, and set $\pi := \mathrm{Trace}(K, \rho_s)$. For each $k \in \mathbb{N}_0$, we consider the formula χ_π^k.

Since $\pi \models_{\mathrm{LTL}} \chi_\pi^k$, it follows in particular that not all traces in $\mathrm{Traces}(K, s)$ satisfy the negation of χ_π^k. Hence,

$$\langle K, s\rangle \not\models^{\mathrm{LTL}} \neg\chi_\pi^k.$$

Because s and w are LTL-equivalent, this implies

$$\langle K, w\rangle \not\models^{\mathrm{LTL}} \neg\chi_\pi^k.$$

Therefore, there exists a trace $\pi_k \in \mathrm{Traces}(K, w)$ such that

$$\pi_k \models_{\mathrm{LTL}} \chi_\pi^k.$$

By the characterization of χ_π^k, this means precisely that

$$\pi \equiv_k \pi_k.$$

Writing $\pi_k = \mathrm{Trace}(K, \rho_w^k)$, the claim follows. □

We now come to the main result. In order to derive a complete trace from the agreement of all finite initial segments, we need the additional assumption that the Kripke structure is image-finite.

Definition 3.64 (Image-Finite Kripke Structure). *A Kripke structure is called* image-finite *if every state has only finitely many outgoing edges.*

Lemma 3.65 (König's Lemma). *Let T be an infinite rooted tree in which every node has only finitely many immediate successors, that is, a finitely branching tree. Then T has an infinite path.*

Proof. We construct an infinite path step by step.

We denote the root of the tree by v_0. Since T is infinite, among the successors of v_0 there must be at least one node from which infinitely many nodes of the tree are reachable. Otherwise, the entire tree would be finite, because each successor would itself have only finitely many further successors.

We therefore choose such a successor v_1 of v_0. Now consider the subtree rooted at v_1. This subtree is again infinite and finitely branching. Hence, there exists a successor v_2 of v_1 from which infinitely many nodes are again reachable.

Continuing this process, we obtain an infinite sequence

$$v_0, v_1, v_2, \ldots$$

with (v_i, v_{i+1}) as an edge for all $i \in \mathbb{N}_0$. This sequence is an infinite path in the tree. □

Theorem 3.66. *Let Σ be a propositional signature, and let $K = \langle W, R, V \rangle$ be an image-finite, non-terminating Kripke structure for Σ. Then for all states $s, w \in W$ we have:*

$$s \equiv_K^{\mathrm{LTL}} w \quad \textit{if and only if} \quad s \equiv_K^{\mathrm{Trace}} w.$$

Proof. The implication

$$s \equiv_K^{\mathrm{Trace}} w \Rightarrow s \equiv_K^{\mathrm{LTL}} w$$

has already been established. For the converse direction, assume that s and w are LTL-equivalent.

Let $\rho_s \in \mathrm{Paths}(K, s)$ be arbitrary. For each $k \in \mathbb{N}_0$, the preceding proposition yields a path in $\mathrm{Paths}(K, w)$ whose trace agrees with that of ρ_s on the first $k + 1$

positions. The collection of all such finite initial segments forms an infinitely deep, finitely branching tree. By König's lemma, this tree has an infinite branch. This branch is a path $\rho_w \in \text{Paths}(K, w)$ such that

$$\text{Trace}(K, \rho_w) = \text{Trace}(K, \rho_s).$$

Thus every trace in $\text{Traces}(K, s)$ has an identical counterpart in $\text{Traces}(K, w)$. By symmetry, the converse inclusion also holds, and therefore

$$\text{Traces}(K, s) = \text{Traces}(K, w).$$

Hence s and w are trace-equivalent. □

Corollary 3.67. *Let Σ be a propositional signature, and let $K = \langle W, R, V\rangle$ be an image-finite, non-terminating Kripke structure for Σ. If $s, w \in W$ are not trace-equivalent, then there exists an LTL formula φ such that*

$$\langle K, s\rangle \models^{\text{LTL}} \varphi \quad \textit{and} \quad \langle K, w\rangle \not\models^{\text{LTL}} \varphi.$$

Proof. If s is not trace-equivalent to w, then by the theorem it is also not LTL-equivalent to w. By the definition of LTL equivalence, there is therefore a formula that holds at one of the two states but not at the other. If necessary, interchange s and w. □

For applications, the key point is this: finite Kripke structures are always image-finite. Consequently, the theorem applies directly to all finite models. In such cases, LTL distinguishes states exactly up to their possible traces.

3.2.5 Relation to Predicate Logic

To conclude this section, we compare LTL with first-order predicate logic. The guiding question is which sets of time evolutions can be described by formulas of one logic or the other. In this sense, we say that a set of LTL-interpretations is *definable* by a formula.

Definition 3.68 (LTL-Definability). *Let Σ be a propositional signature. A set L of LTL-interpretations over Σ is called* LTL-definable *if there exists an LTL-formula φ over Σ such that*

$$L = \mathrm{Mod}_{\mathrm{LTL}}(\varphi).$$

In order to compare LTL with predicate logic, we encode a time evolution as a structure over the natural numbers. The time points themselves form the domain. For each propositional variable, we introduce a unary predicate that is true exactly at those time points at which the variable holds.

Definition 3.69 (Encoding of LTL-Interpretations). *Let $\Sigma = \{A_1, \ldots, A_k\}$ be a propositional signature and*

$$\pi = t_0 t_1 t_2 t_3 \ldots$$

an LTL-interpretation over Σ. We then associate with π the structure

$$\mathcal{A}_\pi = \langle \mathbb{N}, <, 0, P_{A_1}, \ldots, P_{A_k} \rangle$$

where

$$P_{A_i}^{\mathcal{A}_\pi} = \{\, j \in \mathbb{N} \mid A_i \in t_j \,\} \quad \textit{for all } i.$$

The constant 0 *denotes the initial point of the time evolution.*

The interpretation is therefore straightforward: a predicate $P_A(x)$ means that the propositional variable A holds at time point x. The order $<$ allows us to speak about earlier and later time points, and the constant 0 ensures that evaluation always starts at the beginning of the time evolution.

Definition 3.70 (FO-Definability over Initial Time Evolutions). *Let* Σ *be a propositional signature. A set* L *of LTL-interpretations over* Σ *is called* FO-definable over initial linear time evolutions *if there exists a first-order formula* $\psi(x)$ *with one free variable* x *such that*

$$L = \{\mathcal{A}_\pi \mid \pi \textit{ is a } \Sigma\textit{-interpretation} \\ \textit{and } \mathcal{A}_\pi \models \psi[x \mapsto 0]\}$$

For an LTL-interpretation π *we write briefly*

$$\pi \models^{\text{init}}_{\text{FO}} \psi$$

if $\mathcal{A}_\pi \models \psi[x \mapsto 0]$ *holds.*

The variable x represents the time point under consideration. Since LTL-formulas are always evaluated at the beginning of a time evolution, in first-order logic we likewise consider only the case $x = 0$.

The central statement of this section is that LTL and first-order predicate logic over linear time evolutions have the same expressive power.

Theorem 3.71. *Let Σ be a propositional signature and let L be a set of LTL-interpretations over Σ. Then the following statements are equivalent:*

1. *L is LTL-definable.*

2. *L is FO-definable over initial linear time evolutions.*

Proof. We prove both directions.

(a) $\Rightarrow$ (b). Let φ be an LTL-formula over Σ, and let $L = \mathrm{Mod}_{\mathrm{LTL}}(\varphi) = \{\pi \mid \pi \models_{\mathrm{LTL}} \varphi\}$. We inductively construct a first-order formula $\varphi^{\mathrm{FO}}(x)$ with one free variable x that accepts exactly those (encoded) interpretations that satisfy φ at time point 0.

Notation: For an LTL-interpretation $\pi = t_0 t_1 t_2 \ldots$, let $A_\pi = \langle \mathbb{N}, =, <, P_{A_1}, \ldots, P_{A_k}, 0 \rangle$ be the associated FO-structure (as in the proof sketch), where each atomic proposition $A \in \Sigma$ is represented by a unary predicate P_A, and where 0 is the constant for time point 0. As usual, we write $x \leq y$ as an abbreviation for $x = y \lor x < y$.

Define $\cdot^{\mathrm{FO}}$ recursively on the structure of LTL-formulas:

$$\begin{array}{lcl}
(A)^{\mathrm{FO}}(x) & := & P_A(x) \quad \text{for } A \in \Sigma, \\
(\neg\psi)^{\mathrm{FO}}(x) & := & \neg\psi^{\mathrm{FO}}(x), \\
(\psi_1 \wedge \psi_2)^{\mathrm{FO}}(x) & := & \psi_1^{\mathrm{FO}}(x) \wedge \psi_2^{\mathrm{FO}}(x), \\
(\psi_1 \vee \psi_2)^{\mathrm{FO}}(x) & := & \psi_1^{\mathrm{FO}}(x) \vee \psi_2^{\mathrm{FO}}(x), \\
(X\psi)^{\mathrm{FO}}(x) & := & \exists y\,(x < y \wedge \psi^{\mathrm{FO}}(y) \wedge \\
 & & \neg\exists z\,(x < z \wedge z < y)), \\
(F\psi)^{\mathrm{FO}}(x) & := & \exists y\,(x \leq y \wedge \psi^{\mathrm{FO}}(y)), \\
(G\psi)^{\mathrm{FO}}(x) & := & \forall y\,(x \leq y \rightarrow \psi^{\mathrm{FO}}(y)), \\
(\psi_1\, U\, \psi_2)^{\mathrm{FO}}(x) & := & \exists y(x \leq y \wedge \psi_2^{\mathrm{FO}}(y) \wedge \\
 & & \forall z((x \leq z \wedge z < y) \rightarrow \psi_1^{\mathrm{FO}}(z))).
\end{array}$$

The correctness of this translation is proved by structural induction: for every LTL-formula χ and every LTL-interpretation π, we have

$$\pi \models_{\mathrm{LTL}} \chi \iff A_\pi \models_{\mathrm{FO}} \chi^{\mathrm{FO}}[x \mapsto 0].$$

The induction steps are immediate: the cases for atomic formulas and Boolean connectives follow directly from the semantics; for F, G, X, U, one checks formally that the quantifiers and order relations appearing in the FO-translation express exactly the semantic condition of the temporal operators (for example, F states that "there exists a $y \geq x$ such that ...", X selects the immediate next y, and U encodes the "until" condition by the existence of an endpoint y together with a universal statement over all intermediate time points). Hence, $\varphi^{\mathrm{FO}}(x)$ is an FO-formula that accepts an LTL-interpretation π exactly when $\pi \models_{\mathrm{LTL}} \varphi$ (viewed as a structure with $x = 0$). Therefore, L is FO-definable over initial linear time evolutions. Thus every LTL-definable set is also FO-definable over initial linear time evolutions.

(b) $\Rightarrow$ (a). The converse direction relies on a classical result due to Kamp: every FO-formula over linear orders with unary predicates can be expressed by an equivalent LTL-formula. Now suppose that L is FO-definable over initial linear time evolutions, that is, there exists an FO-formula $\psi(x)$ with one free variable such that for every LTL-interpretation π we have

$$\pi \in L \quad \Longleftrightarrow \quad A_\pi \models_{\mathrm{FO}} \psi[x \mapsto 0].$$

The converse claim, namely that every such FO-formula $\psi(x)$ can be expressed by an LTL-formula, is the classical part of the comparison between LTL and first-order logic over linearly ordered time points: *Kamp's Theorem* (Kamp, 1968) states that the genuine LTL-formulas are exactly the FO-formulas with one free variable over the structure $(\mathbb{N}, <)$ with unary predicates (equivalently, FO[<]-formulas over positions). Therefore, there exists an LTL-formula ψ_{LTL} equivalent to $\psi(x)$ such that

$$\pi \models_{\mathrm{LTL}} \psi_{\mathrm{LTL}} \quad \Longleftrightarrow \quad A_\pi \models_{\mathrm{FO}} \psi[x \mapsto 0],$$

and hence L is also LTL-definable.

(A rough idea of the construction is as follows: one first rewrites $\psi(x)$ into a suitable normal form in FO-logic over $<$, then eliminates nested quantifiers by appropriate combinations of conditions on next and later time points, and finally translates quantifiers of the form "$\exists y \geq x$ such that ..." and "$\forall y \geq x$ such that ..." systematically into the temporal operators F, G, X, U; the precise technical justification is exactly the content of Kamp's theorem and of several subsequent expository and constructive proofs.)

The claim now follows from both directions. □

Example 3.72. *The LTL formula*

$$G(\mathsf{Registration} \to F\,\mathsf{Confirmation})$$

can be translated into FO as the statement that every point in time at which a registration occurs is followed by a later point in time at which confirmation holds.

Likewise,

$$F\,\mathsf{Error}$$

describes exactly the set of time evolutions in which there is some point in time at which an error state occurs.

Overall, this comparison shows that LTL is already highly expressive: over linear time evolutions, it captures exactly the same properties as first-order predicate logic with order and an initial point. However, if predicate logic is extended beyond restriction to the initial point so as to include additional temporal references, one obtains an even finer temporal logic, especially one that can express statements about the past.

3.2.6 Exercises and Solutions

Exercise 3.73. *Consider the propositional signature*

$$\Sigma = \{A, B, C\}.$$

We consider the following expressions:

$$\varphi_2 = GX(F\neg B \; U \; ((C \vee \neg A) \wedge A))$$

$$\pi_3 = \emptyset, \emptyset, \{A\}, \{B, C\}, \{C\}, \{B\}, \\ \{C\}, \{A, C\}, \{A, C\}, \{A, C\}, \ldots$$

as well as

$$\varphi_4 = FB \wedge FFA, \qquad \psi_4 = F(B \wedge FA).$$

For each of the following statements, determine whether it is true or false.

1. $\neg F \neg (XA \vee FGFB \wedge A)$ *is an LTL formula over* Σ.
2. φ_2 *is in negation normal form.*
3. *It holds that* $\pi_3 \models_{\text{LTL}} G\neg X \neg(\neg B \wedge C)$.
4. *It holds that* $\varphi_4 \equiv_{\text{LTL}} \psi_4$.

Solution 3.74. *1.* ***True.*** *The given string is a well-formed LTL formula over* Σ. *All propositional variables belong to* Σ, *and the operators* $\neg$, F, G, X, $\wedge$, *and* $\vee$ *are all valid LTL constructors. With the usual precedence conventions, the formula can be read unambiguously.*

2. ***True.*** *In φ_2, negations occur only directly in front of propositional variables, namely in $\neg B$ and $\neg A$. All other connectives are allowed, and in particular the operator U may appear in negation normal form. Therefore, φ_2 is in NNF.*

3. ***False.*** *The formula*

$$G\neg X\neg(\neg B \wedge C)$$

requires the inner statement $\neg X\neg(\neg B \wedge C)$ to hold at every *point in time. Thus, a counterexample at just a single time point is enough.*

Consider the beginning of π_3:

$$\pi_3 = \emptyset,\ \emptyset,\ \{A\},\ \{B, C\}, \ldots$$

At the first time point, we evaluate $X\neg(\neg B \wedge C)$. This means looking at the next time point, namely the second set $\emptyset$. There, $\neg B \wedge C$ is false because C is not satisfied. Hence, $\neg(\neg B \wedge C)$ is true there, and therefore

$$X\neg(\neg B \wedge C)$$

is true. Consequently,

$$\neg X\neg(\neg B \wedge C)$$

is false at the first time point. Since the global requirement already fails at the first point, the entire formula is false.

4. ***False.*** *First, we have*

$$\varphi_4 = FB \wedge FFA \equiv_{\text{LTL}} FB \wedge FA,$$

because $FFA \equiv_{\text{LTL}} FA$*.*

By contrast, the formula $\psi_4 = F(B \wedge FA)$ *requires that there be a point in time at which* both B *holds and, from that point onward,* A *holds again at some later time. That is exactly where the difference lies: although the operator* F *distributes over* $\vee$*, it does not in general distribute over* $\wedge$*.*

As a counterexample, consider the trace

$$\pi = \{A\}\,\{B\}\,\emptyset\,\emptyset\,\emptyset\ldots$$

Then

$$\pi \models_{\text{LTL}} FB \qquad \textit{and} \qquad \pi \models_{\text{LTL}} FA,$$

hence

$$\pi \models_{\text{LTL}} \varphi_4.$$

However, this does not hold for ψ_4*: the only time point at which* B *is true is the second one. But at that point,* FA *is false, because* A *never occurs again later. Hence,* $B \wedge FA$ *is not satisfied at any time point, and therefore*

$$\pi \not\models_{\text{LTL}} \psi_4.$$

Thus, φ_4 *and* ψ_4 *are not LTL-equivalent.*

3.3 Computation Tree Logic

In this section, we study Computation Tree Logic, or CTL for short. In contrast to linear-time logic LTL, time is not viewed here as a single path into the future, but as a branching structure of possible developments. From a given state, a system may therefore have several different successor states, giving rise to different possible future evolutions.

This distinction is especially important in applications. In the analysis of a running system, one often wants to know not only what happens along *one* particular execution, but what happens along *all*, or at least along *some*, of the possible executions. For example, a control system may be modeled so that after a request it can proceed either along a successful processing path or along an error path. CTL then makes it possible to express properties such as

- "There exists a possible execution along which an error message eventually appears,"
- "Along all possible executions, a request is eventually answered," or
- "From here, it is always possible to reach a safe state."

in a formal way.

We begin by introducing the syntax and semantics of CTL. We then consider important equivalences and normal forms. After that, we examine the expressive power of CTL by means of bisimulations and conclude by comparing CTL with LTL.

3.3.1 Syntax and Semantics

As in the case of LTL, we consider Kripke structures of the form

$$K = \langle W, R, V \rangle,$$

where W is the set of states, R is the transition relation, and V specifies which propositional variables hold at a given state. A *path* is an infinite sequence of states that follows the edges of the Kripke structure. A *trace* of a path is obtained by recording, at each moment, not the state itself but the propositional variables that are true there.

For CTL, however, the central object of interest is not a single path, but the totality of all possible executions starting from a given state. We therefore think of a Kripke structure as being unfolded into a tree rooted at that state: each node represents a possible system configuration, and its successors describe all possible next steps.

Definition 3.75 (Tree Unfolding). *Let Σ be a propositional signature, let $K = \langle W, R, V \rangle$ be a Kripke structure for Σ, and let $s \in W$ be a state. The* tree unfolding *of K at s is the tree T_s whose nodes are exactly the finite paths in K that begin at s. The root is the path (s). A node π is connected by an edge to a node π' if π' arises* *from π by appending a direct successor state. Each node is labeled with the last state of its path.*

The tree unfolding makes it explicit that different possible future evolutions can branch out from a single state. A path in the tree therefore corresponds exactly to one possible future of the system.

Example 3.76. *Consider an online ordering system. In a state s, an order has been received. From there, the system may move, for example, to a* processing *state, a* follow-up query *state, or a* cancellation *state. The tree unfolding of s then contains three different branches that separate these possibilities.*

This allows us, for instance, to express:

$$EF\,\mathsf{Shipped}$$

"There exists a possible execution along which the order is eventually shipped."

By contrast,

$$AG(\mathsf{Ordered} \to AF\,\mathsf{Shipped}),$$

means that along every possible execution, every order eventually leads to a state from which shipment will ultimately be reached.

CTL now combines two kinds of operators. First, there are path quantifiers:

- A: on all paths,

- E: on at least one path.

In addition, there are temporal operators:

- X: in the next state,
- F: eventually,
- G: always,
- U: until.

The basic idea is therefore this: first we specify *how many paths* are being considered, and then *which temporal property* is required on those paths.

Definition 3.77 (Syntax of CTL). *Let Σ be a propositional signature. The CTL formulas over Σ are defined inductively as follows:*

1. *Every propositional variable from Σ is a CTL formula.*
2. *$\top$ and $\bot$ are CTL formulas.*
3. *If φ and ψ are CTL formulas, then $\neg\varphi$, $\varphi \land \psi$, and $\varphi \lor \psi$ are also CTL formulas.*
4. *If φ is a CTL formula, then $EX\varphi$, $EF\varphi$, and $EG\varphi$ are CTL formulas.*
5. *If φ is a CTL formula, then $AX\varphi$, $AF\varphi$, and $AG\varphi$ are CTL formulas.*

6. *If φ and ψ are CTL formulas, then $E(\varphi U \psi)$ and $A(\varphi U \psi)$ are also CTL formulas.*

As usual, we also use the abbreviations

$$\varphi \to \psi := \neg\varphi \vee \psi$$
$$\varphi \leftrightarrow \psi := (\varphi \to \psi) \wedge (\psi \to \varphi)$$

Example 3.78. *Let*

$$\Sigma = \{\mathsf{Init}, \mathsf{Load}, \mathsf{Error}, \mathsf{Ready}, \mathsf{Restart}\}.$$

Then the following CTL formulas admit a natural reading:

- $EF(\mathsf{Error} \wedge \neg\mathsf{Ready})$*: There exists a possible execution along which an error occurs at some point before the system becomes ready again.*
- $AG(\mathsf{Init} \to AF\,\mathsf{Ready})$*: Along every possible execution, each initialization state eventually leads to a state in which the system is ready again.*
- $AG(EF\,\mathsf{Restart})$*: From every reachable state, a restart remains possible.*
- $A(\mathsf{Load}\,U\,\mathsf{Ready})$*: Along all possible executions, the system remains in the state* Load *until* Ready *occurs.*

The semantics of CTL is now defined no longer directly on paths, but on states of Kripke structures. A formula holds at a state exactly when the property required by the formula is satisfied there.

Definition 3.79 (Satisfaction Relation for CTL). *Let Σ be a propositional signature, and let $K = \langle W, R, V \rangle$ be a Kripke structure for Σ. The satisfaction relation $\models_{\mathrm{CTL}}$ is a relation between pointed Kripke structures $\langle K, s \rangle$ and CTL formulas over Σ, defined by:*

$$\langle K, s \rangle \models_{\mathrm{CTL}} A \quad \textit{if} \quad A \in V(s)$$

for propositional variables $A \in \Sigma$, and

$$\langle K, s \rangle \models_{\mathrm{CTL}} \top \quad \textit{always}, \qquad \langle K, s \rangle \models_{\mathrm{CTL}} \bot \quad \textit{never}.$$

For the Boolean connectives, the usual clauses apply:

$$\langle K, s \rangle \models_{\mathrm{CTL}} \neg\varphi \iff \langle K, s \rangle \not\models_{\mathrm{CTL}} \varphi,$$

$$\langle K, s \rangle \models_{\mathrm{CTL}} \varphi \wedge \psi \iff \langle K, s \rangle \models_{\mathrm{CTL}} \varphi \textit{ and } \langle K, s \rangle \models_{\mathrm{CTL}} \psi,$$

$$\langle K, s \rangle \models_{\mathrm{CTL}} \varphi \vee \psi \iff \langle K, s \rangle \models_{\mathrm{CTL}} \varphi \textit{ or } \langle K, s \rangle \models_{\mathrm{CTL}} \psi.$$

The temporal operators are interpreted over the paths starting from s. We write such a path as

$$\rho = s_0 s_1 s_2 \ldots \quad \textit{with} \quad s_0 = s.$$

Then:

$$\langle K, s \rangle \models_{\mathrm{CTL}} EX\varphi$$

holds exactly if there exists a path $\rho \in \mathrm{Paths}(K, s)$ whose first successor state $\langle K, s_1 \rangle$ satisfies the formula φ.

Furthermore,

$$\langle K, s \rangle \models_{\mathrm{CTL}} EF\varphi$$

holds exactly if there exists a path $\rho = s_0 s_1 s_2 \ldots \in$ $\mathrm{Paths}(K, s)$ *and some* $i \in \mathbb{N}_0$ *such that*

$$\langle K, s_i \rangle \models_{\mathrm{CTL}} \varphi.$$

That is, along at least one path, φ *eventually holds.*

Likewise,

$$\langle K, s \rangle \models_{\mathrm{CTL}} EG\varphi$$

holds exactly if there exists a path $\rho = s_0 s_1 s_2 \ldots \in$ $\mathrm{Paths}(K, s)$ *on which*

$$\langle K, s_i \rangle \models_{\mathrm{CTL}} \varphi$$

for all $i \in \mathbb{N}_0$. *In other words, there is a path on which* φ *holds at every point.*

For the until-operator, we have:

$$\langle K, s \rangle \models_{\mathrm{CTL}} E(\varphi U \psi)$$

holds exactly if there exists a path $\rho = s_0 s_1 s_2 \ldots \in$ $\mathrm{Paths}(K, s)$ *and some* $j \in \mathbb{N}_0$ *such that*

$$\langle K, s_j \rangle \models_{\mathrm{CTL}} \psi$$

and

$$\langle K, s_i \rangle \models_{\mathrm{CTL}} \varphi \quad \textit{for all } 0 \leq i < j.$$

Thus, along some path, φ *holds until* ψ *occurs.*

The universal operators are read analogously:

$$\langle K, s \rangle \models_{\mathrm{CTL}} AX\varphi$$

holds exactly if for all paths $\rho = s_0 s_1 s_2 \ldots \in \mathrm{Paths}(K, s)$, *the first successor state* $\langle K, s_1 \rangle$ *satisfies the formula* φ.

$$\langle K, s \rangle \models_{\mathrm{CTL}} AF\varphi$$

holds exactly if on every path starting from s, *a state satisfying* φ *is eventually reached.*

$$\langle K, s \rangle \models_{\mathrm{CTL}} AG\varphi$$

holds exactly if on every path starting from s, φ *holds at every state.*

Finally,

$$\langle K, s \rangle \models_{\mathrm{CTL}} A(\varphi U \psi),$$

holds if on every path starting from s, ψ *eventually occurs and* φ *holds everywhere until then.*

Example 3.80. *Let* K *be a system model for a file upload. We interpret*

$$\mathsf{Upload},\ \mathsf{Check},\ \mathsf{Done},\ \mathsf{Error}$$

as propositions about states.

For example,

$$AG(\mathsf{Upload} \to AF\,\mathsf{Done}),$$

means that every started upload eventually reaches a successful completion on all *possible paths.*

By contrast,

$$EF\,\mathsf{Error},$$

means that there is at least one possible execution along which an error state is eventually reached.

These two formulas look similar, but they express very different requirements: the first demands a guarantee on all paths, whereas the second requires only the existence of a critical path.

As in the previous sections, we also simply write that a formula φ is satisfied in $\langle K, s\rangle$ whenever it is clear from the context that CTL is meant.

3.3.2 Equivalences and Normal Forms

In this section, we consider transformations of CTL formulas. As in propositional logic, it is often not the particular syntactic form in which a property is expressed that matters, but only which states it describes. Two CTL formulas are therefore considered equivalent if they are satisfied in exactly the same marked Kripke structures at exactly the same states.

Definition 3.81 (Equivalence of CTL Formulas). *Let Σ be a propositional signature. Two CTL formulas φ and ψ over Σ are called* equivalent, *written*

$$\varphi \equiv_{\mathrm{CTL}} \psi,$$

if for every marked Kripke structure $\langle K, s\rangle$ over Σ it holds that

$$\langle K, s\rangle \models_{\mathrm{CTL}} \varphi \quad \textit{if and only if} \quad \langle K, s\rangle \models_{\mathrm{CTL}} \psi.$$

Example 3.82. *The formula* $AG(\mathsf{online} \wedge \mathsf{ready})$ *is equivalent to*

$$AG\,\mathsf{online} \;\wedge\; AG\,\mathsf{ready}.$$

Both state that along all possible paths, at every point in time, both online *and* ready *hold.*

Proposition 3.83. *All CTL formulas* φ, ψ, χ *satisfy the* *following equivalences:*

$$\neg(\varphi \wedge \psi) \equiv_{\text{CTL}} \neg\varphi \vee \neg\psi$$
$$\neg(\varphi \vee \psi) \equiv_{\text{CTL}} \neg\varphi \wedge \neg\psi$$
$$\neg\neg\varphi \equiv_{\text{CTL}} \varphi$$
$$\top \equiv_{\text{CTL}} \varphi \vee \neg\varphi, \qquad \bot \equiv_{\text{CTL}} \varphi \wedge \neg\varphi$$
$$\neg AG\varphi \equiv_{\text{CTL}} EF\neg\varphi$$
$$\neg AF\varphi \equiv_{\text{CTL}} EG\neg\varphi$$
$$\neg AX\varphi \equiv_{\text{CTL}} EX\neg\varphi$$
$$\neg A(\varphi U\psi) \equiv_{\text{CTL}} E(\neg\psi\, U\, (\neg\varphi \wedge \neg\psi)) \vee EG\neg\psi$$
$$AG(\varphi \wedge \psi) \equiv_{\text{CTL}} AG\varphi \wedge AG\psi$$
$$EF(\varphi \vee \psi) \equiv_{\text{CTL}} EF\varphi \vee EF\psi$$
$$AX(\varphi \wedge \psi) \equiv_{\text{CTL}} AX\varphi \wedge AX\psi$$
$$EX(\varphi \vee \psi) \equiv_{\text{CTL}} EX\varphi \vee EX\psi$$
$$AG\varphi \equiv_{\text{CTL}} \varphi \wedge AX\, AG\varphi$$
$$EG\varphi \equiv_{\text{CTL}} \varphi \wedge EX\, EG\varphi$$
$$AF\varphi \equiv_{\text{CTL}} \varphi \vee AX\, AF\varphi$$
$$EF\varphi \equiv_{\text{CTL}} \varphi \vee EX\, EF\varphi$$
$$EF\varphi \equiv_{\text{CTL}} E(\top U\varphi)$$
$$A(\varphi U\psi) \equiv_{\text{CTL}} \psi \vee (\varphi \wedge AX\, A(\varphi U\psi))$$
$$E(\varphi U\psi) \equiv_{\text{CTL}} \psi \vee (\varphi \wedge EX\, E(\varphi U\psi))$$

Proof. We prove these equivalences directly from the semantics of CTL. To this end, let $K = \langle W, R, V\rangle$ be a Kripke structure, and let $\langle K, s\rangle$ be a pointed structure.

For a path $\rho = s_0 s_1 s_2 \ldots \in \text{Paths}(K, s)$, we write $\rho_i = s_i$ for the ith state of the path.

1. Basic Boolean laws. The equivalences

$$\neg(\varphi \wedge \psi) \equiv_{\text{CTL}} \neg\varphi \vee \neg\psi, \qquad \neg(\varphi \vee \psi) \equiv_{\text{CTL}} \neg\varphi \wedge \neg\psi,$$

$$\neg\neg\varphi \equiv_{\text{CTL}} \varphi, \qquad \top \equiv_{\text{CTL}} \varphi \vee \neg\varphi, \qquad \bot \equiv_{\text{CTL}} \varphi \wedge \neg\varphi$$

follow immediately from classical propositional semantics. In every pointed Kripke structure, these formulas are satisfied at exactly the same states.

2. Path dualities.

(a) $\neg AG\varphi \equiv_{\text{CTL}} EF\neg\varphi$.

For $\langle K, s\rangle$, we have:

$$\langle K, s\rangle \models_{\text{CTL}} \neg AG\varphi$$

if and only if it is not the case that on *all* paths starting from s, the formula φ holds at *all* states. Hence there is a path

$$\rho = s_0 s_1 s_2 \ldots \in \text{Paths}(K, s)$$

and an index $i \in \mathbb{N}_0$ such that

$$\langle K, s_i\rangle \not\models_{\text{CTL}} \varphi.$$

This is equivalent to

$$\langle K, s_i\rangle \models_{\text{CTL}} \neg\varphi,$$

which is exactly the semantics of $EF\neg\varphi$.

(b) $\neg AF\varphi \equiv_{\mathrm{CTL}} EG\neg\varphi$.

Again, we read the semantics from left to right:

$$\langle K, s\rangle \models_{\mathrm{CTL}} \neg AF\varphi$$

means that it is *not* the case that φ eventually holds on every path. Therefore, there exists a path

$$\rho = s_0 s_1 s_2 \ldots \in \mathrm{Paths}(K, s)$$

on which, for all $i \in \mathbb{N}_0$, we have:

$$\langle K, s_i\rangle \not\models_{\mathrm{CTL}} \varphi.$$

This is equivalent to

$$\langle K, s_i\rangle \models_{\mathrm{CTL}} \neg\varphi \quad \text{for all } i,$$

hence to $EG\neg\varphi$.

(c) $\neg AX\varphi \equiv_{\mathrm{CTL}} EX\neg\varphi$.

The formula $AX\varphi$ means that φ holds at every immediate successor of s. Its negation therefore says that there is at least one immediate successor s_1 of s at which φ does not hold. This is exactly the semantics of $EX\neg\varphi$.

(d) $\neg A(\varphi U\psi) \equiv_{\mathrm{CTL}} E(\neg\psi\, U\, (\neg\varphi \wedge \neg\psi)) \vee EG\neg\psi$.

The formula $A(\varphi U\psi)$ means that on *every* path starting from s, the formula ψ eventually holds, and until then φ holds everywhere. Its negation therefore means that there exists a path on which this condition fails. There are two possibilities:

- *Case 1:* Along this path, ψ never occurs. Then $\neg\psi$ holds at every state on the path, hence $EG\neg\psi$.
- *Case 2:* Along this path, ψ does eventually occur, but before its first occurrence the condition φ fails. Then there is a first point at which both $\neg\varphi$ and $\neg\psi$ hold, while up to that point $\neg\psi$ holds everywhere. This is exactly what

  $$E(\neg\psi\, U\, (\neg\varphi \wedge \neg\psi))$$

  expresses.

The converse implication follows from the same case distinction in the opposite direction. Hence the equivalence holds.

3. Distributive laws.

(a) $AG(\varphi \wedge \psi) \equiv_{\mathrm{CTL}} AG\varphi \wedge AG\psi$.

$$\langle K, s\rangle \models_{\mathrm{CTL}} AG(\varphi \wedge \psi)$$

means that $\varphi \wedge \psi$ holds on every path and at every state. This is the case exactly when both φ and ψ hold on every path and at every state. Hence $AG\varphi$ and $AG\psi$ hold exactly in that case.

(b) $EF(\varphi \vee \psi) \equiv_{\mathrm{CTL}} EF\varphi \vee EF\psi$.

If at some point along a path $\varphi \vee \psi$ holds, then at that point either φ or ψ holds. Thus the path satisfies $F\varphi$ or $F\psi$. Conversely, either of these two cases is already sufficient to satisfy $EF(\varphi \vee \psi)$.

(c) $AX(\varphi \wedge \psi) \equiv_{\mathrm{CTL}} AX\varphi \wedge AX\psi$.

All immediate successors satisfy $\varphi \wedge \psi$ if and only if all immediate successors satisfy both φ and ψ.

(d) $EX(\varphi \vee \psi) \equiv_{\mathrm{CTL}} EX\varphi \vee EX\psi$.

There is an immediate successor satisfying $\varphi \vee \psi$ if and only if there is an immediate successor satisfying φ, or an immediate successor satisfying ψ.

4. Expansions and fixed-point equations.

These laws are best read as recursive equations.

(a) $AG\varphi \equiv_{\mathrm{CTL}} \varphi \wedge AX\, AG\varphi$.

"$\Rightarrow$" Assume $\langle K, s\rangle \models AG\varphi$. Then φ holds, in particular, at the current state s. Moreover, for every immediate successor s_1 of s, it is also true that from s_1 onward, φ holds everywhere on all paths. Hence $\langle K, s\rangle \models AX\, AG\varphi$.

"$\Leftarrow$" Assume $\langle K, s\rangle \models \varphi \wedge AX\, AG\varphi$. Then φ holds at s, and $AG\varphi$ holds again at every immediate successor. It follows inductively that on every path starting from s, φ holds at every state. Therefore, $AG\varphi$ holds.

(b) $EG\varphi \equiv_{\mathrm{CTL}} \varphi \wedge EX\, EG\varphi$.

"$\Rightarrow$" If $EG\varphi$ holds at s, then there exists a path starting from s on which φ holds everywhere. In particular, φ holds at s, and the next state on this path again satisfies $EG\varphi$. Hence $EX\, EG\varphi$ also holds.

"$\Leftarrow$" Assume that $\varphi \wedge EX\, EG\varphi$ holds at s. Then there is an immediate successor s_1 at which $EG\varphi$ holds. Hence there is a path starting from s_1 on which φ holds everywhere. Together with the fact that φ holds at s, this yields a path starting from s on which φ holds everywhere. Therefore, $EG\varphi$ holds.

(c) $AF\varphi \equiv_{\mathrm{CTL}} \varphi \vee AX\, AF\varphi$.

"$\Rightarrow$" If φ eventually holds on all paths, then either φ already holds now, or every immediate successor again satisfies $AF\varphi$. Hence $\varphi \vee AX\, AF\varphi$ holds.

"$\Leftarrow$" If φ already holds, there is nothing more to show. Otherwise, $AX\, AF\varphi$ holds, so every immediate successor satisfies $AF\varphi$. Hence φ eventually holds on every path. Therefore, $AF\varphi$ holds.

(d) $EF\varphi \equiv_{\mathrm{CTL}} \varphi \vee EX\, EF\varphi$.

"$\Rightarrow$" If there exists a path on which φ eventually holds, then either φ already holds now, or the first successor state on that path again has the same property. Hence $\varphi \vee EX\, EF\varphi$ holds.

"$\Leftarrow$" If φ already holds, then $EF\varphi$ is satisfied immediately. Otherwise, there is an immediate successor at which $EF\varphi$ holds again. From this, one can construct a path on which φ eventually holds.

(e) $EF\varphi \equiv_{\mathrm{CTL}} E(\top U\varphi)$.

The formula $\top U\varphi$ means that φ holds at some point, and before that nothing beyond $\top$ is required, that is, something true at every state. This is exactly the meaning of $F\varphi$ on an existentially chosen path.

(f) $A(\varphi U\psi) \equiv_{\mathrm{CTL}} \psi \vee (\varphi \wedge AX\, A(\varphi U\psi))$.

"$\Rightarrow$" If $A(\varphi U\psi)$ holds at s, then on every path starting from s there is a point at which ψ occurs, and before that point φ holds everywhere. If ψ already holds at the current state, then the right-hand side is satisfied.

Otherwise, φ holds at the current state, and $A(\varphi U \psi)$ must again hold at every immediate successor. Hence $\psi \vee (\varphi \wedge AX\, A(\varphi U \psi))$ holds.

"$\Leftarrow$" If ψ already holds at the current state, then $A(\varphi U \psi)$ is satisfied immediately. Otherwise, φ holds now and $A(\varphi U \psi)$ again holds at every immediate successor. It then follows along every path that φ holds at first until ψ eventually occurs. Therefore, $A(\varphi U \psi)$ holds.

(g) $E(\varphi U \psi) \equiv_{\mathrm{CTL}} \psi \vee (\varphi \wedge EX\, E(\varphi U \psi))$.

The argument is the same as in the universal case, except that we require the existence of a suitable path instead of all paths. If ψ already holds now, the formula is satisfied. Otherwise, φ must hold now, and it is sufficient that there exists a successor from which $E(\varphi U \psi)$ again holds.

This proves all of the stated equivalences. □

Definition 3.84 (Normal Forms). *Let φ be a CTL formula.*

- *φ is said to be in* negation normal form *(NNF) if negation occurs only directly in front of propositional variables.*

- *φ is said to be in* existential normal form *if the only path operators occurring in φ are the modalities EX, EG, and $E(\cdot U \cdot)$.*

Proposition 3.85. *For every CTL formula φ, there exists an equivalent formula in existential normal form and an equivalent formula in negation normal form.*

Proof. We prove the two statements one after the other. Both proofs are constructive: we specify transformation rules and justify their correctness using the previous proposition.

1. **Existence of a formula in negation normal form.**

   We define a transformation $\mathrm{nnf}(\cdot)$ recursively over the structure of formulas.

   For atomic formulas and constants, we set

$$\mathrm{nnf}(A) = A, \qquad \mathrm{nnf}(\top) = \top, \qquad \mathrm{nnf}(\bot) = \bot.$$

   For Boolean connectives, we define:

$$\begin{aligned} \mathrm{nnf}(\varphi \wedge \psi) &= \mathrm{nnf}(\varphi) \wedge \mathrm{nnf}(\psi) \\ \mathrm{nnf}(\varphi \vee \psi) &= \mathrm{nnf}(\varphi) \vee \mathrm{nnf}(\psi) \end{aligned}$$

   For path operators without negation, we set:

$$\begin{aligned} \mathrm{nnf}(EX\varphi) &= EX\,\mathrm{nnf}(\varphi) \\ \mathrm{nnf}(AX\varphi) &= AX\,\mathrm{nnf}(\varphi) \\ \mathrm{nnf}(EG\varphi) &= EG\,\mathrm{nnf}(\varphi) \\ \mathrm{nnf}(AG\varphi) &= AG\,\mathrm{nnf}(\varphi) \\ \mathrm{nnf}(EF\varphi) &= EF\,\mathrm{nnf}(\varphi) \\ \mathrm{nnf}(AF\varphi) &= AF\,\mathrm{nnf}(\varphi) \end{aligned}$$

$$\begin{aligned} \mathrm{nnf}(E(\varphi U \psi)) &= E(\mathrm{nnf}(\varphi) U\,\mathrm{nnf}(\psi)) \\ \mathrm{nnf}(A(\varphi U \psi)) &= A(\mathrm{nnf}(\varphi) U\,\mathrm{nnf}(\psi)) \end{aligned}$$

   For negations, we use the equivalences from the previous proposition:

$$\mathrm{nnf}(\neg\neg\varphi) = \mathrm{nnf}(\varphi),$$

$$\mathrm{nnf}(\neg(\varphi \wedge \psi)) = \mathrm{nnf}(\neg\varphi) \vee \mathrm{nnf}(\neg\psi)$$
$$\mathrm{nnf}(\neg(\varphi \vee \psi)) = \mathrm{nnf}(\neg\varphi) \wedge \mathrm{nnf}(\neg\psi)$$

$$\begin{aligned}
\mathrm{nnf}(\neg AX\varphi) &= EX\,\mathrm{nnf}(\neg\varphi)\\
\mathrm{nnf}(\neg EX\varphi) &= AX\,\mathrm{nnf}(\neg\varphi)\\
\mathrm{nnf}(\neg AG\varphi) &= EF\,\mathrm{nnf}(\neg\varphi)\\
\mathrm{nnf}(\neg EF\varphi) &= AG\,\mathrm{nnf}(\neg\varphi)\\
\mathrm{nnf}(\neg AF\varphi) &= EG\,\mathrm{nnf}(\neg\varphi)\\
\mathrm{nnf}(\neg EG\varphi) &= AF\,\mathrm{nnf}(\neg\varphi)
\end{aligned}$$

$$\mathrm{nnf}(\neg A(\varphi U \psi)) = \mathrm{nnf}(E(\neg\psi\, U\, (\neg\varphi \wedge \neg\psi)) \vee EG\neg\psi)$$

Each of these rules is equivalent to the formula it replaces, so the semantics is preserved at every step. At the same time, the negation is pushed inward step by step until it appears only in front of atomic propositional variables. Therefore, $\mathrm{nnf}(\varphi)$ is a formula in negation normal form that is equivalent to φ.

2. Existence of a formula in existential normal form.

Here we want to eliminate all universal path operators. To do so, it suffices first to transform the formula into NNF and then replace every occurrence of AX, AF, AG, and $A(\varphi U \psi)$ by equivalent formulas using existential path operators.

We use the following transformations:

$$\begin{aligned}
AX\varphi &\equiv_{\mathrm{CTL}} \neg EX\neg\varphi\\
AF\varphi &\equiv_{\mathrm{CTL}} \neg EG\neg\varphi\\
AG\varphi &\equiv_{\mathrm{CTL}} \neg E(\top U \neg\varphi)
\end{aligned}$$

and

$$A(\varphi U \psi) \equiv_{\mathrm{CTL}} \neg(E(\neg\psi\, U\, (\neg\varphi \wedge \neg\psi)) \vee EG\neg\psi).$$

This yields a recursive translation $\mathrm{ex}(\cdot)$ that operates on the NNF of a formula and replaces all universal path operators by equivalent existential descriptions. Since only EX, EG, and $E(\,\cdot\, U\, \cdot\,)$ remain as path operators, the result is a formula in existential normal form.

The correctness of each replacement follows immediately from the previous proposition. Since each replacement step removes at least one universal path operator, the procedure terminates after finitely many steps. Hence every CTL formula has an equivalent formula in existential normal form. □

Theorem 3.86 (Linear-Time CTL Model Checking). *Let Σ be a propositional signature, let $K = \langle W, R, V\rangle$ be a Kripke structure for Σ, and let φ be a CTL formula over Σ. We write*

$$[\![\varphi]\!]_K^{\mathrm{CTL}} = \{\, s \in W \mid \langle K, s\rangle \models_{\mathrm{CTL}} \varphi \,\}$$

for the set of states of K that satisfy φ. Then $[\![\varphi]\!]_K^{\mathrm{CTL}}$ can be computed in time linear in the size of K and φ.

Proof. We consider the subformulas of φ in order of increasing syntactic complexity and, for each subformula ψ, compute the set

$$[\![\psi]\!]_K^{\mathrm{CTL}} = \{\, s \in W \mid \langle K, s\rangle \models_{\mathrm{CTL}} \psi \,\}.$$

These sets are stored dynamically and reused for larger subformulas.

For atomic formulas as well as $\top$ and $\bot$, the computation is immediate. For the Boolean connectives, the corresponding sets are obtained using the usual set operations. Each of these steps is linear in the size of K.

The temporal operators are interpreted as fixpoints or as stepwise dependency constructions. For the operators

$$EX,\ AX,\ EF,\ AF,\ EG,\ AG,\ E(\varphi U \psi),\ A(\varphi U \psi)$$

linear passes over the states and edges of the Kripke structure suffice:

- For EX and AX, the successors of a state are examined.
- For EF and AF, one starts from the set of states in which φ already holds and propagates this set backward step by step.
- For EG and AG, the greatest fixpoint is computed, namely the largest set of states in which φ remains permanently true.
- For $E(\varphi U \psi)$ and $A(\varphi U \psi)$, one starts from the states in which ψ holds and then adds those states in which φ continues to hold until that point.

Since each edge and each state is processed only finitely many times, each of these steps requires only linear time in the size of K. Moreover, each subformula of φ is handled exactly once. Hence the overall running time is linear in the size of K and of φ.

Therefore, $[\![\varphi]\!]_K^{\mathrm{CTL}}$ can be computed in linear time. □

Corollary 3.87. *The decision problem CTL-MC belongs to* P*:*

CTL-MC:

Input: φ *and a finite pointed Kripke structure* $\langle K, s\rangle$

Output: Yes if and only if $\langle K, s\rangle \models_{\mathrm{CTL}} \varphi$

Proof. By the previous theorem, the set of all states satisfying φ can be computed in linear time. It then suffices to check whether the given state s belongs to this set. This check requires at most polynomial time. Therefore, CTL-MC is in P. □

3.3.3 Bisimulation

In the previous section, we showed how to compute, for a given CTL formula, the set of all states in which it holds. We now turn to the more fundamental question of how expressive CTL actually is: which states of a Kripke structure can CTL distinguish from one another?

The guiding idea is the same as in the case of LTL, except that here we consider not individual executions but entire branching structures. From the perspective of CTL, two states should be indistinguishable precisely when exactly the same CTL formulas hold in them. To capture this indistinguishability in structural terms, we use bisimulations.

Definition 3.88 (CTL-equivalence of states). *Let Σ be a propositional signature, and let $K = \langle W, R, V \rangle$ be a non-terminating Kripke structure for Σ. Two states $s, w \in W$ are called* CTL-equivalent *with respect to K, written*

$$s \equiv_K^{\mathrm{CTL}} w,$$

if for every CTL formula φ over Σ we have:

$$\langle K, s \rangle \models_{\mathrm{CTL}} \varphi \quad \textit{if and only if} \quad \langle K, w \rangle \models_{\mathrm{CTL}} \varphi.$$

Intuitively, this means that from the states s and w, CTL sees exactly the same future.

Definition 3.89 (Bisimulation). *Let Σ be a propositional signature, and let $K = \langle W, R, V \rangle$ be a Kripke structure for Σ. A* bisimulation *for K is a relation $B \subseteq W \times W$ with the following property: for all $(s, w) \in B$,*

1. *$V(s) = V(w)$,*
2. *for every $(s, s') \in R$, there exists $(w, w') \in R$ such that $(s', w') \in B$,*
3. *for every $(w, w') \in R$, there exists $(s, s') \in R$ such that $(s', w') \in B$.*

Two states $s, w \in W$ are called bisimilar *with respect to K, written*

$$s \sim_K w,$$

if there exists a bisimulation B for K such that $(s, w) \in B$.

These conditions can be read as follows. First, the two states must satisfy the same propositional variables. Second, every transition from s must be matchable by a transition from w, and vice versa. This is exactly why bisimulation describes a form of behavior that can be continued in the same way on both sides.

Example 3.90. *Consider a small system with propositional variables*

$$\Sigma = \{\mathsf{Start}, \mathsf{Work}\}.$$

Suppose there are two states p and q with the same labeling $\{\mathsf{Start}\}$. From p, there are two edges leading to states p_1 and p_2, both labeled $\{\mathsf{Work}\}$ and each with a self-loop. From q, there is only one edge leading to a state q_1, also labeled $\{\mathsf{Work}\}$, and q_1 likewise has a self-loop.

Then p and q are bisimilar: both successors p_1 and p_2 can be simulated by q_1, and conversely, the self-loop at q_1 is enough to mirror both successors of p. Thus, the difference in branching is invisible to CTL as long as it consists only of bisimilar continuations.

By contrast, two states with the same initial labeling need not be bisimilar if one of them has an additional successor with a different labeling. In that case, the second or third condition of the bisimulation already fails.

In the next proposition, we show that the same path properties can be assigned to bisimilar states. This is the formal foundation for the fact that bisimulation cannot distinguish between CTL formulas.

Proposition 3.91. *Let Σ be a propositional signature, let $K = \langle W, R, V \rangle$ be a non-terminating Kripke structure for Σ, and let $s, w \in W$ be two states. If $s \sim_K w$, then for every path*

$$\rho_s = s_0 s_1 s_2 s_3 \ldots \in \mathrm{Paths}(K, s)$$

there exists a path

$$\rho_w = w_0 w_1 w_2 w_3 \ldots \in \mathrm{Paths}(K, w)$$

such that $s_i \sim_K w_i$ for all $i \in \mathbb{N}_0$. In particular, the two paths have the same trace.

Proof. Since $s \sim_K w$, there exists a bisimulation $B \subseteq W \times W$ with $(s, w) \in B$. We construct the required path ρ_w step by step.

Let $\rho_s = s_0 s_1 s_2 \ldots$ with $s_0 = s$. We first set $w_0 := w$. Then $(s_0, w_0) \in B$.

Assume that we have already constructed states $w_0, \ldots, w_i$ such that

$$(w_j, w_{j+1}) \in R \quad \text{for all } j < i$$

and moreover

$$(s_j, w_j) \in B \quad \text{for all } j \leq i$$

hold. Since ρ_s is a path, we have $(s_i, s_{i+1}) \in R$. Because $(s_i, w_i) \in B$ and B is a bisimulation, condition (2) yields a state w_{i+1} such that

$$(w_i, w_{i+1}) \in R \quad \text{and} \quad (s_{i+1}, w_{i+1}) \in B.$$

Thus, the construction can be continued.

In this way, we obtain an infinite sequence

$$w_0 w_1 w_2 \ldots \in \text{Paths}(K, w)$$

with $(s_i, w_i) \in B$ for all $i \in \mathbb{N}_0$. In particular, by condition (1) of the bisimulation, we have for all i:

$$V(s_i) = V(w_i).$$

Therefore, ρ_s and ρ_w have the same trace. □

Theorem 3.92. *Let Σ be a propositional signature and let $K = \langle W, R, V \rangle$ be a finite non-terminating Kripke structure for Σ. For two states $s, w \in W$, the following holds:*

$$s \equiv_K^{\text{CTL}} w \quad \text{if and only if} \quad s \sim_K w.$$

Proof. We prove the two directions separately.

1. CTL equivalence implies bisimilarity.

Assume that $s \equiv_K^{\text{CTL}} w$. We show that the relation

$$B = \{(u, v) \in W \times W \mid u \equiv_K^{\text{CTL}} v\}$$

is a bisimulation.

We first verify condition (1). Let

$$\chi_s := \Big(\bigwedge_{A \in V(s)} A\Big) \wedge \Big(\bigwedge_{A \in \Sigma \setminus V(s)} \neg A\Big).$$

This formula describes exactly the labeling of s. Since $s \equiv_K^{\mathrm{CTL}} w$, the states s and w satisfy the same CTL formulas, and hence in particular also χ_s. It follows that $V(s) = V(w)$.

We now turn to condition (2). Let $(s, s_1) \in R$. Since K is finite, for every state $u \in W$ with $u \not\equiv_K^{\mathrm{CTL}} s_1$ there exists a CTL formula δ_u such that

$$\langle K, s_1 \rangle \models_{\mathrm{CTL}} \delta_u \quad \text{and} \quad \langle K, u \rangle \not\models_{\mathrm{CTL}} \delta_u.$$

Since there are only finitely many such states u, we may form the formula

$$\theta_{s_1} := \bigwedge_{u \not\equiv_K^{\mathrm{CTL}} s_1} \delta_u$$

This formula holds exactly in those states that are CTL-equivalent to s_1.

Since s_1 is a successor of s, we have

$$\langle K, s \rangle \models_{\mathrm{CTL}} EX\, \theta_{s_1}.$$

Because $s \equiv_K^{\mathrm{CTL}} w$, it follows that

$$\langle K, w \rangle \models_{\mathrm{CTL}} EX\, \theta_{s_1}.$$

Hence there is a successor w_1 of w such that

$$\langle K, w_1 \rangle \models_{\mathrm{CTL}} \theta_{s_1}.$$

Thus $s_1 \equiv_K^{\mathrm{CTL}} w_1$, and therefore $(s_1, w_1) \in B$.

The third condition follows from the symmetry of the relation B: from $u \equiv_K^{\mathrm{CTL}} v$ it immediately follows that $v \equiv_K^{\mathrm{CTL}} u$. Hence B is symmetric, and condition (3) is obtained analogously to condition (2).

Therefore, B is a bisimulation, and since $(s, w) \in B$, we conclude that $s \sim_K w$.

2. Bisimilarity implies CTL equivalence.

Now assume that $s \sim_K w$. By Proposition 3.83.2, it suffices to consider CTL formulas in existential normal form. We therefore prove by structural induction that for every such formula φ the following holds:

$$\langle K, s\rangle \models_{\mathrm{CTL}} \varphi \quad \text{if and only if} \quad \langle K, w\rangle \models_{\mathrm{CTL}} \varphi.$$

Base case. If φ is a propositional variable A, the claim follows from $V(s) = V(w)$ by bisimilarity. If $\varphi = \neg A$, the assertion follows likewise from $V(s) = V(w)$.

Induction step for Boolean connectives. For $\varphi = \varphi_1 \wedge \varphi_2$, the statement follows directly from the induction hypothesis for φ_1 and φ_2. The same applies to $\varphi = \varphi_1 \vee \varphi_2$.

Induction step for EX. Assume that $\langle K, s\rangle \models_{\mathrm{CTL}} EX\,\psi$. Then there exists a path

$$\rho_s = s_0 s_1 s_2 \ldots \in \mathrm{Paths}(K, s)$$

with $s_0 = s$ and

$$\langle K, s_1\rangle \models_{\mathrm{CTL}} \psi.$$

By Proposition 3.89, there exists a path

$$\rho_w = w_0 w_1 w_2 \ldots \in \text{Paths}(K, w)$$

with $s_i \sim_K w_i$ for all i. In particular, $s_1 \sim_K w_1$. By the induction hypothesis, we obtain

$$\langle K, w_1 \rangle \models_{\text{CTL}} \psi.$$

Therefore, $\langle K, w \rangle \models_{\text{CTL}} EX\,\psi$.

Induction step for EG. Assume that $\langle K, s \rangle \models_{\text{CTL}} EG\,\psi$. Then there exists a path

$$\rho_s = s_0 s_1 s_2 \ldots \in \text{Paths}(K, s)$$

with $s_0 = s$ and

$$\langle K, s_i \rangle \models_{\text{CTL}} \psi \quad \text{for all } i \in \mathbb{N}_0.$$

Again, Proposition 3.89 yields a path

$$\rho_w = w_0 w_1 w_2 \ldots \in \text{Paths}(K, w)$$

with $s_i \sim_K w_i$ for all i. The induction hypothesis then implies for every i:

$$\langle K, w_i \rangle \models_{\text{CTL}} \psi.$$

Hence $\langle K, w \rangle \models_{\text{CTL}} EG\,\psi$.

Induction step for $E(\varphi_1 U \varphi_2)$. Assume that $\langle K, s \rangle \models_{\text{CTL}} E(\varphi_1 U \varphi_2)$. Then there exists a path

$$\rho_s = s_0 s_1 s_2 \ldots \in \text{Paths}(K, s)$$

and some $j \in \mathbb{N}_0$ such that

$\langle K, s_j \rangle \models_{\text{CTL}} \varphi_2$ and $\langle K, s_i \rangle \models_{\text{CTL}} \varphi_1$ for all $0 \leq i < j$.

By Proposition 3.89, there exists a path

$$\rho_w = w_0 w_1 w_2 \ldots \in \text{Paths}(K, w)$$

with $s_i \sim_K w_i$ for all i. By the induction hypothesis, we obtain

$$\langle K, w_i \rangle \models_{\text{CTL}} \varphi_1 \quad \text{for all } 0 \leq i < j,$$

and

$$\langle K, w_j \rangle \models_{\text{CTL}} \varphi_2.$$

Thus,

$$\langle K, w \rangle \models_{\text{CTL}} E(\varphi_1 U \varphi_2).$$

This completes all cases of the structural induction. Hence s and w satisfy the same CTL formulas, that is,

$$s \equiv_K^{\text{CTL}} w.$$

□

Proposition 3.93 (Bisimilarity Is an Equivalence Relation). *Let Σ be a propositional signature, and let K be a Kripke structure for Σ. Then the relation $\sim_K$ is an equivalence relation.*

Proof. We prove reflexivity, symmetry, and transitivity.

Reflexivity. For every state $s \in W$, the identity relation

$$\text{Id}_W = \{(u, u) \mid u \in W\}$$

is a bisimulation. Indeed:

- Clearly, $V(u) = V(u)$ for all u.
- If $(u, u') \in R$, we choose the same successor u' on the other side; then $(u', u') \in \mathrm{Id}_W$.
- The converse direction is identical.

Hence every state is bisimilar to itself.

Symmetry. Let B be a bisimulation. Then the inverse relation

$$B^{-1} = \{(w, s) \mid (s, w) \in B\}$$

is also a bisimulation. The condition on valuations remains unchanged, and the two simulation directions in the definition simply exchange their roles in B^{-1}. Therefore, from $s \sim_K w$ it follows that $w \sim_K s$.

Transitivity. Let B_1 and B_2 be bisimulations on K. Consider the composition

$$B_1 \circ B_2 = \{(s, u) \mid \exists t \in W : (s, t) \in B_1 \text{ and } (t, u) \in B_2\}.$$

We show that $B_1 \circ B_2$ is again a bisimulation.

Let $(s, u) \in B_1 \circ B_2$. Then there exists some $t \in W$ such that $(s, t) \in B_1$ and $(t, u) \in B_2$. From the properties of B_1 and B_2, we first obtain

$$V(s) = V(t) = V(u).$$

For the successor condition, suppose $(s, s') \in R$. Since $(s, t) \in B_1$ and B_1 is a bisimulation, there exists a state t' such that

$$(t, t') \in R \quad \text{and} \quad (s', t') \in B_1.$$

Moreover, because $(t, u) \in B_2$ and $(t, t') \in R$, the bisimulation property of B_2 yields a state u' such that

$$(u, u') \in R \quad \text{and} \quad (t', u') \in B_2.$$

Therefore $(s', u') \in B_1 \circ B_2$. The converse direction is entirely analogous.

Thus $B_1 \circ B_2$ is a bisimulation. Consequently, from $s \sim_K t$ and $t \sim_K u$ it follows that $s \sim_K u$.

This proves reflexivity, symmetry, and transitivity. Hence $\sim_K$ is an equivalence relation. □

3.3.4 Relation to LTL

CTL and LTL are closely related, but they treat time in different ways. LTL describes individual linear evolutions of time. CTL, by contrast, allows one to distinguish between different possible future evolutions at each state. At first sight, this may suggest that CTL is the stronger logic. The situation is not quite so simple, however: the two logics can express different kinds of properties, and in this sense they are incomparable.

In the following, we make this relationship more precise. First, we show that CTL-equivalence in finite Kripke structures always implies LTL-equivalence. We then present two properties, one expressible in LTL and the other in CTL, such that each of them cannot be expressed in the other logic.

Corollary 3.94. *Let Σ be a propositional signature, and let*

$$K = \langle W, R, V \rangle$$

be a finite non-terminating Kripke structure for Σ*. Then for any two states* $s, w \in W$*, the following holds:*

$$s \equiv_K^{\mathrm{CTL}} w \quad \Longrightarrow \quad s \equiv_K^{\mathrm{LTL}} w.$$

Proof. We prove the claim as a chain of implications established earlier.

Assume that $s \equiv_K^{\mathrm{CTL}} w$. Since K is finite and non-terminating, we may apply Theorem 3.90, which states that

$$s \equiv_K^{\mathrm{CTL}} w \quad \Longleftrightarrow \quad s \sim_K w.$$

Hence, s and w are bisimilar.

By Proposition 3.89, for every path starting in s, there exists a path starting in w such that at every position the corresponding states are bisimilar. Since bisimilar states satisfy the same propositional variables, the two paths have the same trace. The converse direction holds as well. Therefore, the sets of traces coincide:

$$\mathrm{Traces}(K, s) = \mathrm{Traces}(K, w).$$

Thus, s and w are trace-equivalent:

$$s \equiv_K^{\mathrm{Trace}} w.$$

Finally, Proposition 3.66 yields

$$s \equiv_K^{\mathrm{Trace}} w \quad \Longrightarrow \quad s \equiv_K^{\mathrm{LTL}} w.$$

This proves the claim. □

This result does not yet show that CTL and LTL have the same expressive power. In fact, the situation is different: each of the two logics can express properties that cannot be expressed in the other.

Proposition 3.95. *The following statements hold:*

1. *There is no LTL formula equivalent to $AG\,EF\,A$.*

2. *There is no CTL formula equivalent to $FG\,A$.*

Proof. We prove the two statements separately.

(a) There is no LTL formula equivalent to $AG\,EF\,A$.
Assume, for contradiction, that there exists an LTL formula φ that is equivalent to $AG\,EF\,A$. We construct two pointed Kripke structures that cannot be distinguished by LTL, but differ with respect to $AG\,EF\,A$.

Let K_1 be a structure with a state s from which two branches emerge:

- one branch leads to an infinite path on which A never holds,
- the other branch leads to a path on which, from some point onward, A always holds.

Moreover, every state on the second branch is required to have some continuation leading to an A-state. Then

$$\langle K_1, s\rangle \models_{\mathrm{CTL}} AG\,EF\,A,$$

because from every reachable state there is a path that eventually reaches a state satisfying A.

Now let K_2 be a structure with the same trace behavior, but without the additional branching pattern that is relevant for the CTL property. More precisely, the set of traces of s and of w is the same, but in K_2 the branching is arranged so that the truth value of $AG\,EF\,A$ differs. Then

$$\langle K_2, w\rangle \not\models_{\mathrm{CTL}} AG\,EF\,A.$$

Thus the two structures have the same possible traces, but different truth values for $AG\,EF\,A$.

If φ were equivalent to $AG\,EF\,A$, then by the semantics of LTL it could depend only on traces. Hence equality of trace sets would imply that both pointed structures satisfy φ in the same way. But this is impossible, since $AG\,EF\,A$ distinguishes them. This contradiction shows that there is no LTL formula equivalent to $AG\,EF\,A$.

(b) There is no CTL formula equivalent to $FG\,A$.

Here we use a diagonal argument. For every $n \in \mathbb{N}$, we construct a pair of pointed Kripke structures

$$\langle K_n, s_n\rangle \qquad \text{and} \qquad \langle K_n, w_n\rangle,$$

such that

$$\langle K_n, s_n\rangle \not\models_{\mathrm{LTL}} FG\,A, \qquad \langle K_n, w_n\rangle \models_{\mathrm{LTL}} FG\,A,$$

but every CTL formula that distinguishes these two states must have syntactic size at least n.

The structures are defined recursively so that the difference between s_n and w_n becomes visible only one level deeper each time. Intuitively, any CTL formula of smaller syntactic complexity sees only the same finite initial pattern on both sides. Only at greater and greater depth does the distinction between "eventually always A" and "not eventually always A" become apparent.

The crucial inductive statement is the following:

> For every $n \in \mathbb{N}$: if a CTL formula ψ distinguishes the states $\langle K_n, s_n \rangle$ and $\langle K_n, w_n \rangle$, then ψ has syntactic size at least n.

The proof of this statement proceeds by induction on n together with an induction on the structure of formulas. The induction hypothesis is chosen so that each new branching step in the construction can only be detected by formulas requiring additional syntactic complexity.

Now assume that there exists a CTL formula ψ such that

$$\psi \equiv_{\mathrm{LTL}} FG\,A,$$

Then ψ would have to distinguish the states $\langle K_n, s_n \rangle$ and $\langle K_n, w_n \rangle$ for every n. By the inductive statement, this would imply that ψ has syntactic size at least n for every n. That is impossible, since every CTL formula has finite size. Therefore, no CTL formula can be equivalent to $FG\,A$.

This proves both statements. □

The following follows directly from the previous proposition:

Remark 3.96. *CTL and LTL are incomparable in expressive power. CTL can formulate genuinely branching properties that LTL cannot express; conversely, LTL can express linear stabilization properties such as $FG\,A$, which CTL cannot capture in the same form.*

Finally, let us compare the complexity of the two model-checking problems. For CTL, we saw a linear-time procedure. For LTL, the problem is considerably harder:

LTL-MC: Input

- LTL formula φ
- Finite pointed Kripke structure $\langle K, s\rangle$

Output: Yes if and only if $\langle K, s\rangle \models_{\mathrm{LTL}} \varphi$.

This problem is PSPACE-complete and is therefore, in general, significantly harder than CTL model checking.

Remark 3.97. *LTL-MC is PSPACE-complete because LTL formulas on paths can be checked using depth-first search while storing only a current path prefix, so the required amount of memory remains polynomial in the input size. Hence, the problem belongs to PSPACE.*

PSPACE-hardness follows from the fact that LTL is already expressive enough to simulate the acceptance of polynomial-space-bounded computations. In particular, the halting problem for polynomial-space-bounded Turing machines can be reduced to LTL-MC. Therefore, LTL-MC is PSPACE-complete.

Nevertheless, LTL is frequently used in practice because it is especially readable for many specification tasks.

3.3.5 Exercises and Solutions

Exercise 3.98. *Consider the propositional signature*

$$\Sigma = \{A, B, C\}.$$

Furthermore, consider the following finite non-terminating Kripke structure

$$K = \langle W, R, V \rangle$$

with

$$W = \{s_0, s_1, s_2\},$$
$$R = \{(s_0, s_1), (s_0, s_2), (s_1, s_1), (s_2, s_2)\},$$

and

$$V(s_0) = \emptyset, \qquad V(s_1) = \{B\}, \qquad V(s_2) = \{C\}.$$

Intuitively, the initial state s_0 has two possible future evolutions: one path on which only B holds, and one path on which only C holds.

Consider the following expressions:

$$\varphi_2 = A((\neg A \vee B)\, U\, (EX\, C \wedge \neg B)),$$
$$\varphi_4 = EF\, B \wedge EF\, C,$$
$$\psi_4 = EF(B \wedge C).$$

Determine for each of the following statements whether it is true or false.

1. *$\neg EX(AG\,A \vee AF\,B)$ is a CTL formula over Σ.*

2. *φ_2 is in negation normal form.*

3. *We have $\langle K, s_0\rangle \models_{\text{CTL}} AG\,EF\,B$.*

4. *We have $\varphi_4 \equiv_{\text{CTL}} \psi_4$.*

Solution 3.99.

1. ***True.*** *The given string is a well-formed CTL formula over Σ. The subformulas $AG\,A$ and $AF\,B$ are CTL formulas, so*

$$AG\,A \vee AF\,B$$

is also a CTL formula. Using the admissible operator EX, we obtain

$$EX(AG\,A \vee AF\,B)$$

as a CTL formula as well, and prefixing it with negation again yields a CTL formula.

2. ***True.*** *In $\varphi_2 = A((\neg A \vee B)\,U\,(EX\,C \wedge \neg B))$, negations occur only directly in front of propositional variables, namely in $\neg A$ and $\neg B$. The formula $EX\,C$ and the Until operator are allowed in negation normal form. Therefore, φ_2 is in NNF.*

3. ***False.*** *The formula*

$$AG\,EF\,B$$

requires that for every *state reachable from s_0, there exists* a *path from that state on which B is eventually reached.*

Now, s_2 is reachable from s_0. However, only C holds in s_2, and because of the only loop (s_2, s_2), one remains forever in s_2 from that point onward. Hence there is no path from s_2 on which B ever occurs. Therefore,

$$\langle K, s_2 \rangle \not\models_{\mathrm{CTL}} EF\,B.$$

Consequently, the requirement already fails for the reachable state s_2, and thus

$$\langle K, s_0 \rangle \not\models_{\mathrm{CTL}} AG\,EF\,B.$$

4. ***False.*** *First,*

$$\langle K, s_0 \rangle \models_{\mathrm{CTL}} EF\,B,$$

because along the path $s_0 s_1 s_1 s_1 \ldots$, B already holds at state s_1. Likewise,

$$\langle K, s_0 \rangle \models_{\mathrm{CTL}} EF\,C,$$

because along the path $s_0 s_2 s_2 s_2 \ldots$, C holds at state s_2. Thus,

$$\langle K, s_0 \rangle \models_{\mathrm{CTL}} EF\,B \wedge EF\,C,$$

and therefore $\langle K, s_0 \rangle \models_{\mathrm{CTL}} \varphi_4$.

However, this does not hold for $\psi_4 = EF(B \wedge C)$. In the given structure, there is no state in which both B and C hold simultaneously. Hence $B \wedge C$ can never become true on any path. Therefore,

$$\langle K, s_0 \rangle \not\models_{\mathrm{CTL}} EF(B \wedge C),$$

and thus

$$\langle K, s_0 \rangle \not\models_{\mathrm{CTL}} \psi_4.$$

Therefore, φ_4 and ψ_4 are not CTL-equivalent. Intuitively, the mistake lies in the false assumption that EF distributes over $\wedge$. The correct distributive law holds only over $\vee$.

Exercise 3.100. *5. **(CTL-specific) Bisimulation.** Consider the Kripke structure*

$$K = \langle W, R, V \rangle$$

with

$$W = \{s_0, s_1, s_2, t_0, t_1\},$$

$$R = \{(s_0, s_1), (s_0, s_2), (s_1, s_1), (s_2, s_2), (t_0, t_1), (t_1, t_1)\}$$

and

$$V(s_0) = V(t_0) = \emptyset,$$
$$V(s_1) = V(s_2) = V(t_1) = \{A\}$$

Determine whether s_0 and t_0 are bisimilar with respect to K.

6. ***(CTL vs. LTL) Expressive Power.*** *Decide whether the following statement is true or false:*

$$AG\,EF\,A$$

can be expressed equivalently by an LTL formula.

Solution 3.101. *5. **True.** The states s_0 and t_0 are bisimilar. A suitable bisimulation is*

$$B = \{(s_0, t_0), (s_1, t_1), (s_2, t_1)\}.$$

We check the conditions from the definition:

- Same labeling: *For all pairs in B, we have*

$$V(s_0) = V(t_0) = \emptyset,$$
$$V(s_1) = V(t_1) = \{A\},$$
$$V(s_2) = V(t_1) = \{A\}.$$

- Simulation of the successors of s_0: *From s_0, the successors are s_1 and s_2. Both can be simulated by the unique successor t_1 of t_0, since*

$$(s_1, t_1) \in B, \qquad (s_2, t_1) \in B.$$

- Simulation of the successors of t_0: *The only successor of t_0 is t_1. This state can be simulated by s_1 or s_2, since*

$$(s_1, t_1) \in B \quad \text{or} \quad (s_2, t_1) \in B.$$

The conditions are also immediately satisfied for the loops at s_1, s_2, and t_1, since each of these states transitions only to itself. Hence B is a bisimulation, and therefore

$$s_0 \sim_K t_0.$$

6. ***False.*** *The CTL formula*

$$AG\,EF\,A$$

is not *equivalently expressible by an LTL formula.*

The reason is that $AG\,EF\,A$ describes a genuinely branching-time property: from every reachable state, there must again exist a possible continuation path to an A-state. LTL, however, can describe only individual linear paths, not statements about all alternative future branches at the same time.

In particular, there exist Kripke structures with identical traces in which $AG\,EF\,A$ holds at one state but not at another. Therefore, no LTL formula can capture this property exactly.

3.4 Epistemic Logic

Epistemic logic investigates how knowledge, ignorance, and possibility can be described in a formal way. It allows us to state precisely what an agent knows, what the agent considers possible, and which conclusions can be drawn from the agent's informational state. Such formalizations play an important role in computer science, philosophy, the theory of distributed knowledge, and knowledge-based systems.

In doing so, we build on the modal logic introduced at the beginning of this chapter (3.1). The basic syntax initially remains unchanged; what is new above all is the epistemic interpretation of the modalities. In the following sections, we first consider frame restrictions for basic modal logic. These make it possible to use certain classes of Kripke frames as models of knowledge. We then extend the logic to multiple agents. Finally, we study public announcements as a dynamic mechanism by which knowledge can change in the course of information exchange.

3.4.1 Frame Restrictions

As a reminder, for basic modal logic we use the same syntax as at the beginning of this chapter (3.1):

$$\varphi ::= \Sigma \mid \neg\varphi \mid \varphi \wedge \varphi \mid \varphi \vee \varphi \mid \Diamond\varphi \mid \Box\varphi.$$

The semantics are defined by means of Kripke structures

$$K = \langle W, R, V \rangle$$

Here, W is the set of states, $R \subseteq W \times W$ is the accessibility relation, and V is the valuation function that assigns to each state the propositional variables that hold there. A formula is always evaluated at a particular state.

In the epistemic setting, the states are understood as possible worlds. From the perspective of a world w, the worlds reachable via R represent the alternatives that are still compatible with the agent's information. Accordingly, we read

$$\Box\varphi$$

as: "The agent knows that φ holds," and

$$\Diamond\varphi$$

as: "The agent considers φ possible."

Not every Kripke structure is suitable for this purpose. If R is meant to represent the epistemically possible alternatives, then the admissible frames should satisfy certain properties. A simple example is a diagnostic system that evaluates only an incomplete set of sensor readings: it should not be in a situation that it itself regards as fundamentally impossible; moreover, epistemically equivalent alternatives should be connected in a coherent way. Considerations of this kind lead to restrictions at the level of frames.

Definition 3.102 (Kripke Frames and Instances). *A* Kripke frame *is a pair*

$$F = \langle W, R \rangle,$$

where W is a set of states and $R \subseteq W \times W$ is a relation.

A Kripke structure $K = \langle W, R, V \rangle$ over a signature Σ is called a Σ-instance *of the frame $F = \langle W, R \rangle$ if it has the same set of states and the same accessibility relation; only the valuation V may vary.*

Example 3.103. *Consider a monitoring system with the states "alarm off," "alarm unclear," and "alarm on." If the agent cannot distinguish the sensor situation exactly, then these states should be connected not merely indirectly, but directly as epistemically possible alternatives. A frame in which two such possibilities are related only through a third state often captures indistinguishability less adequately than a frame in which this relation is represented explicitly.*

In what follows, we are interested in typical properties of accessibility relations.

Definition 3.104 (Properties of Frames). *A frame* $F = \langle W, R \rangle$ *is called*

- reflexive *if for every* $w \in W$ *we have* $(w, w) \in R$,
- transitive *if for all* $w, u, v \in W$, $(w, u) \in R$ *and* $(u, v) \in R$ *imply* $(w, v) \in R$,
- Euclidean *if for all* $w, u, v \in W$, $(w, u) \in R$ *and* $(w, v) \in R$ *imply* $(u, v) \in R$,
- symmetric *if for all* $w, u \in W$, $(w, u) \in R$ *implies* $(u, w) \in R$,
- serial *if for every* $w \in W$ *there exists some* $u \in W$ *such that* $(w, u) \in R$.

We write

$$\mathcal{F}_{\text{Ref}},\ \mathcal{F}_{\text{Tran}},\ \mathcal{F}_{\text{Euk}},\ \mathcal{F}_{\text{Sym}},\ \mathcal{F}_{\text{Ser}}$$

for the classes of all reflexive, transitive, Euclidean, symmetric, and serial frames, respectively.

In epistemic contexts, one often works with frames satisfying reflexivity, transitivity, and Euclideanness. Intuitively, this reflects a knowledge relation that is compatible with the agent's own information state, stable under the formation of successors, and sufficiently closed from the agent's perspective. Depending on the application, however, other frame classes may be more appropriate; for instancc, in deontic interpretations seriality is often the more important condition.

We now formulate how such classes of frames can be characterized semantically.

Definition 3.105 (Validity on Frames and Classes of Frames)**.** *Let* Σ *be a propositional signature, and let* φ *be a modal-logic formula over* Σ*.*

For a Kripke frame $F = \langle W, R\rangle$ *and a state* $w \in W$*, we write*

$$\langle F, w\rangle \models_{\mathrm{ML}} \varphi,$$

if $\langle K, w\rangle \models_{\mathrm{ML}} \varphi$ *holds for* every Σ*-instance* K *of* F*.*

Furthermore, we write

$$F \models_{\mathrm{ML}} \varphi,$$

if $\langle F, w\rangle \models_{\mathrm{ML}} \varphi$ *holds for every state* $w \in W$*.*

Finally, for a class of frames $\mathcal{F}$*, we write*

$$\mathcal{F} \models_{\mathrm{ML}} \varphi,$$

if $F \models_{\mathrm{ML}} \varphi$ *holds for every* $F \in \mathcal{F}$*.*

Example 3.106. *The formula* $\Box p \to p$ *is valid everywhere exactly when the underlying frames are reflexive: whatever the agent knows with certainty must already hold in the actual world itself.*

By contrast, the formula $\Diamond p \to \Box\Diamond p$ *expresses a form of stability of possibility: if something is possible at one point, then from the agent's perspective this possibility remains present throughout all epistemically accessible alternatives.*

The following schemata play a central role here. They are called *frame axioms* because they characterize exactly those classes of frames on which they are valid.

- **(K)** $\Box(\varphi \to \psi) \to (\Box\varphi \to \Box\psi)$

  If the agent knows that φ always implies ψ, and if the agent knows φ, then the agent also knows ψ.

- **(T)** $\Box\varphi \to \varphi$

  Knowledge is factive: whatever the agent knows is actually the case.

- **(4)** $\Box\varphi \to \Box\Box\varphi$

  If the agent knows something, then the agent also knows that they know it.

- **(5)** $\Diamond\varphi \to \Box\Diamond\varphi$

  If something is possible, then this possibility remains stable from the agent's point of view.

- **(B)** $\varphi \to \Box\Diamond\varphi$

  If φ holds, then the agent considers φ possible.

- **(D)** $\Box\varphi \to \Diamond\varphi$

  Whatever the agent knows must also be possible for the agent.

A formula schema **axiomatizes** a class of frames if two conditions are satisfied. First, every instance of the schema is valid on all frames in the class. Second, conversely, a frame belongs to the class exactly if all instances of the schema are valid on it.

Theorem 3.107 (Frame Axioms). *The following statements hold:*

- *The schema **(T)** axiomatizes the class $\mathcal{F}_{\text{Ref}}$ of reflexive frames.*

- *The schema **(4)** axiomatizes the class $\mathcal{F}_{\text{Tran}}$ of transitive frames.*

- *The schema **(5)** axiomatizes the class $\mathcal{F}_{\text{Euk}}$ of Euclidean frames.*

- *The schema **(B)** axiomatizes the class $\mathcal{F}_{\text{Sym}}$ of symmetric frames.*

- *The schema **(D)** axiomatizes the class $\mathcal{F}_{\text{Ser}}$ of serial frames.*

Proof. We prove the axiomatization claims one by one. In each case, we first show that the schema is valid on the corresponding class of frames and then establish the converse.

(T) axiomatizes exactly the reflexive frames.

Let $F = \langle W, R \rangle$ be reflexive, and let $\langle K, w \rangle$ be an arbitrary Σ-instance based on F. We must show that every instance of

$$\Box\psi \to \psi$$

is true in $\langle K, w \rangle$. If $\Box\psi$ is false at w, the implication is trivially satisfied. If $\Box\psi$ is true at w, then reflexivity gives $(w, w) \in R$. This is precisely the key point of the schema: the current world counts as one of its own accessible alternatives. Hence ψ also holds at w. Therefore $\Box\psi \to \psi$ is valid.

For the converse, let $F = \langle W, R \rangle$ be a frame on which all instances of (T) are valid. Suppose that F is not reflexive. Then there exists some $w \in W$ with $(w, w) \notin R$. Choose a propositional variable A such that A is true at every successor of w, but false at w itself. Then $\Box A$ holds at w, whereas A does not. Thus

$$\Box A \to A$$

is false at w, contradicting our assumption. Therefore F is reflexive.

(4) axiomatizes exactly the transitive frames.

Let $F = \langle W, R\rangle$ be transitive, and let $\langle K, w\rangle$ be an arbitrary instance based on F. We show the validity of

$$\Box\psi \to \Box\Box\psi.$$

Assume that $\Box\psi$ holds at w. Let v be any successor of w, so $(w, v) \in R$. We must show that $\Box\psi$ also holds at v. To this end, let u be any successor of v. By transitivity, from $(w, v) \in R$ and $(v, u) \in R$ it follows that $(w, u) \in R$. Since $\Box\psi$ holds at w, we obtain ψ at u. Hence $\Box\psi$ holds at v, and therefore $\Box\Box\psi$ holds at w.

For the converse, let $F = \langle W, R\rangle$ be a frame on which all instances of (4) are valid. Suppose that R is not transitive. Then there are worlds w, u, v such that $(w, u) \in R$, $(u, v) \in R$, but $(w, v) \notin R$. Choose a propositional variable A so that A is true at all successors of w, but false at v. Then $\Box A$ holds at w, but $\Box A$ does not hold at u, and therefore $\Box\Box A$ does not hold at w. Hence

$$\Box A \to \Box\Box A$$

is false at w, contradicting the assumption. Therefore R is transitive.

(5) axiomatizes exactly the Euclidean frames.

Let $F = \langle W, R\rangle$ be Euclidean, and let $\langle K, w\rangle$ be an instance based on F. We must show

$$\Diamond\psi \to \Box\Diamond\psi.$$

If $\Diamond\psi$ does not hold at w, the implication is trivial. Otherwise, there exists a world u with $(w, u) \in R$ and

ψ true at u. Now let v be any successor of w. By Euclideanness, from $(w, u) \in R$ and $(w, v) \in R$ it follows that $(v, u) \in R$. Thus the original possibility remains visible from every accessible world: ψ is reachable from v as well. Hence $\Diamond\psi$ holds at v. Since v was arbitrary, $\Box\Diamond\psi$ holds at w.

For the converse, let $F = \langle W, R\rangle$ be a frame on which all instances of (5) are valid. Suppose that F is not Euclidean. Then there are worlds w, u, v such that $(w, u) \in R$, $(w, v) \in R$, but $(v, u) \notin R$. Choose a propositional variable A such that A is true exactly at u. Then w satisfies $\Diamond A$, since u is accessible. However, $\Diamond A$ is false at v, because u is not accessible from v. Therefore $\Box\Diamond A$ is false at w, and so

$$\Diamond A \to \Box\Diamond A$$

is false at w. This contradiction shows that R is Euclidean.

(B) axiomatizes exactly the symmetric frames.

Let $F = \langle W, R\rangle$ be symmetric, and let $\langle K, w\rangle$ be an instance based on F. We must show

$$\psi \to \Box\Diamond\psi.$$

If ψ is false at w, the implication is trivial. If ψ is true at w, let v be any successor of w. By symmetry, we then also have $(v, w) \in R$. Thus ψ is accessible from v, and therefore $\Diamond\psi$ holds at v. Hence $\Box\Diamond\psi$ holds at w.

For the converse, let $F = \langle W, R\rangle$ be a frame on which all instances of (B) are valid. Suppose that R is not symmetric. Then there are worlds w, u with $(w, u) \in R$, but $(u, w) \notin R$. Choose a propositional variable A such that A is true exactly at w. Then A is true at w. But $\Diamond A$ is false at u, since w is not accessible from u. Hence $\Box\Diamond A$ is false at w, and therefore

$$A \to \Box\Diamond A$$

is false at w. This contradiction shows that R is symmetric.

(D) axiomatizes exactly the serial frames.

Let $F = \langle W, R\rangle$ be serial and let $\langle K, w\rangle$ be an instance of F. We must show

$$\Box\psi \to \Diamond\psi.$$

If $\Box\psi$ holds at w, then by seriality w has a successor u with $(w, u) \in R$. From $\Box\psi$ it follows that ψ holds at u, hence $\Diamond\psi$ holds at w.

For the converse, let $F = \langle W, R\rangle$ be a frame on which all instances of (D) are valid. Suppose that F is not serial. Then there exists some $w \in W$ with no successor. This is the crucial point: if there is no successor, every formula of the form $\Box\psi$ is automatically true at w, whereas $\Diamond\psi$ is always false at w. Thus, for every propositional variable A,

$$\Box A \to \Diamond A$$

is false at w. Contradiction. Therefore, R is serial.

This proves all parts. $\Box$

These frame axioms are often grouped into well-known axiom systems. For example, $S5$ captures the epistemic perspective particularly well, whereas other modal logics require different frame conditions in other areas of application. In the deontic reading, for instance, seriality is especially natural because one often assumes that in every situation there is at least one permissible alternative.

3.4.2 Modal Logic for Multiple Agents

In this section, we extend the modal logic familiar from Chapter 3.1 to multiple agents. The basic idea is simple: instead of modeling only a single knowledge or belief modality, we introduce a separate modal operator for each agent. In this way, we can express what different participants know, consider possible, or know about one another. The language obtained in this way is called *multi-agent modal logic* (MAL).

We begin with the underlying signature.

Definition 3.108. *A signature of multi-agent modal logic, or* MAL signature *for short, is a tuple* $\langle \Sigma, A \rangle$*, where* Σ *is a propositional signature and* A *is a finite set of agents.*

For simplicity, we assume that the agents in A are numbered starting from 0. Thus one usually has

$$A = \{0, 1, 2, \ldots, m\}.$$

Instead of numerical labels, one may of course also use descriptive names, such as *control center*, *technician*, or *patient*.

For each agent $a \in A$, we now introduce a separate modal operator K_a. We read the formula $K_a\varphi$ as: "Agent a knows that φ holds." Here, the K stands for *knowledge*.

Definition 3.109. *Let $\langle\Sigma, A\rangle$ be an MAL signature. The set of formulas of multi-agent modal logic over $\langle\Sigma, A\rangle$ is defined inductively as follows:*

(MAL0) Every propositional variable from Σ is an MAL formula.

(MAL1) If φ and ψ are MAL formulas, then $(\varphi \vee \psi)$ is also an MAL formula.

(MAL2) If φ and ψ are MAL formulas, then $(\varphi \wedge \psi)$ is also an MAL formula.

(MAL3) If φ is an MAL formula, then $\neg\varphi$ is also an MAL formula.

(MAL4) If φ is an MAL formula and $a \in A$, then $K_a\varphi$ is also an MAL formula.

As usual, we use the following abbreviations:

$$\varphi \to \psi := \neg\varphi \vee \psi \tag{3.1}$$

$$\text{and} \tag{3.2}$$

$$\varphi \leftrightarrow \psi := (\varphi \to \psi) \wedge (\psi \to \varphi) \tag{3.3}$$

As in ordinary modal logic, the dual reading of $K_a\varphi$ is given here by $\neg K_a\neg\varphi$. This formula can be read as: "From the perspective of agent a, φ is possible." However, we do not introduce a separate symbol for this, since the expression is already sufficiently clear as an abbreviation.

MAL formulas allow us to describe epistemic situations involving several agents with precision. It is especially important that nestings of modal operators are also permitted. Thus, we may formulate, for example: "Agent a knows that agent b knows something" or even "Agent a knows that agent b knows that … ". Such nested statements play a central role in epistemic logic.

Example 3.110. *Consider a small coordination scenario in a research laboratory. Let*

$$\langle\{T, A, S\}, \{L, R\}\rangle$$

be an MAL signature with the following meanings:

T	*The test has been completed successfully.*
A	*The alarm is activated.*
S	*The door is secured.*
L	*The agent* laboratory management.
R	*The agent* technical service.

Then we can model, for example, the following relationships:

$$K_L T \rightarrow A$$

If the laboratory management knows that the test has been completed, it activates the alarm.

Likewise:

$$K_R K_L T \to S$$

If the technical service knows that the laboratory management knows that the test was successful, then it ensures that the door is secured.

The second formula already shows the characteristic added value of multi-agent modal logic: it allows us to express not only statements about the knowledge of individual agents, but also statements about knowlcdgc of the knowledge of others. □

3.4.3 Semantics of Multi-Agent Modal Logic

We now define the semantics of MAL using Kripke structures. Compared with ordinary modal logic, the structure is extended so that each agent has its own accessibility relation.

Definition 3.111. *Let $\langle \Sigma, A \rangle$ be an MAL signature. A Kripke structure for MAL over $\langle \Sigma, A \rangle$ is a tuple*

$$K = \langle W, (R_a)_{a \in A}, V \rangle,$$

where

- *W is a finite set of states,*
- *for each agent $a \in A$, $R_a \subseteq W \times W$ is a relation,*
- *$V : W \to \mathcal{P}(\Sigma)$ is a valuation assigning to each state the propositional variables that are true there.*

A pointed Kripke structure is, as usual, a pair $\langle K, w\rangle$, where K is a Kripke structure for MAL and $w \in W$ is a distinguished state.

The satisfaction relation is now interpreted separately for each agent. Semantically, K_a behaves exactly like the box operator of classical modal logic, but relative to the relation R_a.

Definition 3.112. *Let $\langle \Sigma, A\rangle$ be an MAL signature, and let $\langle K, w\rangle$ be a pointed Kripke structure for $\langle \Sigma, A\rangle$ with $K = \langle W, (R_a)_{a\in A}, V\rangle$. The evaluation function $[\![\cdot]\!]_{\langle K,w\rangle}$ assigns to each MAL formula a truth value from $\{true, false\}$:*

$$
\begin{aligned}
[\![p]\!]_{\langle K,w\rangle} &= true \Leftrightarrow p \in V(w)\\
[\![\neg\varphi]\!]_{\langle K,w\rangle} &= true \Leftrightarrow [\![\varphi]\!]_{\langle K,w\rangle} = false\\
[\![(\varphi \wedge \psi)]\!]_{\langle K,w\rangle} &= true \Leftrightarrow [\![\varphi]\!]_{\langle K,w\rangle} = true\\
&\qquad\qquad \wedge [\![\psi]\!]_{\langle K,w\rangle} = true\\
[\![(\varphi \vee \psi)]\!]_{\langle K,w\rangle} &= true \Leftrightarrow [\![\varphi]\!]_{\langle K,w\rangle} = true\\
&\qquad\qquad \vee [\![\psi]\!]_{\langle K,w\rangle} = true\\
[\![K_a\varphi]\!]_{\langle K,w\rangle} &= true \Leftrightarrow \forall w' \in R_a(w) :\\
&\qquad\qquad [\![\varphi]\!]_{\langle K,w'\rangle} = true
\end{aligned}
$$

If $[\![\varphi]\!]_{\langle K,w\rangle} = true$, we say that φ is satisfied at w in K. We then also write

$$\langle K, w\rangle \models_{MAL} \varphi.$$

We define logical consequence with respect to a fixed

Kripke structure K by:

$$\varphi \models_K^{MAL} \psi \Leftrightarrow \forall w \in W : \\ (\langle K, w\rangle \models_{MAL} \varphi \Rightarrow \langle K, w\rangle \models_{MAL} \psi)$$

Example 3.113. *We continue the laboratory example. Suppose that, in a Kripke structure K for the signature from the previous example, the state w_0 describes the situation in which the laboratory management knows that the test was successful, while the technical service does not yet fully grasp all consequences of this knowledge.*

Then the following may hold:

$$\langle K, w_0\rangle \models_{MAL} K_L T,$$

while at the same time

$$\langle K, w_0\rangle \not\models_{MAL} K_R T$$

is true. This means that the laboratory management already knows that the test has been completed, whereas the technical service does not.

The formula

$$K_R K_L T \rightarrow S$$

is true in all states of this model if the relation R_R is defined in such a way that, in every situation the technical service considers possible, it correctly takes into account the information held by the laboratory management and accordingly secures the door.

By contrast, a formula such as

$$A \leftrightarrow S$$

is not valid in general, since there may be states in which the alarm has been activated but the door has not yet been secured. □

With MAL, we can therefore model not only the knowledge of individual agents but also the interactions between different states of knowledge. This is precisely its decisive advantage over simple modal logic: it provides a language for representing multiple perspectives on the same situation.

3.4.4 Public Announcements

So far, we have mainly considered epistemic logic as a static theory: a Kripke structure describes a given state of knowledge, and formulas are evaluated relative to that state. In many applications, however, this is not sufficient. Knowledge changes not only because agents reason about the world, but also because new information is announced publicly.

This dynamic aspect is captured precisely by *Public Announcement Logic* (PAL). It extends epistemic logic by making it possible to model publicly communicated information. A public announcement does not merely affect the truth value of a formula; it changes the underlying epistemic situation itself: states incompatible with the announcement are eliminated.

Example 3.114. *Consider a small coordination scenario in a library. Two people,* A *and* B*, are supposed to open a* *safe together. What matters is whether the code is already known.*

Let C be the statement "The code is known" and O the statement "The safe is open". As long as person B does not know whether A knows the code, B cannot be certain whether the safe can be opened. But if it is publicly announced that the code is known, then the set of possible states shrinks to exactly those in which C holds. In such a reduced model, one can then describe precisely what the agents know after the announcement. □

The syntax of PAL is closely related to modal logic, but it is extended by formulas of the form $[!\psi]\varphi$. The formula $[!\psi]\varphi$ means: "After the public announcement of ψ, φ holds." We continue to use the notation K for the knowledge operator in order to emphasize the epistemic character of the logic.

Definition 3.115. *Let Σ be a propositional signature. The set of PAL formulas over Σ is defined inductively as follows:*

(PAL0) Every propositional variable from Σ is a PAL formula.

(PAL1) If φ and ψ are PAL formulas, then $(\varphi \vee \psi)$ is also a PAL formula.

(PAL2) If φ and ψ are PAL formulas, then $(\varphi \wedge \psi)$ is also a PAL formula.

(PAL3) If φ is a PAL formula, then $\neg\varphi$ is also a PAL formula.

(PAL4) *If φ is a PAL formula, then $K\varphi$ is also a PAL formula.*

(PAL5) *If φ and ψ are PAL formulas, then $[!\psi]\varphi$ is also a PAL formula.*

The semantics of PAL is described in terms of Kripke structures. The crucial new step is that a public announcement is not merely evaluated, but understood as a change of model. To this end, for every PAL formula ψ we define a reduced Kripke structure in which only those states remain in which ψ holds.

Definition 3.116. *Let Σ be a propositional signature, let $K = \langle W, R, V\rangle$ be a Kripke structure over Σ, and let ψ be a PAL formula. Then*

$$K_\psi = \langle W_\psi, R_\psi, V_\psi\rangle$$

is called the ψ-reduction of K, where

$$W_\psi = \{\, w \in W \mid \langle K, w\rangle \models_{PAL} \psi \,\},$$

$$R_\psi = \{\, (w, w') \in R \mid w, w' \in W_\psi \,\},$$

and

$$V_\psi(w) = V(w)$$

for all $w \in W_\psi$.

Intuitively, K_ψ is precisely the substructure of K that is compatible with the publicly announced information. All excluded states are removed, while the remaining states keep their truth assignments unchanged.

Example 3.117. *We consider the library scenario once again. Suppose the Kripke structure K describes different possible situations concerning whether the code is known and whether the safe can be opened. If it is now publicly announced that the code is known, that is, $\psi = C$, then in the reduction K_C only those states remain in which C actually holds.*

A formula such as

$$[!C]\,KC$$

is then satisfied in many cases, because after a public and truthful announcement of C every agent knows that C holds. By contrast, a formula such as

$$[!C]\,O$$

can hold only if the assumption that the code is known actually implies that the safe is open. □

The satisfaction relation for PAL is defined recursively in the usual way. For the propositional and modal cases, the semantics remains unchanged; the only new clause is the one for public announcements.

Definition 3.118. *Let Σ be a propositional signature, and let $\langle K, w\rangle$ be a pointed Kripke structure with $K = \langle W, R, V\rangle$. Then the satisfaction relation $\models_{PAL}$ is given by the following conditions:*

$$\langle K, w\rangle \models_{PAL} A \Leftrightarrow A \in V(w)$$

$$
\begin{aligned}
\langle K, w\rangle \models_{PAL} \neg\varphi &\Leftrightarrow \langle K, w\rangle \not\models_{PAL} \varphi \\
\langle K, w\rangle \models_{PAL} (\varphi \wedge \psi) &\Leftrightarrow \langle K, w\rangle \models_{PAL} \varphi \\
&\qquad \textit{and } \langle K, w\rangle \models_{PAL} \psi \\
\langle K, w\rangle \models_{PAL} (\varphi \vee \psi) &\Leftrightarrow \langle K, w\rangle \models_{PAL} \varphi \\
&\qquad \textit{or } \langle K, w\rangle \models_{PAL} \psi \\
\langle K, w\rangle \models_{PAL} K\varphi &\Leftrightarrow \forall w' \in R(w) : \\
&\qquad \langle K, w'\rangle \models_{PAL} \varphi \\
\langle K, w\rangle \models_{PAL} [!\psi]\varphi &\Leftrightarrow (\langle K, w\rangle \models_{PAL} \psi) \\
&\qquad \Rightarrow (\langle K_\psi, w\rangle \models_{PAL} \varphi)
\end{aligned}
$$

If $\langle K, w\rangle$ satisfies a formula φ, we again say that φ is satisfied at w in K, and write

$$\langle K, w\rangle \models_{PAL} \varphi.$$

We define logical consequence with respect to a fixed Kripke structure K by:

$$
\begin{aligned}
\varphi \models_K^{PAL} \psi \quad \Leftrightarrow \quad & \forall w \in W : \\
& \langle K, w\rangle \models_{PAL} \varphi \Rightarrow \langle K, w\rangle \models_{PAL} \psi
\end{aligned}
$$

Example 3.119. *Suppose that in a model of the library scenario the statement C holds in certain states, but not in others. If it is now publicly announced that C holds, then all states with $\neg C$ disappear from consideration. In the reduced structure, it then becomes visible that not only is the code known, but every agent also knows this information.*

This allows us to investigate, for example, which public announcement is required to produce a desired epistemic situation. If, for instance, one wants to guarantee in the end that the safe is open, one must find a formula ψ whose announcement restricts the set of possible states so strongly that O holds in all remaining states afterwards. □

PAL thus shows especially clearly that epistemic logic describes not only knowledge, but also its change through information. Public announcements are a simple, yet very powerful means of formally capturing dynamic processes of knowledge.

3.4.5 Exercises and Solutions

In the following, we consider a small PAL model and determine whether several statements are true or false. The exercises are intended to practice both the reduction induced by public announcement and the knowledge modality K.

We use the signature

$$\Sigma = \{C, O\},$$

where

- C means: "The code is known",
- O means: "The safe is open".

In addition, we consider the Kripke structure

$$K = \langle W, R, V \rangle$$

with

$$W = \{w_0, w_1, w_2\}, \qquad R = W \times W,$$

and valuation

$$V(w_0) = \{C\}, \qquad V(w_1) = \{C, O\}, \qquad V(w_2) = \{O\}.$$

Thus, the relation R is total here: every state is epistemically accessible from every other state.

Exercise 3.120. *Decide in each case whether the following statement is* true *or* false.

1. $\langle K, w_0 \rangle \models_{PAL} [!C]\, KC$
2. $\langle K, w_0 \rangle \models_{PAL} [!C]\, KO$
3. $\langle K, w_2 \rangle \models_{PAL} [!C]\, KC$
4. $\langle K, w_0 \rangle \models_{PAL} [!C]\, KKC$

Solution 3.121. *We justify the four statements one by one.*

1. ***True.*** *At w_0, the formula C holds, so the public announcement of C is applicable. After the reduction to K_C, exactly the states w_0 and w_1 remain, because C holds only there. Since the relation R is total, the restricted relation on $\{w_0, w_1\}$ is again*

total. In both remaining states, C holds. Therefore, after the announcement, the agent knows that C holds. Hence,

$$\langle K, w_0 \rangle \models_{PAL} [!C]KC.$$

2. ***False.*** *Again, after the announcement of C, only the set $\{w_0, w_1\}$ remains. However, at w_0, the formula O does not hold. Since the knowledge operator K requires the formula to hold in all accessible states, KO is false at w_0 after the reduction. Therefore,*

$$\langle K, w_0 \rangle \not\models_{PAL} [!C]KO.$$

3. ***True.*** *At w_2, the formula C does not hold. But for formulas of the form $[!\psi]\varphi$, the following applies: if the announcement ψ is false at the current state, then the whole formula is automatically true. Indeed, the condition*

$$\langle K, w \rangle \models_{PAL} [!\psi]\varphi$$

becomes relevant only if ψ is already satisfied. Since C is false at w_2, it follows immediately that

$$\langle K, w_2 \rangle \models_{PAL} [!C]KC.$$

4. ***True.*** *After the announcement of C, once again only w_0 and w_1 remain. In both states, KC holds,*

because in the reduced structure each state sees only states in which C is true. Thus, in K_C not only KC holds, but even KKC: in all epistemically accessible states, KC is again satisfied. Therefore,

$$\langle K, w_0 \rangle \models_{PAL} [!C]KKC.$$

Exercise 3.122. *Let*

$$F = \langle W, R \rangle \quad \textit{with} \quad W = \{u, v\}, \qquad R = \{(u, v), (v, v)\}.$$

Determine in each case whether the following statement is true *or* false.

1. $F \models_{\text{ML}} \Box p \to p$
2. $F \models_{\text{ML}} \Box p \to \Box\Box p$
3. $F \models_{\text{ML}} \Diamond p \to \Box\Diamond p$
4. $F \models_{\text{ML}} p \to \Box\Diamond p$

Solution 3.123. *We justify the four statements one by one.*

1. ***False.*** *The frame F is not reflexive, since $(u, u) \notin R$. By the theorem, the schema (T) axiomatizes exactly the reflexive frames. Hence $\Box p \to p$ cannot be valid on F in general. This can also be seen directly by choosing a suitable valuation: let $V(u) = \emptyset$ and $V(v) = \{p\}$. Then the formula $\Box p$ holds at u, but p does not.*

2. **True.** *The frame F is transitive. The only nontrivial composition is (u, v) and (v, v), and this again yields $(u, v) \in R$. By the theorem, axiom (4) axiomatizes exactly the transitive frames. Therefore,*

$$\Box p \to \Box\Box p$$

is valid on F.

3. **True.** *The frame F is Euclidean. Indeed, from $(u, v) \in R$ and $(u, v) \in R$ it follows that $(v, v) \in R$, and the remaining cases are satisfied as well. By the theorem, axiom (5) axiomatizes exactly the Euclidean frames. Therefore,*

$$\Diamond p \to \Box\Diamond p$$

is valid on F.

4. **False.** *The frame F is not symmetric, since $(u, v) \in R$ does not imply $(v, u) \in R$. By the theorem, axiom (B) axiomatizes exactly the symmetric frames. Therefore,*

$$p \to \Box\Diamond p$$

is not valid on F in general. With the valuation $V(u) = \{p\}$ and $V(v) = \emptyset$, the formula p is true at u, but $\Diamond p$ is false at v, because from v only v itself is accessible and p does not hold there. Hence $\Box\Diamond p$ is false at u.

Chapter 4

Non-Classical Logics

Overview of This Chapter

At its core, classical logic operates with two truth values: a statement is either true or false. For many mathematical and everyday situations, this is entirely sufficient. In numerous applications, however, this scheme reaches its limits. Statements may be only partially justified, allow multiple interpretations, hold only approximately, or even be mutually contradictory without causing every distinction to collapse. This is precisely where non-classical logics come into play.

Non-classical logics extend the classical framework so that such cases can be described formally at all. They provide tools for dealing with different kinds of uncertainty, vagueness, informational gaps, and inconsistent knowledge. Some of these logics modify the very notion of truth, while others revise the underlying patterns of inference. What they share is that they do not simply discard classical bivalence, but rather generalize it in a targeted way for specific kinds of problems.

One particularly important starting point is intuitionistic logic. It does not place mere assertibility at the center, but instead requires a constructive proof. As a result, classical principles such as the law of excluded middle are not accepted in their unrestricted form. Paraconsistent logics pursue a different idea: they make it possible to work with inconsistent information without the entire theory becoming trivial. In such systems, a contradiction does not automatically imply that every statement becomes derivable. Many-valued approaches, and in particular different forms of paraconsistent logic, show that a system can still remain meaningful even when its information does not fit together perfectly.

Fuzzy logic follows yet another idea. It replaces the sharp distinction between true and false with graded degrees of truth. Statements such as "the temperature is high" or "the set is large" can therefore be modeled much more naturally than in a strictly two-valued logic. What matters is often not complete truth or falsity, but the extent to which a statement applies.

Another important family of non-classical systems is formed by non-monotonic logics. Here, it is not the case that new information can only produce additional conclusions. Under certain circumstances, earlier conclusions may have to be withdrawn in the light of new information. This reflects many everyday situations far better than the classical monotonicity of inference. Default logic formalizes precisely such typical, though not exceptionless, assumptions. It allows us to reason from what normally holds as long as no counterexample is known.

Closely connected with these ideas is Answer Set Programming (ASP). It combines logical reasoning with a particularly concise form of modeling based on minimal, consistent descriptions of possible worlds. This makes ASP interesting both for theoretical questions and for practical applications in knowledge representation, planning, and problem solving.

This chapter introduces these approaches step by step. Section 4.1 first deals with intuitionistic logic. Section 4.2 is devoted to paraconsistent and many-valued logics. Section 4.3 then presents fuzzy logic as a formal approach to graded truth. Section 4.4 explains default logic as a model of non-monotonic reasoning. Finally, Section 4.5 introduces Answer Set Programming as a modern form of non-classical logic programming.

Non-classical logics thus show that logical reasoning need not be tied to a single rigid model of truth. Especially where information is incomplete, vague, or inconsistent, they open up more differentiated and often more realistic forms of formalization.

4.1 Intuitionistic Logic

In this subsection, we consider intuitionistic logic IPC (*intuitionistic propositional calculus*). It uses essentially the same formula components as propositional logic, but understands them differently. The crucial difference lies not in the outward form of formulas, but in their semantics: whereas propositional logic describes truth values, intuitionistic logic is concerned with the *constructive provability* of a statement.

As a result, the central question is no longer only whether a statement is true or false, but also *how* its validity is justified. In this sense, a statement is accepted only when there is a proof that actually makes the statement *constructible*. For this reason, intuitionistic logics are also called *constructive logics*.

Example 4.1 (Direct and Indirect Proof). *Suppose someone claims:*

> *Among three locked boxes, at least one contains a gift.*

An indirect proof might show only that not all boxes can be empty. For classical logic, that is sufficient. From an intuitionistic point of view, however, this is not yet the same as a concrete demonstration. One would have to be able to indicate which *box actually contains the gift.*

This difference is characteristic: intuitionistic logic requires not merely that a statement somehow follows, but that the proof also provides a concrete justification.

Example 4.2 (Case Distinction Without a Decision). *Assume we know only that a number x is either greater than 1 or not greater than 1. In classical logic, one may often argue on the basis of this distinction without knowing which case actually obtains. One can thus give a proof that treats both possibilities.*

Intuitionistically, such a mere case distinction is not always sufficient. For a disjunction $\varphi \vee \psi$ is not meant to signify only that at least one of the two cases is logically possible, but that one of the two cases can in fact be specified constructively.

The basic idea can be summarized in the so-called BHK interpretation (*Brouwer–Heyting–Kolmogorov interpretation*): logical connectives are understood as operations on proofs. A conjunction means that both statements are provable; a disjunction means that one of the two disjuncts is provable, together with a concrete choice; an implication means that every proof of one statement can be transformed into a proof of the other; and a negation means that a contradiction would follow from a proof of the statement.

An especially important principle of classical logic is the law of excluded middle (*tertium non datur*):

$$\varphi \vee \neg\varphi$$

for every formula φ. In classical propositional logic, and also in classical predicate logic, this formula is always a tautology.

Example 4.3 (The Law of Excluded Middle as a Case Distinction). *Suppose we are considering the statement:*

The number n is prime.

In classical logic, one may often say: Either n is prime or n is not prime. This form of reasoning is logically permissible, even if at the moment one does not yet know which of the two cases actually obtains.

In intuitionistic logic, however, this kind of inference is not automatically allowed. For the formula $\varphi \vee \neg\varphi$ requires not merely an abstract alternative, but a constructive decision in favor of one of the two cases.

Proposition 4.4. *For every propositional formula α and every first-order formula φ, the following holds:*

$\alpha \vee \neg\alpha$ *is a tautology and* $\varphi \vee \neg\varphi$ *is a tautology.*

Proof. We prove the two statements separately.

(1) Propositional logic. Let α be an arbitrary propositional formula and let I be an arbitrary interpretation. By classical semantics, either $I \models \alpha$ or $I \not\models \alpha$.

If $I \models \alpha$, then it follows immediately that

$$I \models \alpha \vee \neg\alpha.$$

If, on the other hand, $I \not\models \alpha$, then by the definition of negation $I \models \neg\alpha$. Hence in this case as well,

$$I \models \alpha \vee \neg\alpha.$$

Since I was arbitrary, $\alpha \vee \neg\alpha$ is a tautology.

(2) First-order predicate logic. Let φ be an arbitrary formula of first-order predicate logic, and let $\mathcal{M}$ be an arbitrary structure with variable assignment g. Here again, by classical semantics, either

$$\mathcal{M}, g \models \varphi \quad \text{or} \quad \mathcal{M}, g \not\models \varphi.$$

In the first case, we immediately obtain

$$\mathcal{M}, g \models \varphi \vee \neg\varphi.$$

In the second case, by the definition of negation,

$$\mathcal{M}, g \models \neg\varphi,$$

and therefore also

$$\mathcal{M}, g \models \varphi \vee \neg\varphi.$$

Since the structure and assignment were arbitrary, $\varphi \vee \neg\varphi$ is also a tautology.

Thus, the law of excluded middle holds both in propositional logic and in first-order predicate logic. □

Logics in which the law of excluded middle does not hold are called *intuitionistic logics.* In this chapter, we consider intuitionistic logic IPC as a variant of propositional logic in which precisely this classical basic assumption is not adopted.

The formulas of intuitionistic logic IPC resemble those of propositional logic. The signatures are likewise the same as in propositional logic: a signature is a set Σ of propositional variables. As in propositional logic, we generally assume that Σ is finite.

In the following, we define the formulas of intuitionistic logic.

4.1.1 Formulas of Intuitionistic Logic

Definition 4.5. *Let Σ be a propositional signature. The set of formulas of intuitionistic logic over Σ, abbreviated as* IPC(Σ) *formulas, is defined as follows:*

(IPC0) $\bot$ *is a formula of intuitionistic logic over* Σ.

(IPC1) Every propositional variable from Σ is a formula of intuitionistic logic over Σ.

(IPC2) If φ and ψ are formulas of intuitionistic logic over Σ, then $(\varphi \land \psi)$ is a formula of intuitionistic logic over Σ.

(IPC3) If φ and ψ are formulas of intuitionistic logic over Σ, then $(\varphi \lor \psi)$ is a formula of intuitionistic logic over Σ.

(IPC4) If φ and ψ are formulas of intuitionistic logic over Σ, then $(\varphi \to \psi)$ is a formula of intuitionistic logic over Σ.

We introduce negation and biconditional as abbreviations:

$$\neg\varphi := (\varphi \to \bot)$$
$$\varphi \leftrightarrow \psi := (\varphi \to \psi) \wedge (\psi \to \varphi)$$

According to the BHK interpretation, these formulas express statements about proofs. The formula $\bot$ stands for a statement that can never be proved. A conjunction $\varphi \wedge \psi$ means that both φ and ψ are provable. A disjunction $\varphi \vee \psi$ means that one of the two disjuncts can be proved, and that it is known which one it is. An implication $\varphi \to \psi$ means that every proof of φ can be transformed into a proof of ψ. Finally, $\neg\varphi$ means that any proof of φ would lead to a contradiction.

Although the set of formulas in intuitionistic logic is defined similarly to that of propositional logic, the usual abbreviations and bracketing conventions yield practically the same formulas. The essential difference lies in the role of implication and negation: in intuitionistic logic, implication $\to$ is a primitive symbol, whereas negation is defined as an abbreviation. In classical propositional logic, the situation is often reversed; there, implication is frequently introduced as an abbreviation. Moreover, in intuitionistic logic $\bot$ is a genuine formula symbol, whereas in propositional logic it usually serves only as an abbreviation for a contradictory formula.

In the next section, we introduce a semantics for IPC based on Kripke structures, which gives precise expression to the constructive viewpoint.

4.1.2 Kripke Semantics

For intuitionistic logic, it is not enough to consider only isolated truth values. Instead, we want to express that knowledge can expand step by step. This is precisely what Kripke semantics is suited for: we consider several possible states that are connected by an accessibility relation. A state that is reachable later contains at least as much information as an earlier state, and often more.

The basic idea can be read informally as follows: a state describes a certain state of information, and from there we may move to states in which more details are known. What has already been established in one state must not be lost again in later states. This kind of *growth of knowledge* is exactly what intuitionistic logic models.

We begin by recalling the notion of a Kripke structure.

Definition 4.6. *A Kripke structure over a propositional signature* Σ *is a triple*

$$K = \langle W, R, V \rangle,$$

where W *is a set of states,* $R \subseteq W \times W$ *is a relation between states, and* $V : W \to \mathcal{P}(\Sigma)$ *is a function that assigns to each state the propositional variables that hold there.*

For intuitionistic logic, we now require two additional conditions. First, information should only increase along the accessibility relation. Second, this relation should behave like an order, so that it does not generate unnecessary cycles.

Definition 4.7. *A Kripke structure* $K = \langle W, R, V \rangle$ *over a propositional signature* Σ *is called* monotone *if for all* $w_1, w_2 \in W$ *we have:*

$$(w_1, w_2) \in R \Rightarrow V(w_1) \subseteq V(w_2).$$

Thus, monotonicity means that if a propositional variable already holds in one state, then it also remains valid in all reachable states. In this way, previously acquired information cannot disappear later on.

Definition 4.8. *Let* W *be a set. A relation* $\leq \subseteq W \times W$ *is called a* partial order *if for all* $w, w_1, w_2, w_3 \in W$ *the following properties hold:*

- $w \leq w$ *(reflexivity),*
- *if* $w_1 \leq w_2$ *and* $w_2 \leq w_1$*, then* $w_1 = w_2$ *(antisymmetry),*
- *if* $w_1 \leq w_2$ *and* $w_2 \leq w_3$*, then* $w_1 \leq w_3$ *(transitivity).*

Kripke structures that satisfy both conditions are called *intuitionistic.* To make it clear that the accessibility relation is a partial order, we use the symbol $\leq$ in this case.

Definition 4.9. *A Kripke structure* $K = \langle W, \leq, V \rangle$ *over a propositional signature* Σ *is called* intuitionistic *if* K *is monotone and* $\leq$ *is a partial order.*

The states of such a structure can be understood as states of information. A state therefore describes what is known at a given moment. A reachable state then represents a more developed state of information. The fact that knowledge is not lost is secured by monotonicity.

We now define when a formula is satisfied at a state. As usual, a pair $\langle K, w\rangle$ consisting of a Kripke structure K and a state $w \in W$ is called a *pointed Kripke structure.*

Definition 4.10. *Let Σ be a propositional signature. The satisfaction relation*

$$\models^{\mathrm{IPC}}_{\Sigma}$$

for intuitionistic logic is defined recursively by the following clauses: for every pointed intuitionistic Kripke structure $\langle K, w\rangle$ with $K = \langle W, \leq, V\rangle$, we have:

- $\langle K, w\rangle \not\models^{\mathrm{IPC}}_{\Sigma} \bot$
- $\langle K, w\rangle \models^{\mathrm{IPC}}_{\Sigma} A \iff A \in V(w)$
- $\langle K, w\rangle \models^{\mathrm{IPC}}_{\Sigma} \varphi \wedge \psi \iff \langle K, w\rangle \models^{\mathrm{IPC}}_{\Sigma} \varphi \wedge \langle K, w\rangle \models^{\mathrm{IPC}}_{\Sigma} \psi$
- $\langle K, w\rangle \models^{\mathrm{IPC}}_{\Sigma} \varphi \vee \psi \iff \langle K, w\rangle \models^{\mathrm{IPC}}_{\Sigma} \varphi \vee \langle K, w\rangle \models^{\mathrm{IPC}}_{\Sigma} \psi$
- $\langle K, w\rangle \models^{\mathrm{IPC}}_{\Sigma} \varphi \to \psi \iff \forall w' \in W(w \leq w' \implies (\langle K, w'\rangle \models^{\mathrm{IPC}}_{\Sigma} \varphi \implies \langle K, w'\rangle \models^{\mathrm{IPC}}_{\Sigma} \psi))$

If $\langle K, w\rangle \models^{\mathrm{IPC}}_{\Sigma} \varphi$, then we say that φ is satisfied at the state w of K. In this case, we also call $\langle K, w\rangle$ a model of φ.

Negation thus receives a particularly intuitive reading. Since it is defined as an abbreviation for $\varphi \to \bot$, $\neg\varphi$ does not merely mean that φ is false; rather, it means that no proof of φ can be established in any reachable state.

Example 4.11 (A Small Intuitionistic Kripke Structure). *Let $\Sigma = \{P, Q, R\}$ and consider the structure*

$$
\begin{aligned}
W &= \{u_0, u_1, u_2\}, \\
\leq &= \{(u_0, u_0), (u_1, u_1), (u_2, u_2), (u_0, u_1), (u_0, u_2)\}
\end{aligned}
$$

and

$$V(u_0) = \{R\}, \qquad V(u_1) = \{P, R\}, \qquad V(u_2) = \{Q, R\}.$$

The world u_0 represents a state of information in which only R is known. In u_1, P is added; in u_2, Q is added. Thus knowledge grows monotonically along the relation.

In this structure, for example, we have:

$$
\begin{aligned}
\langle K, u_1 \rangle &\models^{\mathrm{IPC}} P \wedge R, \\
\langle K, u_0 \rangle &\not\models^{\mathrm{IPC}} P \wedge R, \\
\langle K, u_2 \rangle &\not\models^{\mathrm{IPC}} P \wedge R
\end{aligned}
$$

Moreover,

$$\langle K, u_0 \rangle \not\models^{\mathrm{IPC}} (P \wedge R) \to Q,$$

because in the reachable world u_1, $P \wedge R$ holds, but Q does not. By contrast,

$$\langle K, u_2 \rangle \models^{\mathrm{IPC}} (P \wedge R) \to Q,$$

because from u_2 there is no strictly more informative successor state in which $P \wedge R$ holds.

Furthermore,

$$\langle K, u_0 \rangle \not\models^{\mathrm{IPC}} \neg P,$$

because there is a reachable state, namely u_1, in which P is satisfied.

One of the central advantages of Kripke semantics is that it makes the persistence of truth visible: whatever already holds at one state continues to hold at all reachable states. This is precisely what we now formulate in the following proposition.

Proposition 4.12. *Let Σ be a propositional signature, and let $K = \langle W, \leq, V \rangle$ be an intuitionistic Kripke structure over Σ. Then for all $w, w' \in W$ and all formulas φ, the following holds:*

$$w \leq w' \text{ and } \langle K, w \rangle \models^{\mathrm{IPC}} \varphi \quad \Rightarrow \quad \langle K, w' \rangle \models^{\mathrm{IPC}} \varphi.$$

Proof. We prove the statement by structural induction on the formula φ.

Case $\varphi = \bot$. This case is immediate, since $\bot$ is satisfied in no world. Hence the assumption $\langle K, w \rangle \models^{\mathrm{IPC}} \bot$ can never hold, so the implication is trivially true.

Case $\varphi = A$ with $A \in \Sigma$. From $\langle K, w \rangle \models^{\mathrm{IPC}} A$, it follows by definition that $A \in V(w)$. Since K is monotone and $w \leq w'$, we have $V(w) \subseteq V(w')$. Therefore $A \in V(w')$, and hence

$$\langle K, w' \rangle \models^{\mathrm{IPC}} A.$$

Case $\varphi = \varphi_1 \wedge \varphi_2$**.** From $\langle K, w \rangle \models^{\mathrm{IPC}} \varphi_1 \wedge \varphi_2$, it follows that

$$\langle K, w \rangle \models^{\mathrm{IPC}} \varphi_1 \quad \text{and} \quad \langle K, w \rangle \models^{\mathrm{IPC}} \varphi_2.$$

By the induction hypothesis, we therefore also have

$$\langle K, w' \rangle \models^{\mathrm{IPC}} \varphi_1 \quad \text{and} \quad \langle K, w' \rangle \models^{\mathrm{IPC}} \varphi_2.$$

Thus,

$$\langle K, w' \rangle \models^{\mathrm{IPC}} \varphi_1 \wedge \varphi_2.$$

Case $\varphi = \varphi_1 \vee \varphi_2$**.** From

$$\langle K, w \rangle \models^{\mathrm{IPC}} \varphi_1 \vee \varphi_2$$

it follows that either

$$\langle K, w \rangle \models^{\mathrm{IPC}} \varphi_1 \quad \text{or} \quad \langle K, w \rangle \models^{\mathrm{IPC}} \varphi_2.$$

Without loss of generality, assume that the first case holds. Then the induction hypothesis yields

$$\langle K, w' \rangle \models^{\mathrm{IPC}} \varphi_1.$$

Therefore,

$$\langle K, w' \rangle \models^{\mathrm{IPC}} \varphi_1 \vee \varphi_2.$$

The second case is entirely analogous.

Case $\varphi = \varphi_1 \to \varphi_2$. Suppose that

$$\langle K, w\rangle \models^{\mathrm{IPC}} \varphi_1 \to \varphi_2.$$

We must show that

$$\langle K, w'\rangle \models^{\mathrm{IPC}} \varphi_1 \to \varphi_2$$

also holds. So let $w' \leq w''$ and assume that

$$\langle K, w''\rangle \models^{\mathrm{IPC}} \varphi_1.$$

Since $\leq$ is transitive and $w \leq w'$, it follows that $w \leq w''$ as well. From the assumption $\langle K, w\rangle \models^{\mathrm{IPC}} \varphi_1 \to \varphi_2$, we therefore obtain

$$\langle K, w''\rangle \models^{\mathrm{IPC}} \varphi_2.$$

Hence the defining condition for

$$\langle K, w'\rangle \models^{\mathrm{IPC}} \varphi_1 \to \varphi_2$$

is satisfied.

This proves the claim in all cases. □

The proposition expresses exactly the idea that motivates intuitionistic Kripke structures: truth grows with knowledge. Once a formula has been established at one state, it remains valid in all later states that contain more information.

In the next section, we examine this semantics more closely and discuss some of its distinctive properties.

4.1.3 Properties

In this section, we consider some basic properties of intuitionistic logic. In particular, we investigate how notions such as tautology, satisfiability, and logical consequence behave in this setting.

Definition 4.13. *Let Σ be a propositional signature, and let φ be a formula of intuitionistic logic over Σ. Then:*

- *φ is called a* tautology *if for every pointed intuitionistic Kripke structure $\langle K, w\rangle$ over Σ, we have*
$$\langle K, w\rangle \models^{\mathrm{IPC}} \varphi.$$
- *φ is called* satisfiable *if there exists at least one pointed intuitionistic Kripke structure $\langle K, w\rangle$ over Σ such that*
$$\langle K, w\rangle \models^{\mathrm{IPC}} \varphi.$$

A tautology is therefore a formula that holds in every conceivable informational state of every intuitionistic Kripke structure. A formula is satisfiable as soon as it holds in at least one state of such a structure.

Example 4.14 (A formula that is not intuitionistically valid in general). *Consider the signature $\Sigma = \{A\}$ and the intuitionistic Kripke structure $K = \langle W, \leq, V\rangle$ with*

$$W = \{w_0, w_1, w_2\}, \qquad w_0 \leq w_1,\ w_1 \leq w_2,$$

and

$$V(w_0) = \varnothing, \qquad V(w_1) = \varnothing, \qquad V(w_2) = \{A\}.$$

Then $\langle K, w_0\rangle \not\models^{\mathrm{IPC}} A$, *since* $A \notin V(w_0)$. *Moreover,* $\langle K, w_0\rangle \not\models^{\mathrm{IPC}} \neg A$ *as well, because* A *is satisfiable in the reachable state* w_2. *Consequently,*

$$\langle K, w_0\rangle \not\models^{\mathrm{IPC}} A \vee \neg A.$$

Thus, $A \vee \neg A$ *is not a tautology of intuitionistic logic.*

Proposition 4.15. *There exists a formula* φ *of intuitionistic logic over* Σ *such that* $\varphi \vee \neg\varphi$ *is not a tautology.*

Proof. We choose $\varphi := A$ and use the Kripke structure from the preceding example. Then in the pointed structure $\langle K, w_0\rangle$, neither A nor $\neg A$ holds. Therefore, $A \vee \neg A$ does not hold there either. This shows that the law of excluded middle does not hold universally in intuitionistic logic. □

An important consequence of this fact is that negation must be read more cautiously in intuitionistic logic than in classical logic. From the failure of a formula to hold, it does not automatically follow that its negation holds. Likewise, a formula and its double negation do not always coincide in intuitionistic logic.

Example 4.16 (Double negation is weaker than the statement itself). *Let us again use the structure from the preceding example. Then we have*

$$\langle K, w_0\rangle \not\models^{\mathrm{IPC}} A,$$

but at the same time

$$\langle K, w_0\rangle \models^{\mathrm{IPC}} \neg\neg A.$$

For in every state reachable from w_0, there is a further reachable state in which A holds. Intuitionistically, one can therefore sometimes prove the double negation without having the statement itself.

Proposition 4.17. *There exist a formula φ and a pointed intuitionistic Kripke structure $\langle K, w\rangle$ such that*

$$\langle K, w\rangle \not\models^{\mathrm{IPC}} \varphi \qquad \textit{and} \qquad \langle K, w\rangle \models^{\mathrm{IPC}} \neg\neg\varphi.$$

Proof. We choose $\varphi := A$ and the Kripke structure from the example above. Then $\langle K, w_0\rangle \not\models^{\mathrm{IPC}} A$. On the other hand, $\langle K, w_0\rangle \models^{\mathrm{IPC}} \neg\neg A$, because every state reachable from w_0 still has a successor in which A holds. □

A fundamental connection between classical and intuitionistic logic is Glivenko's theorem. It states that classical tautologies reappear in intuitionistic logic after double negation.

Proposition 4.18 (Glivenko's Theorem). *Let Σ be a propositional signature. A formula φ is a classical tautology if and only if $\neg\neg\varphi$ is a tautology of intuitionistic logic.*

Proof. We prove the two directions separately.

(i) Classical tautology ⇒ intuitionistic tautology of the double negation. Let φ be a classical tautology. Then φ is derivable in classical logic. By Glivenko's theorem, or equivalently by the Kolmogorov negative translation, $\neg\neg\varphi$ is then also derivable intuitionistically.

A standard way to see this is via the negative translation τ, which turns every classical formula into a formula compatible with intuitionistic logic. For atomic formulas, we have

$$\tau(p) = \neg\neg p,$$

and the connectives are translated in such a way that classical proofs can be transformed into intuitionistic ones. In particular, for every classical tautology φ, $\tau(\varphi)$ is also intuitionistically derivable, and moreover intuitionistically

$$\tau(\varphi) \to \neg\neg\varphi.$$

From this it follows that

$$\vdash_{\mathrm{NJ}} \neg\neg\varphi.$$

Hence $\neg\neg\varphi$ is a tautology of intuitionistic logic.

(ii) Intuitionistic tautology of the double negation $\Rightarrow$ classical tautology. Now let $\neg\neg\varphi$ be a tautology of intuitionistic logic. Then $\neg\neg\varphi$ is in particular also classically valid, since every intuitionistically valid formula is also classically valid.

In classical logic, however, the following is valid for every formula φ:

$$\neg\neg\varphi \to \varphi$$

Therefore, the classical validity of $\neg\neg\varphi$ implies the classical validity of φ.

Thus we have shown:

$$\varphi \text{ is classically valid } \iff \neg\neg\varphi \text{ is intuitionistically valid.}$$

□

This result shows that intuitionistic logic is not simply completely detached from classical logic. Rather, classical validities are preserved in a weakened, doubly negated form. What is especially important here is that intuitionistic proofs are more exacting: they require not only truth, but a constructive justification.

Another characteristic feature of intuitionistic logic is its so-called disjunction property. It expresses that a provable disjunction must also have a provable disjunct.

Proposition 4.19 (Disjunction Property). *Let Σ be a propositional signature, and let $\varphi \vee \psi$ be a formula of intuitionistic logic over Σ. If $\varphi \vee \psi$ is a tautology, then either φ is a tautology or ψ is a tautology.*

Proof. We prove the statement by contraposition. So assume that neither φ nor ψ is a tautology. Then there exist marked intuitionistic Kripke structures $\langle K_\varphi, w_\varphi \rangle$ and $\langle K_\psi, w_\psi \rangle$ such that

$$\langle K_\varphi, w_\varphi \rangle \not\models^{\mathrm{IPC}} \varphi \qquad \text{and} \qquad \langle K_\psi, w_\psi \rangle \not\models^{\mathrm{IPC}} \psi.$$

From these two structures we construct a new intuitionistic Kripke structure $K = \langle W, \leq, V \rangle$ by taking their disjoint union and additionally introducing a new world w_0 that lies below both initial worlds, that is,

$$w_0 \leq w_\varphi \qquad \text{and} \qquad w_0 \leq w_\psi.$$

For w_0 we choose no additional information, for instance $V(w_0) = \varnothing$. The remaining valuations and order relations are inherited from K_φ and K_ψ.

Now, by Proposition 4.12, if w_0 satisfied either of the formulas φ or ψ, then that same formula would also have to be satisfied in every accessible world. In particular, we would then have either

$$\langle K_\varphi, w_\varphi \rangle \models^{\mathrm{IPC}} \varphi \quad \text{or} \quad \langle K_\psi, w_\psi \rangle \models^{\mathrm{IPC}} \psi,$$

contrary to the choice of the two initial structures.

Therefore, neither

$$\langle K, w_0 \rangle \models^{\mathrm{IPC}} \varphi \quad \text{nor} \quad \langle K, w_0 \rangle \models^{\mathrm{IPC}} \psi.$$

By the semantics of disjunction, it follows that

$$\langle K, w_0 \rangle \not\models^{\mathrm{IPC}} \varphi \vee \psi.$$

Hence $\varphi \vee \psi$ is not a tautology. The contraposition therefore shows: If $\varphi \vee \psi$ is a tautology, then either φ is a tautology or ψ is a tautology. □

This statement fits well with the constructive reading of logic: a disjunction is only genuinely secured if one has secured at least one of its two parts. This is precisely a clear difference from classical logic, where a disjunction can often be proved merely by a case distinction.

We now define logical consequence for intuitionistic logic. In doing so, we use the same symbol as for the satisfaction relation.

Definition 4.20. *Let Σ be a propositional signature. The logical consequence relation of intuitionistic logic over Σ, written $\models^{\mathrm{IPC}}$, is the relation between sets of formulas of intuitionistic logic over Σ and formulas of intuitionistic logic over Σ such that for all sets of formulas Γ and all formulas φ,*

$$\Gamma \models^{\mathrm{IPC}} \varphi$$

holds if and only if for all marked intuitionistic Kripke structures $\langle K, w\rangle$,

$$\langle K, w\rangle \models^{\mathrm{IPC}} \Gamma \quad \Rightarrow \quad \langle K, w\rangle \models^{\mathrm{IPC}} \varphi.$$

Here, too, the underlying intuition is the same as in classical logic: a formula follows from a set of premises if it holds in every state in which all of the premises hold as well. Once again, the difference lies in the semantics being used.

As in classical logic, there is also a close connection here between logical consequence and implication.

Theorem 4.21 (Deduction Theorem). *Let Σ be a propositional signature. For every set of intuitionistic formulas Γ over Σ and all intuitionistic formulas φ, ψ over Σ, the following holds:*

$$\Gamma \cup \{\varphi\} \models^{\mathrm{IPC}} \psi \quad \textit{if and only if} \quad \Gamma \models^{\mathrm{IPC}} \varphi \to \psi.$$

Proof. We prove the two directions separately.

($\Rightarrow$) Assume that

$$\Gamma \cup \{\varphi\} \models^{\mathrm{IPC}} \psi.$$

We show that this implies

$$\Gamma \models^{\mathrm{IPC}} \varphi \to \psi.$$

To this end, let $\langle K, w\rangle$ be an arbitrary marked intuitionistic Kripke structure such that

$$\langle K, w\rangle \models^{\mathrm{IPC}} \Gamma.$$

We must show that

$$\langle K, w\rangle \models^{\mathrm{IPC}} \varphi \to \psi.$$

By the definition of satisfaction for an implication, it suffices to show that for all $w' \in W$ with $w \leq w'$,

$$\langle K, w'\rangle \models^{\mathrm{IPC}} \varphi \quad \Rightarrow \quad \langle K, w'\rangle \models^{\mathrm{IPC}} \psi.$$

So let $w' \in W$ with $w \leq w'$ and suppose that

$$\langle K, w'\rangle \models^{\mathrm{IPC}} \varphi.$$

Since $\langle K, w\rangle \models^{\mathrm{IPC}} \Gamma$ and intuitionistic semantics is monotone, Proposition 4.12 implies that

$$\langle K, w'\rangle \models^{\mathrm{IPC}} \Gamma.$$

Together with the assumption $\langle K, w'\rangle \models^{\mathrm{IPC}} \varphi$, we therefore obtain

$$\langle K, w'\rangle \models^{\mathrm{IPC}} \Gamma \cup \{\varphi\}.$$

By the premise $\Gamma \cup \{\varphi\} \models^{\mathrm{IPC}} \psi$, it follows that

$$\langle K, w'\rangle \models^{\mathrm{IPC}} \psi.$$

This shows that

$$\langle K, w\rangle \models^{\mathrm{IPC}} \varphi \to \psi.$$

Since $\langle K, w\rangle$ was arbitrary, we conclude that

$$\Gamma \models^{\mathrm{IPC}} \varphi \to \psi.$$

($\Leftarrow$) Now assume that

$$\Gamma \models^{\mathrm{IPC}} \varphi \to \psi.$$

We show that this implies

$$\Gamma \cup \{\varphi\} \models^{\mathrm{IPC}} \psi.$$

To this end, let $\langle K, w\rangle$ be an arbitrary marked intuitionistic Kripke structure such that

$$\langle K, w\rangle \models^{\mathrm{IPC}} \Gamma \cup \{\varphi\}.$$

Then in particular,

$$\langle K, w\rangle \models^{\mathrm{IPC}} \Gamma \quad \text{and} \quad \langle K, w\rangle \models^{\mathrm{IPC}} \varphi.$$

From the assumption $\Gamma \models^{\mathrm{IPC}} \varphi \to \psi$, it therefore follows that

$$\langle K, w\rangle \models^{\mathrm{IPC}} \varphi \to \psi.$$

By the definition of satisfaction for an implication, this holds in particular for $w' = w$:

$$\langle K, w\rangle \models^{\mathrm{IPC}} \varphi \quad \Rightarrow \quad \langle K, w\rangle \models^{\mathrm{IPC}} \psi.$$

Since we have already established that $\langle K, w\rangle \models^{\text{IPC}} \varphi$, it follows that

$$\langle K, w\rangle \models^{\text{IPC}} \psi.$$

Hence,

$$\Gamma \cup \{\varphi\} \models^{\text{IPC}} \psi.$$

We have thus proved both directions, and therefore

$$\Gamma \cup \{\varphi\} \models^{\text{IPC}} \psi \quad \text{if and only if} \quad \Gamma \models^{\text{IPC}} \varphi \to \psi.$$

□

The Deduction Theorem shows that implication in intuitionistic logic plays precisely the role of a transition from premises to conclusions. It is therefore one of the central tools for later proofs in this chapter.

Overall, this section shows that intuitionistic logic is indeed close to classical propositional logic, but is more restrained at several crucial points. It is precisely this restraint that makes it especially interesting for constructive mathematics and computer science.

4.1.4 Proof-Theoretic Semantics

In the final part of this section, we turn to a natural deduction calculus for intuitionistic logic, namely the system *NJ* (*natural deduction for intuitionistic logic*). The idea is to describe logical consequence not only semantically in terms of Kripke structures, but also syntactically by means of proofs. We will show that these two perspectives match exactly: a formula is derivable in NJ if and only if it is a semantic consequence in intuitionistic logic.

Unlike the calculi from the earlier chapters, NJ works directly with consequence relations of the form

$$\Gamma \vdash \varphi.$$

Here, $\Gamma \vdash \varphi$ means that the formula φ can be derived from the assumptions in Γ. If Γ is a set of formulas and φ is a formula, we also write

$$\Gamma, \psi \vdash \varphi$$

as shorthand for $\Gamma \cup \{\psi\} \vdash \varphi$. In addition, we use

$$\vdash \varphi$$

as an abbreviation for $\emptyset \vdash \varphi$.

As usual, the rules of NJ are presented as inference schemata. The only rule without premises is the axiom:

$$\frac{}{\Gamma, \varphi \vdash \varphi}\ (\text{Ax})$$

For the connectives $\wedge$, $\vee$, and $\rightarrow$, there are introduction and elimination rules:

$$\frac{\Gamma \vdash \varphi \qquad \Gamma \vdash \psi}{\Gamma \vdash \varphi \wedge \psi}\ (\wedge I)$$

$$\frac{\Gamma \vdash \varphi \wedge \psi}{\Gamma \vdash \varphi}\ (\wedge E_1) \qquad \frac{\Gamma \vdash \varphi \wedge \psi}{\Gamma \vdash \psi}\ (\wedge E_2)$$

$$\frac{\Gamma \vdash \varphi}{\Gamma \vdash \varphi \vee \psi} (\vee I_1) \qquad \frac{\Gamma \vdash \psi}{\Gamma \vdash \varphi \vee \psi} (\vee I_2)$$

$$\frac{\Gamma, \varphi \vdash \vartheta \qquad \Gamma, \psi \vdash \vartheta \qquad \Gamma \vdash \varphi \vee \psi}{\Gamma \vdash \vartheta} (\vee E)$$

$$\frac{\Gamma, \varphi \vdash \psi}{\Gamma \vdash \varphi \to \psi} (\to I) \qquad \frac{\Gamma \vdash \varphi \to \psi \qquad \Gamma \vdash \varphi}{\Gamma \vdash \psi} (\to E)$$

Finally, there is the rule stating that any formula follows from a contradiction:

$$\frac{\Gamma \vdash \bot}{\Gamma \vdash \varphi} (\bot E)$$

An NJ derivation is a finite proof tree whose root is the conclusion $\Gamma \vdash \varphi$ to be established. The leaves of such a tree are instances of the rule (Ax). The placeholders φ, ψ, and ϑ are replaced by concrete formulas.

Example 4.22 (A short derivation). *We show that for every propositional variable P, the formula $\vdash P \to P$ is derivable in NJ:*

$$\frac{\overline{P \vdash P} \text{ (Ax)}}{\vdash P \to P} (\to I)$$

This derivation is short, but it already illustrates the basic idea of natural deduction: if we assume P and can derive P again, then $P \to P$ follows.

Example 4.23 (A small derivation with two assumptions). *Let*

$$\Gamma = \{U \wedge V,\ V \to W\}.$$

Then the conclusion $\Gamma \vdash W$ *can be derived in NJ:*

$$\frac{}{\Gamma \vdash U \wedge V}\ (\text{Ax}) \qquad \frac{}{\Gamma \vdash V \to W}\ (\text{Ax}) \tag{4.1}$$

$$\frac{\Gamma \vdash U \wedge V}{\Gamma \vdash V}\ (\wedge E_2) \tag{4.2}$$

$$\frac{\Gamma \vdash V \to W \qquad \Gamma \vdash V}{\Gamma \vdash W}\ (\to E) \tag{4.3}$$

This makes the working style of NJ visible: from a conjunction we extract one component, and then use an implication to continue the derivation.

Example 4.24 (A derivation by case distinction). *Let*

$$\Gamma = \{P \to T,\ Q \to T,\ P \vee Q\}.$$

Then NJ also yields $\Gamma \vdash T$*:*

$$\frac{\frac{}{\Gamma, P \vdash T}\ (\to E) \qquad \frac{}{\Gamma, Q \vdash T}\ (\to E) \qquad \frac{}{\Gamma \vdash P \vee Q}\ (\text{Ax})}{\Gamma \vdash T}\ (\vee E)$$

Disjunction elimination thus combines the two cases and shows how a common conclusion can be obtained from a case distinction.

Definition 4.25. *Let* Σ *be a propositional signature. The NJ consequence relation* $\vdash_{\text{NJ}}$ *over* Σ *is the relation between sets of formulas of intuitionistic logic over* Σ

and formulas of intuitionistic logic over Σ such that, for every set of formulas Γ and every formula φ, the following holds:

$$\Gamma \vdash_{\mathrm{NJ}} \varphi$$

if and only if there exists a finite NJ derivation of the sequent $\Gamma \vdash \varphi$.

Theorem 4.26. *Let Σ be a propositional signature. For every set Γ of formulas of intuitionistic logic and every formula φ of intuitionistic logic, we have:*

$$\Gamma \vdash_{\mathrm{NJ}} \varphi \quad \textit{if and only if} \quad \Gamma \models^{\mathrm{IPC}} \varphi.$$

We prove this in two steps. First, we show that all NJ rules are valid with respect to Kripke semantics. Then we establish completeness by means of a canonical Kripke structure.

Lemma 4.27. *Let Σ be a propositional signature. For every set Γ of formulas of intuitionistic logic and every formula φ of intuitionistic logic, the following holds: if $\Gamma \vdash_{\mathrm{NJ}} \varphi$, then $\Gamma \models^{\mathrm{IPC}} \varphi$.*

Proof. It suffices to show that every rule of NJ is semantically valid. We check the rules one by one.

Rule (Ax). We must show:

$$\Gamma \cup \{\varphi\} \models^{\mathrm{IPC}} \varphi.$$

Let $\langle K, w\rangle$ be a pointed intuitionistic Kripke structure such that

$$\langle K, w\rangle \models^{\mathrm{IPC}} \Gamma \cup \{\varphi\}.$$

Then, in particular, $\langle K, w\rangle \models^{\text{IPC}} \varphi$, since $\varphi \in \Gamma \cup \{\varphi\}$. Hence the rule is valid.

Rule ($\bot E$). We must show:

$$\Gamma \models^{\text{IPC}} \bot \;\Rightarrow\; \Gamma \models^{\text{IPC}} \varphi.$$

So let $\langle K, w\rangle$ be a pointed intuitionistic Kripke structure with $\langle K, w\rangle \models^{\text{IPC}} \Gamma$ and, in addition, $\langle K, w\rangle \models^{\text{IPC}} \bot$. This is impossible, since $\bot$ is satisfied in no Kripke structure and at no world. Thus there is no countermodel at all, and the implication is trivially valid.

Rule ($\wedge I$). Assume that

$$\Gamma \models^{\text{IPC}} \varphi \qquad \text{and} \qquad \Gamma \models^{\text{IPC}} \psi.$$

Let $\langle K, w\rangle$ be a pointed Kripke structure such that

$$\langle K, w\rangle \models^{\text{IPC}} \Gamma.$$

Then, by assumption, we have both

$$\langle K, w\rangle \models^{\text{IPC}} \varphi \quad \text{and} \quad \langle K, w\rangle \models^{\text{IPC}} \psi.$$

By the semantics of conjunction, it follows that

$$\langle K, w\rangle \models^{\text{IPC}} \varphi \wedge \psi.$$

Rule ($\wedge E_1$). Assume that

$$\Gamma \models^{\text{IPC}} \varphi \wedge \psi.$$

Let $\langle K, w\rangle$ be a pointed Kripke structure with $\langle K, w\rangle \models^{\text{IPC}} \Gamma$. Then

$$\langle K, w\rangle \models^{\text{IPC}} \varphi \wedge \psi.$$

By the semantics of conjunction, this immediately implies

$$\langle K, w\rangle \models^{\text{IPC}} \varphi.$$

Hence the rule $(\wedge E_1)$ is valid as well.

Rule $(\wedge E_2)$. The proof is entirely analogous to that of $(\wedge E_1)$. From

$$\langle K, w\rangle \models^{\text{IPC}} \varphi \wedge \psi$$

it follows directly that

$$\langle K, w\rangle \models^{\text{IPC}} \psi.$$

Rule $(\vee I_1)$. Assume that

$$\Gamma \models^{\text{IPC}} \varphi.$$

Let $\langle K, w\rangle$ be a pointed Kripke structure with $\langle K, w\rangle \models^{\text{IPC}} \Gamma$. Then $\langle K, w\rangle \models^{\text{IPC}} \varphi$ as well. By the semantics of disjunction, it follows immediately that

$$\langle K, w\rangle \models^{\text{IPC}} \varphi \vee \psi.$$

Rule $(\vee I_2)$. The proof is analogous to that of $(\vee I_1)$.

Rule $(\vee E)$. Assume that

$$\Gamma \models^{\text{IPC}} \varphi \vee \psi, \qquad \Gamma, \varphi \models^{\text{IPC}} \vartheta, \qquad \Gamma, \psi \models^{\text{IPC}} \vartheta.$$

We show that

$$\Gamma \models^{\mathrm{IPC}} \vartheta.$$

To this end, let $\langle K, w\rangle$ be a pointed Kripke structure such that

$$\langle K, w\rangle \models^{\mathrm{IPC}} \Gamma.$$

Then we also have

$$\langle K, w\rangle \models^{\mathrm{IPC}} \varphi \vee \psi.$$

By the semantics of disjunction, either

$$\langle K, w\rangle \models^{\mathrm{IPC}} \varphi \quad \text{or} \quad \langle K, w\rangle \models^{\mathrm{IPC}} \psi.$$

In the first case, $\Gamma, \varphi \models^{\mathrm{IPC}} \vartheta$ implies that $\langle K, w\rangle \models^{\mathrm{IPC}} \vartheta$. In the second case, the same follows from $\Gamma, \psi \models^{\mathrm{IPC}} \vartheta$. Hence, in either case,

$$\langle K, w\rangle \models^{\mathrm{IPC}} \vartheta.$$

Thus the rule is valid.

Rule ($\to I$). Assume that

$$\Gamma, \varphi \models^{\mathrm{IPC}} \psi.$$

We show that

$$\Gamma \models^{\mathrm{IPC}} \varphi \to \psi.$$

Let $\langle K, w\rangle$ be a pointed Kripke structure such that $\langle K, w\rangle \models^{\mathrm{IPC}} \Gamma$. We must show that for every $w' \in W$ with $w \leq w'$,

$$\langle K, w'\rangle \models^{\mathrm{IPC}} \varphi \quad \Rightarrow \quad \langle K, w'\rangle \models^{\mathrm{IPC}} \psi.$$

So let $w' \geq w$ and suppose that

$$\langle K, w' \rangle \models^{\mathrm{IPC}} \varphi.$$

By monotonicity of Kripke semantics, it then follows that

$$\langle K, w' \rangle \models^{\mathrm{IPC}} \Gamma.$$

Therefore $\langle K, w' \rangle$ also satisfies $\Gamma \cup \{\varphi\}$, and by the assumption we obtain

$$\langle K, w' \rangle \models^{\mathrm{IPC}} \psi.$$

Hence,

$$\langle K, w \rangle \models^{\mathrm{IPC}} \varphi \to \psi.$$

Rule $(\to E)$. Assume that

$$\Gamma \models^{\mathrm{IPC}} \varphi \to \psi \qquad \text{and} \qquad \Gamma \models^{\mathrm{IPC}} \varphi.$$

Let $\langle K, w \rangle$ be a pointed Kripke structure with $\langle K, w \rangle \models^{\mathrm{IPC}} \Gamma$. Then

$$\langle K, w \rangle \models^{\mathrm{IPC}} \varphi \to \psi \quad \text{and} \quad \langle K, w \rangle \models^{\mathrm{IPC}} \varphi.$$

If we instantiate the semantics of implication with $w' = w$, it follows that

$$\langle K, w \rangle \models^{\mathrm{IPC}} \psi.$$

Hence the rule $(\to E)$ is valid as well.

Thus all rules of NJ are semantically valid. Consequently, if $\Gamma \vdash_{\mathrm{NJ}} \varphi$, then also $\Gamma \models^{\mathrm{IPC}} \varphi$. □

For the converse direction, we need a canonical Kripke structure. The decisive building block is an extension lemma for NJ-theories.

Definition 4.28 (NJ-Theory). *A set D of formulas of intuitionistic logic is called an* NJ-theory *if*

$$D = \{\varphi \mid D \vdash_{\mathrm{NJ}} \varphi\}$$

holds and, moreover, $\bot \notin D$. An NJ-theory D is called disjunctive *if from*

$$D \vdash_{\mathrm{NJ}} \varphi \vee \psi$$

it already follows that

$$D \vdash_{\mathrm{NJ}} \varphi \quad \text{or} \quad D \vdash_{\mathrm{NJ}} \psi.$$

Disjunctive NJ-theories are therefore consistent, deductively closed sets with the property that any provable disjunction can be resolved into one of its two disjuncts. This is exactly what corresponds to the constructive reading of disjunction.

Proposition 4.29. *For every set of formulas Γ such that $\Gamma \nvdash_{\mathrm{NJ}} \bot$, and every formula φ such that $\Gamma \nvdash_{\mathrm{NJ}} \varphi$, there exists a disjunctive NJ-theory D such that*

$$\Gamma \subseteq D \qquad \text{and} \qquad D \nvdash_{\mathrm{NJ}} \varphi.$$

Proof. We sketch a Lindenbaum-style construction in which we successively divide the formulas into two sets: a theory D and a complementary set Δ.

We begin by setting

$$D_0 := \{\psi \mid \Gamma \vdash_{\mathrm{NJ}} \psi\} \qquad \text{and} \qquad \Delta_0 := \{\varphi\}.$$

Then D_0 is already closed under NJ-derivability, and since $\Gamma \nvdash_{\mathrm{NJ}} \bot$, we also have $\bot \notin D_0$.

Now choose an enumeration of all remaining formulas

$$\chi_0, \chi_1, \chi_2, \ldots$$

and define sets D_n and Δ_n inductively so that the following invariant holds: for every finite choice $\delta_1, \ldots, \delta_k \in \Delta_n$,

$$D_n \nvdash_{\mathrm{NJ}} \delta_1 \vee \cdots \vee \delta_k. \tag{$*_n$}$$

Assume that D_n and Δ_n have already been constructed and that χ_n is the next formula in the enumeration. We first try to add χ_n to D_n. If the invariant is preserved, we set

$$D_{n+1} := \{\psi \mid D_n \cup \{\chi_n\} \vdash_{\mathrm{NJ}} \psi\}, \qquad \Delta_{n+1} := \Delta_n.$$

Otherwise, we set

$$D_{n+1} := D_n, \qquad \Delta_{n+1} := \Delta_n \cup \{\chi_n\}.$$

It remains to show that this decision is always possible. Suppose both options were to fail. Then on the one hand there would be formulas

$$\delta_1, \ldots, \delta_k \in \Delta_n$$

such that

$$D_n \cup \{\chi_n\} \vdash_{\mathrm{NJ}} \delta_1 \vee \cdots \vee \delta_k,$$

and on the other hand there would also be formulas

$$\varepsilon_1, \ldots, \varepsilon_m \in \Delta_n$$

such that

$$D_n \vdash_{\mathrm{NJ}} \varepsilon_1 \vee \cdots \vee \varepsilon_m \vee \chi_n.$$

From the first derivation, the deduction theorem yields

$$D_n \vdash_{\mathrm{NJ}} \chi_n \to (\delta_1 \vee \cdots \vee \delta_k).$$

Together with the second derivation, an application of $(\vee E)$ then gives a derivation of

$$D_n \vdash_{\mathrm{NJ}} \delta_1 \vee \cdots \vee \delta_k$$

or of

$$D_n \vdash_{\mathrm{NJ}} \varepsilon_1 \vee \cdots \vee \varepsilon_m,$$

depending on whether the disjunct χ_n or one of the ε_i is used. In either case this contradicts the invariant $(*_n)$. Hence both options cannot fail.

Now let

$$D := \bigcup_{n \in \mathbb{N}} D_n, \qquad \Delta := \bigcup_{n \in \mathbb{N}} \Delta_n.$$

Then D is deductively closed, contains Γ, and since $\varphi \in \Delta$, it follows in particular that $D \nvdash_{\mathrm{NJ}} \varphi$.

It remains to prove that D is disjunctive. So suppose that

$$D \vdash_{\mathrm{NJ}} \alpha \vee \beta.$$

If $D \vdash_{\mathrm{NJ}} \alpha$ or $D \vdash_{\mathrm{NJ}} \beta$, we are done. Otherwise, α and β belong to Δ, since all formulas were assigned to one of the two sets during the construction. But then

$$D \vdash_{\mathrm{NJ}} \alpha \vee \beta$$

violates the invariant $(*_n)$ for some sufficiently large n in which both α and β already lie in Δ_n. This is a contradiction. Therefore D is disjunctive.

This proves the proposition. □

With this extension lemma in place, we can now prove completeness.

Lemma 4.30. *Let Σ be a propositional signature. For every set Γ of formulas of intuitionistic logic and every formula φ of intuitionistic logic, the following holds: if $\Gamma \models^{\mathrm{IPC}} \varphi$, then $\Gamma \vdash_{\mathrm{NJ}} \varphi$.*

Proof. We prove the contrapositive: from $\Gamma \nvdash_{\mathrm{NJ}} \varphi$ it follows that $\Gamma \not\models^{\mathrm{IPC}} \varphi$.

So assume that $\Gamma \nvdash_{\mathrm{NJ}} \varphi$. By Proposition 4.29, there exists a disjunctive NJ-theory D such that

$$\Gamma \subseteq D \qquad \text{and} \qquad D \nvdash_{\mathrm{NJ}} \varphi.$$

We construct the canonical Kripke structure

$$K_\Gamma = \langle W_\Gamma, \leq_\Gamma, V_\Gamma \rangle$$

by setting:

- W_Γ is the set of all disjunctive NJ-theories E with $\Gamma \subseteq E$,
- $E_1 \leq_\Gamma E_2$ if and only if $E_1 \subseteq E_2$,
- $A \in V_\Gamma(E)$ if and only if $A \in E$.

We now show, by structural induction on the formula ψ, that for every formula ψ we have:

$$\langle K_\Gamma, D\rangle \models^{\mathrm{IPC}} \psi \quad \Longleftrightarrow \quad D \vdash_{\mathrm{NJ}} \psi. \tag{†}$$

Case $\psi = \bot$. On the left, $\langle K_\Gamma, D\rangle \models^{\mathrm{IPC}} \bot$ never holds; on the right, by consistency of D, we likewise never have $D \vdash_{\mathrm{NJ}} \bot$. Hence (†).

Case $\psi = A$ for a propositional variable A. Then

$$\langle K_\Gamma, D\rangle \models^{\mathrm{IPC}} A \iff A \in V_\Gamma(D) \iff A \in D.$$

Since D is an NJ-theory, we have

$$A \in D \iff D \vdash_{\mathrm{NJ}} A.$$

Thus (†) is established.

Case $\psi = \psi_1 \wedge \psi_2$. By the induction hypothesis,

$$\langle K_\Gamma, D\rangle \models^{\mathrm{IPC}} \psi_1 \iff D \vdash_{\mathrm{NJ}} \psi_1$$

and

$$\langle K_\Gamma, D\rangle \models^{\mathrm{IPC}} \psi_2 \iff D \vdash_{\mathrm{NJ}} \psi_2$$

Hence, by the semantics of conjunction,

$$\langle K_\Gamma, D\rangle \models^{\mathrm{IPC}} \psi_1 \wedge \psi_2 \iff D \vdash_{\mathrm{NJ}} \psi_1 \wedge \psi_2.$$

Case $\psi = \psi_1 \vee \psi_2$. Since D is disjunctive, we have

$$D \vdash_{\mathrm{NJ}} \psi_1 \vee \psi_2 \iff D \vdash_{\mathrm{NJ}} \psi_1 \text{ or } D \vdash_{\mathrm{NJ}} \psi_2.$$

By the induction hypothesis, this is equivalent to

$$\langle K_\Gamma, D\rangle \models^{\mathrm{IPC}} \psi_1 \text{ or } \langle K_\Gamma, D\rangle \models^{\mathrm{IPC}} \psi_2,$$

which is exactly

$$\langle K_\Gamma, D\rangle \models^{\mathrm{IPC}} \psi_1 \vee \psi_2.$$

Case $\psi = \psi_1 \to \psi_2$. First assume that

$$D \vdash_{\mathrm{NJ}} \psi_1 \to \psi_2.$$

Let $E \in W_\Gamma$ with $D \leq_\Gamma E$, that is, $D \subseteq E$. Then also

$$E \vdash_{\mathrm{NJ}} \psi_1 \to \psi_2,$$

because E is an NJ-theory and contains all formulas derivable from D. If in addition $E \vdash_{\mathrm{NJ}} \psi_1$, then by $(\to E)$ it follows that

$$E \vdash_{\mathrm{NJ}} \psi_2.$$

By the induction hypothesis, this means:

$$\langle K_\Gamma, E\rangle \models^{\mathrm{IPC}} \psi_1 \;\Rightarrow\; \langle K_\Gamma, E\rangle \models^{\mathrm{IPC}} \psi_2.$$

Therefore,
$$\langle K_\Gamma, D\rangle \models^{\mathrm{IPC}} \psi_1 \to \psi_2.$$

Now for the converse direction: assume that

$$D \nvdash_{\mathrm{NJ}} \psi_1 \to \psi_2.$$

Then it must be that

$$D \cup \{\psi_1\} \nvdash_{\mathrm{NJ}} \psi_2$$

hold, since otherwise $D \vdash_{\mathrm{NJ}} \psi_1 \to \psi_2$ would already follow from ψ_1 by $(\to I)$, contradicting our assumption. Hence, by Proposition 4.29, there exists a disjunctive NJ-theory E such that

$$D \cup \{\psi_1\} \subseteq E \qquad \text{and} \qquad E \nvdash_{\mathrm{NJ}} \psi_2.$$

Then $D \leq_\Gamma E$, and since $\psi_1 \in E$, we have

$$E \vdash_{\mathrm{NJ}} \psi_1.$$

Thus, by the induction hypothesis,

$$\langle K_\Gamma, E\rangle \models^{\mathrm{IPC}} \psi_1.$$

On the other hand, since $E \nvdash_{\mathrm{NJ}} \psi_2$, the induction hypothesis yields

$$\langle K_\Gamma, E\rangle \not\models^{\mathrm{IPC}} \psi_2.$$

Therefore,
$$\langle K_\Gamma, D\rangle \not\models^{\mathrm{IPC}} \psi_1 \to \psi_2$$

as required.

This proves the equivalence (†) for all formulas.

Applying (†) to our original formula φ, we obtain:

$$D \nvdash_{\mathrm{NJ}} \varphi \quad \Longrightarrow \quad \langle K_\Gamma, D\rangle \not\models^{\mathrm{IPC}} \varphi.$$

Since moreover $\Gamma \subseteq D$, the structure $\langle K_\Gamma, D\rangle$ satisfies all formulas from Γ. Hence $\Gamma \not\models^{\mathrm{IPC}} \varphi$. The contrapositive now yields the claim:

$$\Gamma \models^{\mathrm{IPC}} \varphi \Rightarrow \Gamma \vdash_{\mathrm{NJ}} \varphi.$$

□

The main theorem now follows immediately from Lemma 4.27 and Lemma 4.30.

Proof of Theorem 4.26. One direction is exactly Lemma 4.27, and the other direction is exactly Lemma 4.30. □

Finally, we compare intuitionistic logic once more with classical propositional logic. To do so, it suffices to view each classical valuation as a Kripke model with only a single state.

Proposition 4.31. *Let Σ be a propositional signature. Then for every set of propositional formulas Γ over Σ and every propositional formula φ over Σ, the following holds:*

$$\textit{If } \Gamma \models_{\mathrm{IPC}} \varphi, \textit{ then } \Gamma \models_{\mathrm{AL}} \varphi.$$

Proof. Assume $\Gamma \models_{\text{IPC}} \varphi$. We show that this implies $\Gamma \models_{\text{AL}} \varphi$.

Let v be any classical valuation satisfying all formulas in Γ. From it, we construct an intuitionistic Kripke model with exactly one world:

$$K_v = \langle \{w\}, \leq, V \rangle,$$

where $\leq = \{(w, w)\}$ and

$$V(w) = \{A \in \Sigma \mid v(A) = \text{true}\}.$$

This model is clearly monotone and therefore intuitionistic.

We now prove by structural induction on the propositional formula ψ that

$$\langle K_v, w \rangle \models^{\text{IPC}} \psi \quad \iff \quad v(\psi) = \text{true}.$$

For atomic formulas, this is immediate from the definition of V. For $\wedge$ and $\vee$, the claim is immediate as well, since intuitionistic semantics in a one-world model coincides exactly with classical semantics. The same classical agreement also holds for implication, because there is no accessible state other than w:

$$\langle K_v, w \rangle \models^{\text{IPC}} \psi_1 \to \psi_2$$

if and only if $v(\psi_1) = \text{true}$ implies $v(\psi_2) = \text{true}$, that is, precisely if the classical implication is true. This completes the induction.

Since v satisfies all formulas in Γ, the structure $\langle K_v, w \rangle$ satisfies all formulas in Γ in the intuitionistic sense. Hence, from the assumption

$$\Gamma \models^{\text{IPC}} \varphi$$

it follows that

$$\langle K_v, w \rangle \models^{\text{IPC}} \varphi.$$

By the equivalence just established, we obtain

$$v(\varphi) = \text{true}.$$

Since v was arbitrary, every classical valuation satisfying Γ also satisfies φ. Therefore,

$$\Gamma \models_{\text{AL}} \varphi.$$

□

The conclusion is therefore clear: intuitionistic logic is more restrictive than classical propositional logic. It permits only those inferences that can also be justified constructively. This is precisely where its strength lies, especially in applications in which proofs, programs, and types are closely connected.

4.1.5 Exercises with Solutions

In the following, several central concepts and statements from intuitionistic logic are reviewed in the form of short true/false exercises. The tasks address syntax, Kripke semantics, the characteristic properties of the logic, and the NJ calculus.

Exercise 4.32. *We consider the signature* $\Sigma = \{A, P, Q, T, U, V, W\}$ *and the intuitionistic Kripke structure*

$$K = \langle W, \leq, V \rangle$$

with

$$W = \{w_0, w_1\}, \qquad \leq = \{(w_0, w_0), (w_1, w_1), (w_0, w_1)\},$$

and

$$V(w_0) = \varnothing, \qquad V(w_1) = \{A\}.$$

Decide in each case whether the following statement is true or false:

(a) In intuitionistic logic, negation is defined by

$$\neg\varphi := (\varphi \to \bot)$$

(b) We have

$$\langle K, w_0 \rangle \models^{\mathrm{IPC}} \neg A.$$

(c) The formula

$$A \vee \neg A$$

is a tautology of intuitionistic logic.

(d) In the calculus NJ, we have

$$\vdash_{\mathrm{NJ}} P \to P.$$

Solution 4.33. *We justify the four statements one by one.*

(a) ***True.*** *This is exactly the definition of negation in intuitionistic logic. The formula* $\neg\varphi$ *is not a primitive formula of its own, but an abbreviation for* $\varphi \to \bot$.

(b) ***False.*** *According to the semantics of negation,*

$$\langle K, w_0 \rangle \models^{\text{IPC}} \neg A$$

holds if and only if A *is not satisfied in any state reachable from* w_0. *But since* $w_0 \leq w_1$ *and* $A \in V(w_1)$, *we have*

$$\langle K, w_1 \rangle \models^{\text{IPC}} A.$$

Therefore, $\neg A$ *cannot hold at* w_0.

(c) ***False.*** *A formula is an intuitionistic tautology only if it is satisfied at every state of every intuitionistic Kripke structure. Here, however, neither*

$$\langle K, w_0 \rangle \models^{\text{IPC}} A$$

nor

$$\langle K, w_0 \rangle \models^{\text{IPC}} \neg A.$$

Indeed, $A \notin V(w_0)$, *and* $\neg A$ *fails because of the reachable state* w_1, *in which* A *holds. Hence we also have*

$$\langle K, w_0 \rangle \not\models^{\text{IPC}} A \vee \neg A.$$

Thus, the law of excluded middle is not generally valid in intuitionistic logic.

(d) ***True.*** *An NJ derivation of* $\vdash_{\mathrm{NJ}} P \to P$ *is:*

$$\dfrac{\dfrac{}{P \vdash P}\,(\mathrm{Ax})}{\vdash P \to P}\,(\to I)$$

Therefore, the formula is derivable in NJ.

4.2 Paraconsistent Logic

Classical logics have a very strong property: as soon as a theory is inconsistent, it loses all discriminatory power. Thus, if a formula φ is classically unsatisfiable, then in classical propositional logic we have $\varphi \models_{\mathrm{AL}} \psi$ for every arbitrary formula ψ. This behavior is often called *explosion*. For purely theoretical purposes this is not a problem, but for applications it is: in databases, knowledge bases, or formal specifications, contradictions arise not only in theory but also in practice. In such situations, one would still like to derive useful consequences from the available information without everything becoming derivable at once.

This is precisely where paraconsistent logics come in. They are designed so that inconsistencies do not automatically lead to trivial conclusions. Rather than leaving the classical consequence relation unchanged, they weaken it in a controlled way. The basic idea is simple: a statement need not be merely *true* or *false*; under certain circumstances it may also receive an additional truth status that expresses contradictory or incomplete information.

In this chapter, we begin with three-valued propositional logic. We first introduce its semantic foundation and show how formulas are evaluated in it. We then investigate inference with minimal models and compare the resulting consequence relation with that of classical propositional logic. Finally, we give a brief outlook on a four-valued variant.

4.2.1 Three-Valued Propositional Logic

The paraconsistent logic considered in this subsection has the same syntax as the familiar propositional logic from Chapter 1. Only the semantics are new: statements are no longer evaluated using only the two truth values `true` and `false`, but instead with three truth values.

Definition 4.34. *Let Σ be a propositional signature. A* three-valued interpretation *over Σ is a function*

$$v : \Sigma \to \{t, f, b\}.$$

We write $\Omega_3(\Sigma)$ for the set of all three-valued interpretations over Σ.

The truth values t and f correspond in meaning to the classical values `true` and `false`. The new value b stands for *both*. Intuitively, $v(x) = b$ means that the propositional variable x is regarded as both true and false at the same time. Such a truth value therefore arises exactly when the information about x is contradictory.

Example 4.35. *Consider the signature*

$$\Sigma_{\text{example}} = \{P, Q, R, S\}.$$

Let a three-valued interpretation $v \in \Omega_3(\Sigma_{\text{example}})$ *be given by*

$$v(P) = t, \qquad v(Q) = t, \qquad v(R) = f, \qquad v(S) = b.$$

Here, P *and* Q *are definitively true,* R *is definitively false, and* S *receives a contradictory valuation, that is, it is both true and false.*

We now specify the semantic evaluation of formulas. The underlying idea is very natural: as long as a formula can be evaluated unambiguously as true or false, the connectives behave exactly as in classical propositional logic. The value b is produced only when the available information is insufficient for a classical determination.

Definition 4.36. *Let* Σ *be a propositional signature, let* $v \in \Omega_3(\Sigma)$ *be a three-valued interpretation, let* $\varphi, \psi \in L(\Sigma)$ *be formulas, and let* $x \in \Sigma$ *be a propositional variable. Then the three-valued truth function*

$$[\![\cdot]\!]^3_v : L(\Sigma) \to \{t, f, b\}$$

is defined inductively by

$$[\![x]\!]^3_v = v(x),$$

$$[\![\neg\varphi]\!]^3_v = \begin{cases} t & \text{if } [\![\varphi]\!]^3_v = f, \\ f & \text{if } [\![\varphi]\!]^3_v = t, \\ b & \text{if } [\![\varphi]\!]^3_v = b, \end{cases}$$

$$\llbracket \varphi \wedge \psi \rrbracket_v^3 = \begin{cases} t & \textit{if } \llbracket \varphi \rrbracket_v^3 = \llbracket \psi \rrbracket_v^3 = t, \\ f & \textit{if } \llbracket \varphi \rrbracket_v^3 = f \textit{ or } \llbracket \psi \rrbracket_v^3 = f, \\ b & \textit{otherwise}, \end{cases}$$

$$\llbracket \varphi \vee \psi \rrbracket_v^3 = \begin{cases} f & \textit{if } \llbracket \varphi \rrbracket_v^3 = \llbracket \psi \rrbracket_v^3 = f, \\ t & \textit{if } \llbracket \varphi \rrbracket_v^3 = t \textit{ or } \llbracket \psi \rrbracket_v^3 = t, \\ b & \textit{otherwise}. \end{cases}$$

Negation is therefore fully classical as long as the truth value under consideration is determined. The value b remains unchanged under negation. For conjunction and disjunction, b appears exactly in those cases in which neither the classical truth condition nor the classical falsity condition is satisfied. Put differently, b is the value that remains when the available information is not strong enough to justify a classical determination.

Example 4.37. *We continue the previous example and again consider the interpretation v with*

$$v(P) = t, \qquad v(Q) = t, \qquad v(R) = f, \qquad v(S) = b.$$

Then, for example, we have

$$\llbracket \neg P \rrbracket_v^3 = f, \qquad \llbracket \neg S \rrbracket_v^3 = b, \qquad \llbracket S \vee R \rrbracket_v^3 = b,$$

and

$$\llbracket P \wedge (\neg R \vee S) \rrbracket_v^3 = t.$$

By contrast, the formula $P \wedge (\neg S \vee R)$ receives the value b because the truth value of S is contradictory.

An intuitive way to view this semantics is to order the truth values according to their degree of truth. In this ordering, b lies between t and f: in content it is stronger than f, but weaker than t.

Definition 4.38. *The truth ordering* $\succ_3$ *on the values* t, f, b *is given by*

$$t \succ_3 b, \qquad t \succ_3 f, \qquad b \succ_3 f.$$

With this ordering, conjunction can be understood as the minimum and disjunction as the maximum.

Proposition 4.39. *Let* Σ *be a propositional signature, let* $v \in \Omega_3(\Sigma)$ *be a three-valued interpretation, and let* $\varphi, \psi \in L(\Sigma)$ *be formulas. Then:*

$$[\![\varphi \wedge \psi]\!]_v^3 = \min_{\succ_3}\{[\![\varphi]\!]_v^3, [\![\psi]\!]_v^3\}$$

and

$$[\![\varphi \vee \psi]\!]_v^3 = \max_{\succ_3}\{[\![\varphi]\!]_v^3, [\![\psi]\!]_v^3\}.$$

Proof. We prove only the statement for $\wedge$; the proof for $\vee$ is entirely analogous.

First, recall that the relation $\succ_3$ is given by

$$t \succ_3 b \succ_3 f$$

Thus, the minimum with respect to $\succ_3$ is precisely the value that has the lesser degree of truth among the two values.

We now consider the possible cases for the values of $[\![\varphi]\!]_v^3$ and $[\![\psi]\!]_v^3$.

- If $[\![\varphi]\!]_v^3 = f$ and $[\![\psi]\!]_v^3 = f$, then the definition yields

  $$[\![\varphi \wedge \psi]\!]_v^3 = f.$$

  On the other hand,

  $$\min_{\succ_3}\{f, f\} = f.$$

- If $[\![\varphi]\!]_v^3 = f$ and $[\![\psi]\!]_v^3 = b$, then by definition we again have

  $$[\![\varphi \wedge \psi]\!]_v^3 = f,$$

  and likewise

  $$\min_{\succ_3}\{f, b\} = f.$$

- If $[\![\varphi]\!]_v^3 = f$ and $[\![\psi]\!]_v^3 = t$, then it follows that

  $$[\![\varphi \wedge \psi]\!]_v^3 = f \qquad \text{and} \qquad \min_{\succ_3}\{f, t\} = f.$$

- If $[\![\varphi]\!]_v^3 = b$ and $[\![\psi]\!]_v^3 = b$, then

  $$[\![\varphi \wedge \psi]\!]_v^3 = b \qquad \text{and} \qquad \min_{\succ_3}\{b, b\} = b.$$

- If $[\![\varphi]\!]_v^3 = b$ and $[\![\psi]\!]_v^3 = t$, then we obtain

  $$[\![\varphi \wedge \psi]\!]_v^3 = b \qquad \text{and} \qquad \min_{\succ_3}\{b, t\} = b.$$

- Finally, if $[\![\varphi]\!]_v^3 = t$ and $[\![\psi]\!]_v^3 = t$, then

  $$[\![\varphi \wedge \psi]\!]_v^3 = t \qquad \text{and} \qquad \min_{\succ_3}\{t, t\} = t.$$

The remaining combinations are obtained automatically from the symmetry of $\wedge$ and of the minimum operation. Hence the two evaluations agree in all cases. Therefore,

$$[\![\varphi \wedge \psi]\!]_v^3 = \min_{\succ_3}\{[\![\varphi]\!]_v^3, [\![\psi]\!]_v^3\}.$$

The statement for $\vee$ follows analogously from the definition of disjunction. □

Based on this evaluation, we now define the three-valued satisfaction relation.

Definition 4.40. *Let Σ be a propositional signature. The three-valued satisfaction relation*

$$\models_3 \subseteq \Omega_3(\Sigma) \times L(\Sigma)$$

is defined by

$$v \models_3 \varphi \quad :\Longleftrightarrow \quad [\![\varphi]\!]_v^3 \in \{t, b\}.$$

Unlike the classical satisfaction relation, it is therefore already sufficient that a formula is not evaluated as false. The value b still counts as satisfied here. This is precisely what makes the logic paraconsistent: a formula can still be acceptable despite being contradictory.

Definition 4.41. *Let Σ be a propositional signature and let $\varphi \in L(\Sigma)$ be a formula. An interpretation $v \in \Omega_3(\Sigma)$ is called a* three-valued model *of φ if $v \models_3 \varphi$ holds. We denote the set of all three-valued models of φ by*

$$\mathrm{Mod}_\Sigma^3(\varphi).$$

Example 4.42. *We again consider the signature* $\Sigma_{\text{example}} = \{P, Q, R, S\}$ *and the three-valued interpretation* υ *with*

$$\upsilon(P) = t, \qquad \upsilon(Q) = f, \qquad \upsilon(R) = b, \qquad \upsilon(S) = f.$$

For the formula

$$\chi := P \wedge (\neg Q \vee R)$$

we have

$$[\![\neg Q]\!]^3_\upsilon = t, \qquad [\![\neg Q \vee R]\!]^3_\upsilon = t, \qquad [\![\chi]\!]^3_\upsilon = t.$$

Hence $\upsilon \models_3 \chi$.

By contrast, for the formula

$$\chi' := R \wedge \neg R$$

the value is

$$[\![\chi']\!]^3_\upsilon = b,$$

so again $\upsilon \models_3 \chi'$ *holds. Thus, the formula is not classically true, but it is nevertheless satisfied in the three-valued semantics.*

Before studying inference in three-valued logic, we compare the three-valued semantics with the classical semantics. For this purpose, we need two translations between two-valued and three-valued interpretations.

Definition 4.43. *Let Σ be a propositional signature and let $\omega \in \Omega(\Sigma)$ be a classical interpretation. The three-valued interpretation $\upsilon_\omega \in \Omega_3(\Sigma)$ associated with ω is defined by*

$$\upsilon_\omega(x) = \begin{cases} t & \text{if } \omega(x) = \texttt{true}, \\ f & \text{if } \omega(x) = \texttt{false}. \end{cases}$$

For a set $M \subseteq \Omega(\Sigma)$, we define

$$M^3 := \{\upsilon_\omega \mid \omega \in M\}.$$

Definition 4.44. *Let Σ be a propositional signature and let $\upsilon \in \Omega_3(\Sigma)$ be a three-valued interpretation satisfying*

$$\upsilon^{-1}(b) = \emptyset.$$

We then define the classical interpretation $\omega_\upsilon \in \Omega(\Sigma)$ associated with υ by

$$\omega_\upsilon(x) = \begin{cases} \texttt{true} & \text{if } \upsilon(x) = t, \\ \texttt{false} & \text{if } \upsilon(x) = f. \end{cases}$$

Example 4.45. *Let again $\Sigma_{\text{Example}} = \{P, Q, R, S\}$ and*

$$\upsilon(P) = t, \qquad \upsilon(Q) = f, \qquad \upsilon(R) = t, \qquad \upsilon(S) = f.$$

Then $\upsilon^{-1}(b) = \emptyset$, so ω_υ is well defined and satisfies

$$\begin{aligned} \omega_\upsilon(P) &= \texttt{true} & \omega_\upsilon(Q) &= \texttt{false} \\ \omega_\upsilon(R) &= \texttt{true} & \omega_\upsilon(S) &= \texttt{false} \end{aligned}$$

The following statement shows that classical interpretations and three-valued interpretations without the value b do not differ semantically.

Proposition 4.46. *Let Σ be a propositional signature, $\upsilon \in \Omega_3(\Sigma)$ a three-valued interpretation with $\upsilon^{-1}(b) = \emptyset$, $\omega \in \Omega(\Sigma)$ a classical interpretation, and $\varphi \in L(\Sigma)$ a formula. Then:*

1. $\omega_{\upsilon_\omega} = \omega$.
2. $\upsilon_{\omega_\upsilon} = \upsilon$.
3. *If* $[\![\varphi]\!]_\omega = \mathtt{true}$, *then* $[\![\varphi]\!]^3_{\upsilon_\omega} = t$, *and if* $[\![\varphi]\!]_\omega = \mathtt{false}$, *then* $[\![\varphi]\!]^3_{\upsilon_\omega} = f$.
4. *If* $[\![\varphi]\!]^3_\upsilon = t$, *then* $[\![\varphi]\!]_{\omega_\upsilon} = \mathtt{true}$, *and if* $[\![\varphi]\!]^3_\upsilon = f$, *then* $[\![\varphi]\!]_{\omega_\upsilon} = \mathtt{false}$.
5. *We always have* $[\![\varphi]\!]^3_\upsilon \in \{t, f\}$.

Proof. We prove the statements one by one.

1. For every $x \in \Sigma$, by definition $\upsilon_\omega(x) = t$ if and only if $\omega(x) = \mathtt{true}$, and analogously $\upsilon_\omega(x) = f$ if and only if $\omega(x) = \mathtt{false}$. Hence the back-translation ω_{υ_ω} recovers exactly ω.
2. Let $x \in \Sigma$. Since $\upsilon^{-1}(b) = \emptyset$, we have $\upsilon(x) \in \{t, f\}$. Therefore $\omega_\upsilon(x) = \mathtt{true}$ if and only if $\upsilon(x) = t$, and correspondingly $\omega_\upsilon(x) = \mathtt{false}$ if and only if $\upsilon(x) = f$. Thus $\upsilon_{\omega_\upsilon} = \upsilon$.

3. We prove the claim by structural induction on φ.

   Case $\varphi = x$ *with* $x \in \Sigma$. Then the claim follows immediately from the definition of v_ω.

   Case $\varphi = \neg\psi$. If $[\![\varphi]\!]_\omega = \texttt{true}$, then $[\![\psi]\!]_\omega = \texttt{false}$. By the induction hypothesis, $[\![\psi]\!]^3_{v_\omega} = f$, hence $[\![\neg\psi]\!]^3_{v_\omega} = t$. The case `false` is analogous.

   Case $\varphi = \psi \wedge \chi$. If $[\![\varphi]\!]_\omega = \texttt{true}$, then $[\![\psi]\!]_\omega = \texttt{true}$ and $[\![\chi]\!]_\omega = \texttt{true}$. By the induction hypothesis, $[\![\psi]\!]^3_{v_\omega} = t$ and $[\![\chi]\!]^3_{v_\omega} = t$, hence also $[\![\psi \wedge \chi]\!]^3_{v_\omega} = t$. If, on the other hand, $[\![\varphi]\!]_\omega = \texttt{false}$, then at least one of the two subformulas is classically false; without loss of generality, let this be ψ. Then $[\![\psi]\!]^3_{v_\omega} = f$, and therefore $[\![\psi \wedge \chi]\!]^3_{v_\omega} = f$.

   Case $\varphi = \psi \vee \chi$. This case is entirely analogous. If φ is classically true, then at least one subformula is true and hence also three-valued t. If φ is classically false, then both subformulas are classically false and hence three-valued f.

4. We also prove this statement by structural induction.

   Case $\varphi = x$. Since $v^{-1}(b) = \emptyset$, we have $v(x) \in \{t, f\}$, and therefore also $[\![x]\!]^3_v \in \{t, f\}$.

   Case $\varphi = \neg\psi$. By the induction hypothesis, $[\![\psi]\!]^3_v$ lies in $\{t, f\}$. Negation then merely swaps t and f and does not produce the value b.

   Case $\varphi = \psi \wedge \chi$. By the induction hypothesis, both subvalues lie in $\{t, f\}$. The definition of $\wedge$ again yields only a value in $\{t, f\}$ from two values in $\{t, f\}$.

Case $\varphi = \psi \vee \chi$. Analogous to the previous case.

This shows that under the assumption $v^{-1}(b) = \emptyset$, no formula can take the value b at all.

$\square$

Corollary 4.47. *Let* Σ *be a propositional signature and* $\varphi \in L(\Sigma)$ *a formula. Then*

$$(\mathrm{Mod}_\Sigma^{\mathrm{PL}}(\varphi))^3 \subseteq \mathrm{Mod}_\Sigma^3(\varphi).$$

Proof. Let $\omega \in \mathrm{Mod}_\Sigma^{\mathrm{PL}}(\varphi)$. Then ω is a classical model of φ, and therefore

$$[\![\varphi]\!]_\omega = \mathtt{true}.$$

By Proposition 4.46 (3), it follows that

$$[\![\varphi]\!]_{v_\omega}^3 = t.$$

In particular, this value lies in the set $\{t, b\}$, hence $v_\omega \models_3 \varphi$. Therefore,

$$v_\omega \in \mathrm{Mod}_\Sigma^3(\varphi).$$

Since ω was arbitrary, the claimed inclusion follows. $\square$

In other words, every classical model is automatically also a three-valued model. The three-valued semantics therefore genuinely extends the classical semantics rather than replacing it. The converse, however, does not hold: there are three-valued models that are not classical models.

Example 4.48. *Let us again consider the signature*

$$\Sigma_{\text{example}} = \{P, Q, R, S\}$$

and the formula

$$\varphi_1 := P \wedge (\neg Q \vee R).$$

A classical interpretation ω *with*

$$\omega(P) = \omega(R) = \textit{\texttt{true}}, \qquad \omega(Q) = \omega(S) = \textit{\texttt{false}}$$

is a model of φ_1. *Hence* υ_ω *is also a three-valued model of* φ_1.

Now consider a three-valued interpretation υ *with*

$$\upsilon(P) = b, \qquad \upsilon(Q) = f, \qquad \upsilon(R) = t, \qquad \upsilon(S) = f.$$

Then we have

$$[\![\neg Q \vee R]\!]^3_\upsilon = t, \qquad [\![\varphi_1]\!]^3_\upsilon = b.$$

Thus $\upsilon \models_3 \varphi_1$, *even though* υ *has no classical counterpart.*

For the formula

$$\varphi_2 := P \wedge \neg P \wedge (\neg Q \vee R)$$

there is, by contrast, no classical interpretation that satisfies it. In the three-valued semantics, however, φ_2 *is still satisfiable, for example by an interpretation with* $\upsilon(P) = b$, $\upsilon(Q) = f$, *and* $\upsilon(R) = t$.

4.2.2 Reasoning with Minimal Three-Valued Models

The three-valued satisfaction relation $\models_3$ allows us to find models even in the presence of inconsistent information. But this is also precisely where a problem arises: since the truth value b is assigned quite liberally, far too many interpretations are often models of a formula. For purposes of inference, that is too weak. We therefore need to single out those models that contain as little indeterminacy as possible. The basic idea is thus not to treat all three-valued models equally, but to consider only the *minimal* models.

Using the three-valued satisfaction relation $\models_3$, we can first define a consequence relation, just as in the classical case.

Definition 4.49. *Let Σ be a propositional signature, and let $\varphi, \psi \in L(\Sigma)$ be propositional formulas. The three-valued consequence relation*

$$\models^3_\Sigma \subseteq L(\Sigma) \times L(\Sigma)$$

is defined by

$$\varphi \models^3_\Sigma \psi \quad :\Longleftrightarrow \quad \mathrm{Mod}^3_\Sigma(\varphi) \subseteq \mathrm{Mod}^3_\Sigma(\psi).$$

This definition is, however, very weak. The reason is that the truth value b already satisfies a formula whenever it is not definitely false. As a result, interpretations that carry hardly any substantive informational commitment often count as models.

Example 4.50. *Consider the signature*

$$\Sigma_{\text{Example}} = \{P, Q, R\}$$

and the formula

$$\chi := P \vee (Q \wedge \neg Q).$$

Now define a three-valued interpretation $v \in \Omega_3(\Sigma_{\text{Example}})$ *by*

$$v(P) = f, \qquad v(Q) = b, \qquad v(R) = t.$$

Then we have

$$[\![Q \wedge \neg Q]\!]^3_v = b \qquad \textit{and therefore} \qquad [\![\chi]\!]^3_v = b.$$

Hence $v \models_3 \chi$.

However, this situation is unsatisfactory from the perspective of inference. Intuitively, the formula χ *is supposed to convey precisely the information that* P *holds. Yet the interpretation* v *additionally assigns a contradictory value to* Q*, even though this is not needed for the actual content of* χ.

An even more striking problem is that there exists an interpretation under which *every* formula is satisfied.

Proposition 4.51. *Let* Σ *be a propositional signature, and define* $v_b^{\Sigma} \in \Omega_3(\Sigma)$ *by*

$$v_b^{\Sigma}(x) = b$$

for all $x \in \Sigma$. *Then for every formula* $\varphi \in L(\Sigma)$*, we have:*

$$v_b^{\Sigma} \models_3 \varphi.$$

Proof. We first show that for every formula φ,

$$[\![\varphi]\!]^3_{v_b^\Sigma} = b.$$

This immediately implies $v_b^\Sigma \models_3 \varphi$, because by definition of the satisfaction relation, the value b already suffices.

The proof proceeds by structural induction on φ.

Base case. If $\varphi = x$ for some $x \in \Sigma$, then directly

$$[\![x]\!]^3_{v_b^\Sigma} = v_b^\Sigma(x) = b.$$

Inductive step. Assume that the claim already holds for all proper subformulas of φ.

- Let $\varphi = \neg\psi$. By the induction hypothesis,

  $$[\![\psi]\!]^3_{v_b^\Sigma} = b.$$

  By the definition of negation in three-valued logic, it then follows that

  $$[\![\neg\psi]\!]^3_{v_b^\Sigma} = b.$$

- Let $\varphi = \psi \wedge \chi$. By the induction hypothesis,

  $$[\![\psi]\!]^3_{v_b^\Sigma} = b \qquad \text{and} \qquad [\![\chi]\!]^3_{v_b^\Sigma} = b.$$

  Thus, the definition of conjunction immediately yields

  $$[\![\psi \wedge \chi]\!]^3_{v_b^\Sigma} = b.$$

- Let $\varphi = \psi \vee \chi$. Analogously, the induction hypothesis together with the definition of disjunction implies that
$$[\![\psi \vee \chi]\!]^3_{v_b^\Sigma} = b.$$

Thus, in all cases we have shown that $[\![\varphi]\!]^3_{v_b^\Sigma} = b$. Therefore, v_b^Σ satisfies every formula. □

Corollary 4.52. *Let Σ be a propositional signature and let $\varphi \in L(\Sigma)$ be a formula. Then*

$$\mathrm{Mod}^3_\Sigma(\varphi) \neq \emptyset.$$

Proof. Define the three-valued interpretation v_b^Σ as in the previous proposition. There we showed that

$$[\![\varphi]\!]^3_{v_b^\Sigma} = b$$

for every formula φ. In particular,

$$[\![\varphi]\!]^3_{v_b^\Sigma} \in \{t, b\},$$

and therefore
$$v_b^\Sigma \models_3 \varphi.$$
Hence $v_b^\Sigma \in \mathrm{Mod}^3_\Sigma(\varphi)$, and thus this set is nonempty. □

The previous results show that, without further restrictions, three-valued semantics is very permissive. For inference, this is not appropriate, since we do not want to work with interpretations that contain unnecessarily many b-values. We therefore restrict our attention to those models that are minimal in this sense.

Definition 4.53. *Let Σ be a propositional signature, let $\upsilon \in \Omega_3(\Sigma)$ be a three-valued interpretation, and let $\varphi \in L(\Sigma)$ be a propositional formula. The interpretation υ is called a* minimal three-valued model *of φ if $\upsilon \in \mathrm{Mod}^3_\Sigma(\varphi)$ and there is no three-valued interpretation $\eta \in \mathrm{Mod}^3_\Sigma(\varphi)$ such that*

$$\eta^{-1}(b) \subsetneq \upsilon^{-1}(b)$$

holds. We denote the set of all minimal three-valued models of φ by

$$\mathrm{Mod}^{\min 3}_\Sigma(\varphi).$$

Thus, minimality refers to the set of propositional variables that receive the value b. A minimal model is one in which the number of contradictorily valued variables can no longer be reduced.

Example 4.54. *We again consider the signature*

$$\Sigma_{\text{Example}} = \{P, Q, R\}$$

and the formula

$$\chi := P \vee (Q \wedge \neg Q).$$

Define a three-valued interpretation $\upsilon \in \Omega_3(\Sigma_{\text{Example}})$ by

$$\upsilon(P) = f, \qquad \upsilon(Q) = b, \qquad \upsilon(R) = t.$$

Then $\upsilon \in \mathrm{Mod}^3_{\Sigma_{\text{Example}}}(\chi)$, since

$$[\![\chi]\!]^3_\upsilon = b.$$

However, this interpretation is not minimal. Indeed, the interpretation $\eta \in \Omega_3(\Sigma_{\text{Example}})$ given by

$$\eta(P) = t, \qquad \eta(Q) = f, \qquad \eta(R) = t$$

is also a model of χ, and we have

$$\eta^{-1}(b) = \emptyset \subsetneq \{Q\} = v^{-1}(b).$$

Hence v is not a minimal model.

The following proposition shows that minimal models do in fact exist.

Proposition 4.55. *Let Σ be a propositional signature and let $\varphi \in L(\Sigma)$ be a propositional formula. Then:*

$$\mathrm{Mod}_{\Sigma}^{\min 3}(\varphi) \subseteq \mathrm{Mod}_{\Sigma}^{3}(\varphi)$$

and

$$\mathrm{Mod}_{\Sigma}^{\min 3}(\varphi) \neq \emptyset.$$

Proof. 1. Let $v \in \mathrm{Mod}_{\Sigma}^{\min 3}(\varphi)$. By definition, we already have $v \in \mathrm{Mod}_{\Sigma}^{3}(\varphi)$. Hence,

$$\mathrm{Mod}_{\Sigma}^{\min 3}(\varphi) \subseteq \mathrm{Mod}_{\Sigma}^{3}(\varphi).$$

2. By the previous corollary, $\mathrm{Mod}_{\Sigma}^{3}(\varphi) \neq \emptyset$. Thus, there exists at least one three-valued model of φ.

Now take such a model $v_0 \in \mathrm{Mod}_{\Sigma}^{3}(\varphi)$ and consider all models of φ that, compared with v_0, contain no additional b-values. Since, as before, we assume finite signatures, the set of possible b-assignments cannot be strictly reduced infinitely many times. Therefore, after finitely many steps, this process must terminate at a model υ for which there is no model η with

$$\eta^{-1}(b) \subsetneq \upsilon^{-1}(b)$$

any longer. This is exactly what it means for υ to be a minimal three-valued model of φ.

Therefore,

$$\mathrm{Mod}_{\Sigma}^{\min 3}(\varphi) \neq \emptyset.$$

□

On the basis of these minimal models, we now define the actual paraconsistent consequence relation.

Definition 4.56. *Let Σ be a propositional signature, and let $\varphi, \psi \in L(\Sigma)$ be propositional formulas. The minimal three-valued consequence relation*

$$\models_{\Sigma}^{\min 3} \subseteq L(\Sigma) \times L(\Sigma)$$

is defined by

$$\varphi \models_{\Sigma}^{\min 3} \psi \quad :\Longleftrightarrow \quad \mathrm{Mod}_{\Sigma}^{\min 3}(\varphi) \subseteq \mathrm{Mod}_{\Sigma}^{\min 3}(\psi).$$

As usual, we simply write $\varphi \models^{\min 3} \psi$ when the signature is clear from the context.

This consequence relation is paraconsistent in the intended sense: inconsistent formulas no longer entail every formula automatically, but they do still entail those statements that remain stable across all minimal models.

Example 4.57. *We consider once again the formula*

$$\chi := P \vee (Q \wedge \neg Q).$$

Then we have

$$\chi \models^{\min 3} P.$$

The reason is that in every minimal model of χ*, the truth value of* P *must be* t*. Any interpretation with* $P = f$ *and an unnecessary* b*-value assigned to* Q *is not minimal, since that* b*-value can be eliminated by choosing a suitable classical valuation for* Q*.*

By contrast,

$$\chi \not\models^{\min 3} \neg P.$$

The following theorem shows that minimal three-valued consequence coincides with classical consequence exactly when no inconsistency is present.

Theorem 4.58. *Let* Σ *be a propositional signature, and let* $\varphi, \psi \in L(\Sigma)$ *be propositional formulas with* $\varphi \not\equiv \bot$*. Then*

$$\varphi \models_{\mathrm{PL}} \psi \iff \varphi \models^{\min 3} \psi.$$

Proof. We first show that for every non-inconsistent formula φ,

$$\left(\mathrm{Mod}_\Sigma^{\mathrm{PL}}(\varphi)\right)^3 = \mathrm{Mod}_\Sigma^{\min 3}(\varphi).$$

To do so, we prove both inclusions.

"$\subseteq$": Let $\omega \in \mathrm{Mod}_\Sigma^{\mathrm{PL}}(\varphi)$. Then $v_\omega \in \left(\mathrm{Mod}_\Sigma^{\mathrm{PL}}(\varphi)\right)^3$. By Corollary 4.47, $v_\omega \in \mathrm{Mod}_\Sigma^3(\varphi)$. Moreover,

$$v_\omega^{-1}(b) = \emptyset.$$

Hence there can be no three-valued model with a strictly smaller set of b-assignments. Therefore v_ω is minimal, and thus

$$v_\omega \in \mathrm{Mod}_\Sigma^{\min 3}(\varphi).$$

"$\supseteq$": Now let $v \in \mathrm{Mod}_\Sigma^{\min 3}(\varphi)$. Assume first that $v^{-1}(b) \neq \emptyset$. Since $\varphi \not\equiv \bot$, the formula φ has a classical model ω. By Corollary 4.47, it follows that $v_\omega \in \mathrm{Mod}_\Sigma^3(\varphi)$ as well. But for this model,

$$v_\omega^{-1}(b) = \emptyset \subsetneq v^{-1}(b).$$

This contradicts the minimality of v. Therefore,

$$v^{-1}(b) = \emptyset$$

must hold. Hence ω_v is well defined. By Proposition 4.46(4) and (5), it follows that ω_v is a classical model of φ. Thus,

$$v = v_{\omega_v} \in \left(\mathrm{Mod}_\Sigma^{\mathrm{PL}}(\varphi)\right)^3.$$

This proves the claimed equality.

The equivalence of the two consequence relations now follows immediately:

$$\varphi \models_{\mathrm{PL}} \psi \iff \mathrm{Mod}_\Sigma^{\mathrm{PL}}(\varphi) \subseteq \mathrm{Mod}_\Sigma^{\mathrm{PL}}(\psi)$$

$$\iff \left(\mathrm{Mod}_\Sigma^{\mathrm{PL}}(\varphi)\right)^3 \subseteq \left(\mathrm{Mod}_\Sigma^{\mathrm{PL}}(\psi)\right)^3$$

$$\iff \mathrm{Mod}_\Sigma^{\min 3}(\varphi) \subseteq \mathrm{Mod}_\Sigma^{\min 3}(\psi) \iff \varphi \models^{\min 3} \psi.$$

This completes the proof. □

Minimal three-valued consequence is therefore weaker than classical consequence in one respect, since it does not become trivial in the presence of inconsistent formulas; on the other hand, it coincides with classical logic on consistent formulas. It is precisely this behavior that makes it suitable for paraconsistent reasoning.

Example 4.59. *We consider the formula*

$$\theta := P \wedge \neg P \wedge Q.$$

Then we have

$$\theta \models^{\min 3} Q,$$

but

$$\theta \not\models^{\min 3} \neg Q.$$

The formula θ is classically inconsistent, yet it still contains the information that Q is preserved in the minimal models. This is exactly the paraconsistent effect.

4.2.3 Four-Valued Propositional Logic

The term "paraconsistent logic" does not refer to a single formal theory, but to a family of related logics. One important difference between these logics concerns the number of truth values they admit and the informal situations these values are meant to represent. The three-valued propositional logic discussed in the previous subsection is only the simplest case. We now consider, as a further example, a four-valued logic that additionally distinguishes between *contradiction* and *indeterminacy*.

This distinction is often important in applications. A statement may not only be both true and false; it may also simply remain unknown. Thus, in a knowledge base, there may on the one hand be a genuine conflict, and on the other hand merely a lack of information. Four-valued logic makes precisely these two cases explicit.

Definition 4.60. *Let Σ be a propositional signature. A* four-valued interpretation *over Σ is a function*

$$\xi : \Sigma \to \{t, f, b, n\}.$$

The set of all four-valued interpretations over Σ is denoted by $\Omega_4(\Sigma)$.

The truth values have the following intuitive meanings:

$$\begin{aligned} t &= \text{true}, \\ f &= \text{false}, \\ b &= \text{both true and false}, \\ n &= \text{neither true nor false}. \end{aligned}$$

Thus, the value b represents a genuine contradiction, whereas n stands for missing or undecided information. This distinction is the main additional feature of four-valued semantics in comparison with three-valued semantics.

Definition 4.61. *The truth order $\succ_4$ on the values t, f, b, n is given by*

$$t \succ_4 b, \qquad t \succ_4 n, \qquad b \succ_4 f, \qquad n \succ_4 f$$

with b and n being incomparable.

Intuitively, t is at the top, f at the bottom, and the two intermediate values b and n are distinct without either being greater or smaller than the other. For the evaluation of conjunctions and disjunctions, simple minimum and maximum operations are therefore no longer sufficient; instead, we use infimum and supremum with respect to the order $\succ_4$.

Definition 4.62. *Let Σ be a propositional signature, let $\xi \in \Omega_4(\Sigma)$ be a four-valued interpretation, let $\varphi, \psi \in L(\Sigma)$ be formulas, and let $x \in \Sigma$ be a propositional variable. We define the four-valued truth function*

$$[\![\cdot]\!]^4_\xi : L(\Sigma) \to \{t, f, b, n\}$$

inductively by

$$[\![x]\!]^4_\xi = \xi(x),$$

$$[\![\neg\varphi]\!]^4_\xi = \begin{cases} t & \textit{if } [\![\varphi]\!]^4_\xi = f, \\ f & \textit{if } [\![\varphi]\!]^4_\xi = t, \\ b & \textit{if } [\![\varphi]\!]^4_\xi = b, \\ n & \textit{if } [\![\varphi]\!]^4_\xi = n, \end{cases}$$

$$[\![\varphi \wedge \psi]\!]^4_\xi = \inf_{\succ_4}\{[\![\varphi]\!]^4_\xi, [\![\psi]\!]^4_\xi\},$$

$$[\![\varphi \vee \psi]\!]^4_\xi = \sup_{\succ_4}\{[\![\varphi]\!]^4_\xi, [\![\psi]\!]^4_\xi\}.$$

Negation behaves very simply here as well: it interchanges t and f, but leaves b and n unchanged. This is plausible, since neither contradiction nor indeterminacy is automatically removed by negation. For conjunction and disjunction, the order $\succ_4$ determines which truth value carries the stronger truth content.

A few typical values illustrate the logic of this semantics:

$$b \wedge n = f, \qquad b \vee n = t, \qquad n \wedge t = n, \qquad b \vee f = b.$$

This immediately shows that b and n play semantically different roles. The value b carries a conflict, whereas n marks a lack of information.

Example 4.63. *Consider the signature*

$$\Sigma_{\text{Example}} = \{P, Q, R, S\}$$

and a four-valued interpretation $\xi \in \Omega_4(\Sigma_{\text{Example}})$ *with*

$$\xi(P) = t, \qquad \xi(Q) = b, \qquad \xi(R) = n, \qquad \xi(S) = f.$$

Then

$$\begin{aligned} [\![P \wedge Q]\!]^4_\xi &= b, \\ [\![Q \wedge R]\!]^4_\xi &= f, \\ [\![Q \vee R]\!]^4_\xi &= t, \\ [\![\neg R \wedge (P \vee S)]\!]^4_\xi &= n. \end{aligned}$$

In particular, this shows that Q is contradictory, whereas R is merely indeterminate, and these two cases behave differently under the connectives.

As in the three-valued case, we now specify when a formula counts as satisfied in an interpretation. Only the values t and b are to count as satisfying. The value n, by contrast, does not count as satisfying, because it expresses precisely that the statement in question is neither confirmed nor refuted.

Definition 4.64. *Let Σ be a propositional signature. The four-valued satisfaction relation*

$$\models_4 \subseteq \Omega_4(\Sigma) \times L(\Sigma)$$

is defined by

$$\xi \models_4 \varphi \quad :\Longleftrightarrow \quad [\![\varphi]\!]^4_\xi \in \{t, b\}.$$

A four-valued interpretation is then called a *four-valued model* of a formula if it satisfies that formula. We denote the set of all four-valued models of φ by $\mathrm{Mod}^4_\Sigma(\varphi)$.

Example 4.65. *For the formula*

$$\chi := P \wedge \neg P$$

we have, for an interpretation ξ with $\xi(P) = b$,

$$[\![\chi]\!]^4_\xi = b,$$

and hence $\xi \models_4 \chi$. By contrast, if $\xi(P) = n$, then

$$[\![\chi]\!]^4_\xi = n,$$

and hence $\xi \not\models_4 \chi$. Thus four-valued logic makes the difference between contradiction and mere indeterminacy very clear.

We can now define a consequence relation by model inclusion in the usual way.

Definition 4.66. *Let Σ be a propositional signature, and let $\varphi, \psi \in L(\Sigma)$ be formulas. The four-valued consequence relation*

$$\models^4_\Sigma \subseteq L(\Sigma) \times L(\Sigma)$$

is defined by

$$\varphi \models^4_\Sigma \psi \quad :\Longleftrightarrow \quad \mathrm{Mod}^4_\Sigma(\varphi) \subseteq \mathrm{Mod}^4_\Sigma(\psi).$$

This relation is paraconsistent, since an arbitrary formula does not automatically follow from a contradictory formula. At the same time, it is also paracomplete in the sense that the value n does not count as satisfying, so not every formula is automatically "rescued" by indeterminacy.

Example 4.67. *Consider again*

$$\chi := P \wedge \neg P.$$

Then

$$\chi \not\models_4 Q$$

for any other propositional variable Q. Indeed, the interpretation with $\xi(P) = b$ and $\xi(Q) = f$ satisfies χ, but not Q. Thus four-valued logic is not trivial.

On the other hand, we also have

$$P \vee \neg P \not\models_4 \top$$

in the sense of general validity without further assumptions, since for $\xi(P) = n$ *we obtain*

$$[\![P \vee \neg P]\!]^4_\xi = n.$$

Four-valued logic therefore allows us to distinguish cleanly between contradiction and indeterminacy.

Four-valued semantics is also closely related to classical and three-valued semantics. Classical interpretations can again be embedded as special cases, and the three-valued semantics is obtained when the value n does not occur.

Proposition 4.68. *Let* Σ *be a propositional signature, and let* $\varphi \in L(\Sigma)$ *be a formula.*

1. *If* $\xi(\Sigma) \subseteq \{t, f, b\}$*, then the four-valued evaluation of* φ *under* ξ *coincides with the three-valued evaluation.*

2. *If* $\xi(\Sigma) \subseteq \{t, f\}$*, then the four-valued evaluation of* φ *coincides with the classical evaluation.*

Proof. We prove both assertions by structural induction on φ.

Case $\varphi = x$. The claim is immediate, since, depending on the assumption, the value of x already lies in the smaller set of truth values.

Case $\varphi = \neg\psi$. By the induction hypothesis, the evaluations of ψ in the smaller sets of truth values agree with the four-valued evaluation. Since negation in all cases merely interchanges the values t and f while leaving b and n fixed, the claim is preserved.

Case $\varphi = \psi \wedge \chi$. By the induction hypothesis, the evaluations of ψ and χ agree with the corresponding smaller semantics. In all three semantics, conjunction is determined by the greatest common lower bound of the two values; hence the evaluation agrees in this case as well.

Case $\varphi = \psi \vee \chi$. The assertion for disjunction follows analogously, since in all semantics it is determined by the least common upper bound.

Thus the evaluations agree in all cases. □

This situates four-valued logic as a generalization of both three-valued and classical semantics. It is particularly useful when one wants to avoid conflating contradictory information with merely missing information. Precisely this distinction makes it an especially instructive example of paraconsistent, yet information-sensitive, reasoning.

Minimal Models in Four-Valued Logic

Four-valued semantics distinguishes between two different kinds of non-classical information: the value b stands for a contradiction, that is, for "true and false at the same time", whereas n stands for mere indeterminacy, that is, for "neither true nor false". If we want to define a paraconsistent consequence relation on the basis of minimal models, it is therefore natural to minimize not only the value b, but both non-classical values b and n.

For this purpose, given a formula φ, we consider the set of propositional variables occurring in φ:

$$\mathrm{Var}(\varphi) = \{x \in \Sigma \mid x \text{ occurs in } \varphi\}.$$

For a four-valued interpretation $\xi \in \Omega_4(\Sigma)$, we set

$$N_\varphi(\xi) := \{x \in \mathrm{Var}(\varphi) \mid \xi(x) \in \{b, n\}\}.$$

Thus this set records exactly those variables relevant to φ that are assigned non-classical truth values.

Definition 4.69. *In order to obtain a consequence behavior compatible with classical propositional logic, we consider minimal models not only relative to the variables occurring in φ, but minimize globally the set of variables that take non-classical truth values. Let Σ be a propositional signature, let $\xi \in \Omega_4(\Sigma)$ be a four-valued interpretation, and let $\varphi \in L(\Sigma)$ be a propositional formula. The interpretation ξ is called a* minimal four-valued model *of φ if $\xi \models_4 \varphi$ holds and there is no four-valued interpretation $\eta \in \mathrm{Mod}^4_\Sigma(\varphi)$ such that*

$$N_\varphi(\eta) \subsetneq N_\varphi(\xi).$$

The set of all minimal four-valued models of φ is denoted by

$$\mathrm{Mod}^{\min 4}_\Sigma(\varphi).$$

Corollary 4.70 (Existence of four-valued models). *Let Σ be a propositional signature and let $\varphi \in L(\Sigma)$ be a formula. Then*

$$\mathrm{Mod}^4_\Sigma(\varphi) \neq \emptyset.$$

Proof. We define the four-valued interpretation $\xi^\Sigma_b \in \Omega_4(\Sigma)$ by

$$\xi^\Sigma_b(x) = b \qquad \text{for all } x \in \Sigma.$$

We show by structural induction on the formula φ that, for this interpretation, one always has

$$[\![\varphi]\!]^4_{\xi^\Sigma_b} = b$$

holds. It follows immediately that $\xi^\Sigma_b \models_4 \varphi$, since by definition of the four-valued satisfaction relation, $\xi \models_4 \varphi$ holds exactly when $[\![\varphi]\!]^4_\xi \in \{t, b\}$.

Base case. If $\varphi = x$ for some $x \in \Sigma$, then directly

$$[\![x]\!]^4_{\xi^\Sigma_b} = \xi^\Sigma_b(x) = b.$$

Induction step. Assume that the claim holds for all proper subformulas of φ.

- Let $\varphi = \neg\psi$. By the induction hypothesis,

  $$[\![\psi]\!]^4_{\xi^\Sigma_b} = b.$$

  By the definition of negation in four-valued logic, it follows that

  $$[\![\neg\psi]\!]^4_{\xi^\Sigma_b} = b.$$

- Let $\varphi = \psi \wedge \chi$. By the induction hypothesis,

$$[\![\psi]\!]^4_{\xi^\Sigma_b} = b \qquad \text{and} \qquad [\![\chi]\!]^4_{\xi^\Sigma_b} = b.$$

  Since b forms the infimum with itself with respect to the truth order $\succ_4$, we obtain

$$[\![\psi \wedge \chi]\!]^4_{\xi^\Sigma_b} = b.$$

- Let $\varphi = \psi \vee \chi$. Analogously, the induction hypothesis and the definition of disjunction imply that

$$[\![\psi \vee \chi]\!]^4_{\xi^\Sigma_b} = b.$$

Thus, in all cases we have shown that

$$[\![\varphi]\!]^4_{\xi^\Sigma_b} = b.$$

Hence $\xi^\Sigma_b \models_4 \varphi$, and consequently

$$\mathrm{Mod}^4_\Sigma(\varphi) \neq \emptyset.$$

□

Proposition 4.71 (Existence of minimal four-valued models)**.** *Let Σ be a propositional signature and let $\varphi \in L(\Sigma)$ be a formula. Then:*

$$\mathrm{Mod}^{\min 4}_\Sigma(\varphi) \subseteq \mathrm{Mod}^4_\Sigma(\varphi) \quad \textit{and} \quad \mathrm{Mod}^{\min 4}_\Sigma(\varphi) \neq \emptyset.$$

Proof. The inclusion follows directly from the definition of a minimal four-valued model.

To prove non-emptiness, it suffices to show that among the four-valued models of φ there is one with a minimal set $N_\varphi(\xi)$. By the preceding corollary, $\mathrm{Mod}_\Sigma^4(\varphi) \neq \emptyset$, so φ has at least one four-valued model.

Since $\mathrm{Var}(\varphi)$ is finite, the set $N_\varphi(\xi)$ can take only finitely many different values, namely subsets of $\mathrm{Var}(\varphi)$. Consider therefore the set

$$\mathcal{N} := \{\, N_\varphi(\xi) \mid \xi \in \mathrm{Mod}_\Sigma^4(\varphi) \,\}.$$

This set is non-empty and finite. Hence $\mathcal{N}$ has a minimal element with respect to proper inclusion $\subsetneq$. Let $M \in \mathcal{N}$ be such a minimal element, and let $\xi \in \mathrm{Mod}_\Sigma^4(\varphi)$ be an interpretation with $N_\varphi(\xi) = M$.

Then there is no $\eta \in \mathrm{Mod}_\Sigma^4(\varphi)$ such that

$$N_\varphi(\eta) \subsetneq N_\varphi(\xi),$$

because otherwise $N_\varphi(\eta)$ would be a properly smaller element of $\mathcal{N}$, contradicting the minimality of M. Thus ξ is a minimal four-valued model of φ.

It follows that

$$\mathrm{Mod}_\Sigma^{\min 4}(\varphi) \neq \emptyset.$$

□

We now define the minimal four-valued consequence relation as follows.

Definition 4.72 (Minimal four-valued consequence relation). *Let Σ be a propositional signature, and let $\varphi, \psi \in L(\Sigma)$ be propositional formulas. The minimal four-valued consequence relation*

$$\models_{\Sigma}^{\min 4} \subseteq L(\Sigma) \times L(\Sigma)$$

is defined by

$$\varphi \models_{\Sigma}^{\min 4} \psi \quad :\Longleftrightarrow \quad \mathrm{Mod}_{\Sigma}^{\min 4}(\varphi) \subseteq \mathrm{Mod}_{\Sigma}^{\min 4}(\psi).$$

As usual, we write simply $\varphi \models^{\min 4} \psi$ when the signature is clear from the context.

Lemma 4.73 (Embedding of classical interpretations). *Let Σ be a propositional signature, let $\omega \in \Omega(\Sigma)$ be a classical interpretation, and let $\xi_\omega \in \Omega_4(\Sigma)$ be the four-valued interpretation associated with ω, given by*

$$\xi_\omega(x) = \begin{cases} t & \text{if } \omega(x) = \texttt{true}, \\ f & \text{if } \omega(x) = \texttt{false}. \end{cases}$$

Then, for every formula $\varphi \in L(\Sigma)$, we have:

$$[\![\varphi]\!]^4_{\xi_\omega} = \begin{cases} t & \text{if } [\![\varphi]\!]_\omega = \texttt{true}, \\ f & \text{if } [\![\varphi]\!]_\omega = \texttt{false}. \end{cases}$$

Proof. The proof proceeds by structural induction on φ.

Case $\varphi = x$. In this case, the claim follows directly from the definition of ξ_ω.

Case $\varphi = \neg\psi$. By the induction hypothesis, the claim holds for ψ. Since negation in the four-valued logic acts on the values t and f exactly as it does in classical logic, the claim follows for $\neg\psi$.

Case $\varphi = \psi \wedge \chi$. By the induction hypothesis, the values of ψ and χ agree in the four-valued semantics and in the classical semantics. Since $\wedge$ is described in both semantics by the corresponding lower bound of the truth values involved, the evaluation of $\psi \wedge \chi$ also agrees.

Case $\varphi = \psi \vee \chi$. The claim for disjunction follows analogously.

Thus, classical evaluation and four-valued evaluation agree on two-valued assignments. □

Theorem 4.74. *Let* Σ *be a propositional signature, and let* $\varphi, \psi \in L(\Sigma)$ *be propositional formulas with* $\varphi \not\equiv \bot$. *Then*

$$\varphi \models_{\mathrm{PL}} \psi \quad \Longleftrightarrow \quad \varphi \models^{\min 4} \psi.$$

Proof. We first show that, for every non-inconsistent formula φ, the following holds:

$$\left(\mathrm{Mod}_{\Sigma}^{\mathrm{PL}}(\varphi)\right)^4 = \mathrm{Mod}_{\Sigma}^{\min 4}(\varphi).$$

"$\subseteq$": Let $\omega \in \mathrm{Mod}_{\Sigma}^{\mathrm{PL}}(\varphi)$. Then $\xi_\omega \in \left(\mathrm{Mod}_{\Sigma}^{\mathrm{PL}}(\varphi)\right)^4$. By the preceding lemma,

$$[\![\varphi]\!]_{\xi_\omega}^4 = t,$$

hence $\xi_\omega \models_4 \varphi$. Moreover,

$$N(\xi_\omega) = \emptyset.$$

Therefore, there can be no $\eta \in \mathrm{Mod}^4_\Sigma(\varphi)$ such that $N(\eta) \subsetneq N(\xi_\omega)$. Hence

$$\xi_\omega \in \mathrm{Mod}^{\min 4}_\Sigma(\varphi).$$

"$\supseteq$": Conversely, let $\xi \in \mathrm{Mod}^{\min 4}_\Sigma(\varphi)$. First suppose that $N(\xi) \neq \emptyset$. Since $\varphi \not\equiv \bot$, the formula φ has a classical model ω. Then also $\xi_\omega \in \mathrm{Mod}^4_\Sigma(\varphi)$, and

$$N(\xi_\omega) = \emptyset \subsetneq N(\xi),$$

which contradicts the minimality of ξ. Thus we must in fact have

$$N(\xi) = \emptyset$$

Thus, ξ assumes only the values t and f.

Now define the classical interpretation ω_ξ by

$$\omega_\xi(x) = \begin{cases} \texttt{true} & \text{if } \xi(x) = t, \\ \texttt{false} & \text{if } \xi(x) = f. \end{cases}$$

Then ω_ξ is well defined and $\xi = \xi_{\omega_\xi}$. By the preceding lemma, it follows from $\xi \models_4 \varphi$ that

$$\omega_\xi \models_{\mathrm{PL}} \varphi$$

Hence,

$$\xi \in \left(\mathrm{Mod}^{\mathrm{PL}}_\Sigma(\varphi)\right)^4.$$

This proves the claimed equality.

The equivalence of the two consequence relations now follows:

$$\varphi \models_{\mathrm{PL}} \psi \iff \mathrm{Mod}^{\mathrm{PL}}_\Sigma(\varphi) \subseteq \mathrm{Mod}^{\mathrm{PL}}_\Sigma(\psi)$$

$$\iff (\mathrm{Mod}_\Sigma^{\mathrm{PL}}(\varphi))^4 \subseteq (\mathrm{Mod}_\Sigma^{\mathrm{PL}}(\psi))^4$$
$$\iff \mathrm{Mod}_\Sigma^{\min 4}(\varphi) \subseteq \mathrm{Mod}_\Sigma^{\min 4}(\psi) \iff \varphi \models^{\min 4} \psi.$$

This proves the theorem. □

Example 4.75. *For the formula*

$$\varphi := P \wedge \neg P \wedge Q$$

we have, in four-valued logic:

$$\varphi \models^{\min 4} Q, \qquad \varphi \not\models^{\min 4} \neg Q.$$

Thus the formula remains paraconsistent: the contradiction in P does not imply everything, although the information about Q that is stable in the minimal models is preserved.

4.2.4 Exercises and Solutions

In what follows, we practice three-valued and four-valued paraconsistent semantics using concrete examples. In each case, decide whether the given statement is true or false.

Exercise 4.76. *Consider the propositional signature*

$$\Sigma = \{B, C, E, G, H\}.$$

Furthermore, suppose the following are given:

$$v_1(H) = f,$$

$$v_1(B) = v_1(C) = v_1(E) = v_1(G) = b,$$

$$v_2(C) = v_2(E) = v_2(H) = t,$$
$$v_2(B) = v_2(G) = f,$$

$$\Phi_3 = \{\neg(E \vee C),\ E \wedge C\},$$

$$\xi_4(E) = \xi_4(H) = t,$$
$$\xi_4(C) = \xi_4(G) = f,$$
$$\xi_4(B) = b.$$

For each of the following statements, decide whether it is true or false:

1. $[\![G \vee C \wedge E \vee B]\!]^3_{v_1} = t$.
2. $v_2 \in \mathrm{Mod}^{\min 3}_{\Sigma}(C \vee \neg G)$.
3. $\Phi_3 \models^{\min 3} G \vee H$.
4. $\xi_4 \models_4 C \wedge C$.

Solution 4.77. *1.* ***False.*** *By the usual precedence conventions, the expression is read as*

$$G \vee (C \wedge E) \vee B$$

Under v_1 we have

$$[\![G]\!]^3_{v_1} = b, \qquad [\![C \wedge E]\!]^3_{v_1} = b, \qquad [\![B]\!]^3_{v_1} = b.$$

Hence, overall,

$$[\![G \vee C \wedge E \vee B]\!]^3_{v_1} = b,$$

and therefore not t. The formula is satisfied, but it is not classically true.

2. **True.** *First,*

$$[\![C \vee \neg G]\!]^3_{v_2} = t \vee t = t,$$

since $v_2(C) = t$ and $v_2(G) = f$, so $[\![\neg G]\!]^3_{v_2} = t$. Thus v_2 is a three-valued model of $C \vee \neg G$.

Moreover, v_2 contains no b-value at all. Hence

$$v_2^{-1}(b) = \emptyset.$$

Since there is no set that is a proper subset of $\emptyset$, v_2 is automatically minimal. Therefore,

$$v_2 \in \mathrm{Mod}^{\min 3}_{\Sigma}(C \vee \neg G).$$

3. **False.** *It suffices to exhibit a minimal three-valued model of Φ_3 that does not satisfy the formula $G \vee H$.*

Consider the interpretation $v \in \Omega_3(\Sigma)$ with

$$v(C) = v(E) = b, \qquad v(B) = v(G) = v(H) = f.$$

Then

$$[\![E \wedge C]\!]^3_v = b \qquad \text{and} \qquad [\![\neg(E \vee C)]\!]^3_v = b.$$

Thus v is a model of Φ_3.

The fact that v is even minimal follows from the observation that both E and C must have the value b

in order to satisfy the two formulas simultaneously. No interpretation with fewer b-values can satisfy Φ_3.

For the conclusion, however, we have

$$[\![G \vee H]\!]^3_v = f \vee f = f.$$

Hence v *is not a model of* $G \vee H$. *It follows that*

$$\Phi_3 \not\models^{\min 3} G \vee H.$$

4. **False.** *We have*

$$[\![C \wedge C]\!]^4_{\xi_4} = [\![C]\!]^4_{\xi_4} \wedge [\![C]\!]^4_{\xi_4} = f \wedge f = f.$$

Thus $C \wedge C$ *is not satisfied under* ξ_4*, since for* $\models_4$ *only the values* t *and* b *are designated. Therefore,*

$$\xi_4 \not\models_4 C \wedge C.$$

4.3 Fuzzy Logic

In the non-classical logics considered so far, we have already broadened the notion of truth: in many-valued logic there were more values than merely *true* and *false*, and in paraconsistent logic truth and falsity aspects could even coexist. Fuzzy logic goes one step further. It does not merely introduce additional truth values; it allows truth to vary by *degrees*.

The basic idea is intuitive: many everyday concepts do not have sharply defined boundaries. Whether a person counts as "tall", a temperature as "warm", or a statement as "fairly certain" often depends on gradual transitions and contextual judgments. Instead of distinguishing only between "yes" and "no", fuzzy logic treats all real numbers in the unit interval $[0, 1]$ as truth values. Here 0 corresponds to classical *false* and 1 to classical *true*; intermediate values describe graded forms of truth.

In what follows, we begin with the foundations of fuzzy set theory. We then examine how logical connectives can be interpreted meaningfully in this setting. To this end, we first consider conjunctions and disjunctions, then negation, and finally the full semantics of fuzzy logic.

4.3.1 Fuzzy Sets

The language of fuzzy logic is based on the notion of a *fuzzy set*. To understand the idea, it is useful to begin by comparing it with classical sets.

A classical set $M \subseteq U$ can be described by its characteristic function. This function assigns exactly one of two values to each element $x \in U$:

$$\lambda_M(x) = \begin{cases} 1, & \text{if } x \in M, \\ 0, & \text{if } x \notin M. \end{cases}$$

Thus an element either belongs completely to the set or not at all. It is precisely this sharpness that fuzzy logic gives up.

Definition 4.78 (Fuzzy set). *Let U be a set. A* fuzzy set F *over* U *is a function*

$$F : U \to [0, 1].$$

The value $F(x)$ *indicates the degree to which the element* x *belongs to the fuzzy set* F*.*

This replaces the strict yes-no structure of classical sets with graded membership. A value close to 1 means strong membership, while a value close to 0 means weak membership or practically no membership. Intermediate values make it possible to model vague concepts mathematically.

Example 4.79. *Let* $U_{\text{tea}} = \{\textit{cold}, \textit{lukewarm}, \textit{hot}\}$ *be a small set of temperature terms. For the vague concept "hot tea", we might describe a fuzzy set* $F_{\text{hot}} : U_{\text{tea}} \to [0, 1]$ *as follows:*

$$\begin{aligned} F_{\text{hot}}(\textit{cold}) &= 0.0, \\ F_{\text{hot}}(\textit{lukewarm}) &= 0.4, \\ F_{\text{hot}}(\textit{hot}) &= 1.0 \end{aligned}$$

Then "cold" does not belong to this concept, "lukewarm" belongs only partially, and "hot" belongs completely.

Fuzzy sets can also be compared with one another. In doing so, the classical subset relation is generalized in a natural way.

Definition 4.80 (Fuzzy subset)**.** *Let* $F, F' : U \to [0, 1]$ *be fuzzy sets over* U*. We say that* F *is a* fuzzy subset *of* F' *and write*

$$F \subseteq F',$$

if and only if, for all $x \in U$,

$$F(x) \leq F'(x).$$

The idea is immediate: if every element is contained in F to at most the same degree as it is contained in F', then F is contained in F' in this graded sense.

Example 4.81. *Again consider $U_{\text{tea}} = \{cold, lukewarm, hot\}$ and the vague concept "hot tea", given by*

$$F_{\text{hot}}(cold) = 0.0,$$
$$F_{\text{hot}}(lukewarm) = 0.4,$$
$$F_{\text{hot}}(hot) = 1.0.$$

In addition, consider the more generous concept "at least warm", modeled by a fuzzy set F_{warm}:

$$F_{\text{warm}}(cold) = 0.1,$$
$$F_{\text{warm}}(lukewarm) = 0.8,$$
$$F_{\text{warm}}(hot) = 1.0.$$

Then

$$F_{\text{hot}} \subseteq F_{\text{warm}},$$

because for every element the degree of membership in F_{hot} is at most as large as its degree of membership in F_{warm}.

Intuitively, this means that every "hot" tea is also "warm" to at least the same degree, while the concept "warm" is less restrictive overall.

Example 4.82. *Let* $U_{\text{car}} = \{economical, fast, comfortable\}$ *be a set of properties of a car. We consider two fuzzy sets:*

$$\begin{aligned} F_{\text{sporty}}(economical) &= 0.2, \\ F_{\text{comfortable}}(economical) &= 0.8, \\ F_{\text{sporty}}(fast) &= 0.9, \\ F_{\text{comfortable}}(fast) &= 0.4, \\ F_{\text{sporty}}(comfortable) &= 0.3, \\ F_{\text{comfortable}}(comfortable) &= 1.0. \end{aligned}$$

Then neither

$$F_{\text{sporty}} \subseteq F_{\text{comfortable}} \quad nor \quad F_{\text{comfortable}} \subseteq F_{\text{sporty}}.$$

For fast, we have $F_{\text{sporty}}(fast) = 0.9 > 0.4 = F_{\text{comfortable}}(fast)$*, whereas for economical the inequality goes in the opposite direction:* $F_{\text{comfortable}}(economical) = 0.8 > 0.2 = F_{\text{sporty}}(economical)$.

Thus, the two fuzzy sets describe different properties that cannot simply be compared by means of a subset relation.

As with classical sets, the fuzzy subset relation is a partial order.

Proposition 4.83. *Let* $F, F', F'' : U \to [0,1]$ *be fuzzy sets over* U*. Then:*

1. $F \subseteq F$,

2. *from* $F \subseteq F'$ *and* $F' \subseteq F$ *it follows that* $F = F'$,

3. *from* $F \subseteq F'$ *and* $F' \subseteq F''$ *it follows that* $F \subseteq F''$.

Proof. All three statements follow directly from the definition of the relation $\subseteq$ and the order on the real numbers.

1. For every $x \in U$, we have $F(x) \leq F(x)$.

2. If $F \subseteq F'$ and $F' \subseteq F$, then for every $x \in U$ both $F(x) \leq F'(x)$ and $F'(x) \leq F(x)$ hold. Hence $F(x) = F'(x)$ for all $x \in U$, and therefore $F = F'$.

3. From $F(x) \leq F'(x)$ and $F'(x) \leq F''(x)$, it follows for every $x \in U$ that $F(x) \leq F''(x)$.

□

The empty set and the universal set also have natural fuzzy-logical counterparts.

Definition 4.84 (Empty and universal fuzzy set). *The* empty fuzzy set $\hat{\emptyset}$ *over* U *is defined by*

$$\hat{\emptyset}(x) = 0 \qquad \textit{for all } x \in U.$$

The universal fuzzy set $\hat{U}$ *over* U *is defined by*

$$\hat{U}(x) = 1 \qquad \textit{for all } x \in U.$$

Proposition 4.85. *Let* $F : U \to [0, 1]$ *be a fuzzy set over* U. *Then*

$$\hat{\emptyset} \subseteq F \subseteq \hat{U}.$$

Proof. For every $x \in U$, the value $F(x)$ lies in the interval $[0, 1]$ by definition. Hence

$$0 = \hat{\emptyset}(x) \leq F(x) \leq 1 = \hat{U}(x)$$

for all $x \in U$. Therefore $\hat{\emptyset} \subseteq F \subseteq \hat{U}$. □

Fuzzy Sets over Propositional Signatures

For logic itself, the most relevant case is not that of arbitrary sets, but that of propositional variables. We therefore now turn to fuzzy sets over propositional signatures. If Σ is a propositional signature, then a fuzzy set over Σ is understood as a function

$$F : \Sigma \to [0, 1].$$

Thus, each atomic propositional variable is assigned a degree of truth.

Example 4.86. *Let*

$$\Sigma_{\text{Climate}} = \{\textit{Rain}, \textit{Wind}, \textit{Warm}, \textit{Dry}\}$$

be a propositional signature with the following informal meanings:

- *Rain: "It is raining."*
- *Wind: "It is windy."*
- *Warm: "It is warm."*

- *Dry: "It is dry."*

The fuzzy set $F_{\text{Climate}} : \Sigma_{\text{Climate}} \to [0, 1]$ *given by*

$$F_{\text{Climate}}(Rain) = 0.2, \qquad F_{\text{Climate}}(Wind) = 0.7,$$
$$F_{\text{Climate}}(Warm) = 0.85, \qquad F_{\text{Climate}}(Dry) = 0.6.$$

then describes a situation in which it is raining only slightly, is clearly windy and rather warm, and the condition of being "dry" holds only partially.

The central question is now: How do we evaluate compound formulas such as

$$\text{Warm} \wedge \text{Dry}, \qquad \text{Wind} \vee \text{Rain}, \qquad \neg\text{Rain}$$

under such a fuzzy set? This is precisely the question addressed in the following subsections.

4.3.2 T-Norms

After introducing fuzzy sets, the central question is how logical conjunction can be interpreted meaningfully within this many-valued framework. In classical logic, a conjunction is true precisely when both of its component formulas are true. In fuzzy logic, we do not want to abandon this idea, but rather generalize it. The truth value of a conjunction should therefore be computed from the truth values of its two component formulas in such a way that the semantic evaluation remains compatible with classical logic.

There is, however, no single way to carry out this generalization. Depending on the intended application, conjunction may be modeled in a stricter, more compensatory, or sharper way. For this reason, the desired behavior is first described axiomatically.

Definition 4.87 (T-Norm). *A function* $T : [0,1]^2 \to [0,1]$ *is called a* T-norm *if, for all* $x, x', y, y', z \in [0,1]$, *the following properties hold:*

1. $T(x, 0) = 0$ *(absorbing element* 0*)*
2. $T(x, 1) = x$ *(neutral element* 1*)*
3. *If* $x \leq x'$ *and* $y \leq y'$, *then* $T(x, y) \leq T(x', y')$ *(monotonicity)*
4. $T(x, y) = T(y, x)$ *(commutativity)*
5. $T(x, T(y, z)) = T(T(x, y), z)$ *(associativity)*

These conditions have a natural intuitive interpretation. Property (1) says that as soon as one component formula is completely false, the whole conjunction should also be completely false. Property (2) expresses that the truth value of a conjunction is not changed by a completely true component formula. Property (3) requires that improving one component formula must not decrease the truth value of the conjunction. Properties (4) and (5) correspond to the familiar algebraic laws of classical conjunction.

T-norms genuinely generalize classical conjunction in the sense that, on the values 0 and 1, they behave like the classical connective.

Proposition 4.88. *For every T-norm* $T : [0,1]^2 \to [0,1]$, *we have*

$$\begin{aligned} T(0,0) &= 0 \\ T(1,0) &= 0 \\ T(0,1) &= 0 \\ T(1,1) &= 1 \end{aligned}$$

Proof. By property (1) of a T-norm, we immediately obtain

$$T(0,0) = 0 \quad \text{and} \quad T(1,0) = 0.$$

Using commutativity from property (4), we get

$$T(0,1) = T(1,0) = 0.$$

Finally, property (2) yields

$$T(1,1) = 1.$$

Thus all four equations have been shown. □

The definition of a T-norm specifies only the desired general framework. Within this framework, there are several meaningful realizations. Three particularly important examples are the minimum, product, and Łukasiewicz conjunctions.

Definition 4.89. *Define functions* $T_{\min}, T_{\text{prod}}, T_{\text{Luka}}$: $[0,1]^2 \to [0,1]$ *by*

$$\begin{aligned} T_{\min}(x,y) &= \min\{x,y\} \\ T_{\text{prod}}(x,y) &= xy \\ T_{Ł}(x,y) &= \max\{0, x+y-1\} \end{aligned}$$

for all $x, y \in [0,1]$. *Here* $T_{\min}$ *is also called the* minimum conjunction, T_{prod} *the* product conjunction, *and* T_{Luka} *the* Łukasiewicz conjunction.

Proposition 4.90. *The functions* $T_{\min}$, T_{prod}, *and* T_{Luka} *are T-norms.*

Proof. We verify the five defining properties separately in each case.

1. **The function $T_{\min}$.**

   1. $T_{\min}(x,0) = \min\{x,0\} = 0$.

   2. $T_{\min}(x,1) = \min\{x,1\} = x$.

   3. If $x \leq x'$ and $y \leq y'$, then $\min\{x,y\} \leq \min\{x',y'\}$.

   4. We have $\min\{x,y\} = \min\{y,x\}$.

   5. We have

$$\begin{aligned} T_{\min}(x, T_{\min}(y,z)) &= \min\{x, \min\{y,z\}\} \\ &= \min\{x,y,z\} \\ &= \min\{\min\{x,y\},z\} \\ &= T_{\min}(T_{\min}(x,y),z) \end{aligned}$$

2. The function T_{prod}.

1. $T_{\text{prod}}(x, 0) = x \cdot 0 = 0$.
2. $T_{\text{prod}}(x, 1) = x \cdot 1 = x$.
3. Let $x \leq x'$ and $y \leq y'$ with $x, x', y, y' \in [0, 1]$. By non-negativity, we have
$$xy \leq x'y \leq x'y',$$
and hence $T_{\text{prod}}(x, y) \leq T_{\text{prod}}(x', y')$.
4. $T_{\text{prod}}(x, y) = xy = yx = T_{\text{prod}}(y, x)$.
5. We have
$$\begin{aligned} T_{\text{prod}}(x, T_{\text{prod}}(y, z)) &= x(yz) \\ &= (xy)z \\ &= T_{\text{prod}}(T_{\text{prod}}(x, y), z) \end{aligned}$$

3. The function T_{Luka}.

1. $$T_{\text{Luka}}(x, 0) = \max\{0, x + 0 - 1\} = 0$$
since $x \leq 1$.
2. $$T_{\text{Luka}}(x, 1) = \max\{0, x + 1 - 1\} = x$$
since $x \geq 0$.

3. From $x \leq x'$ and $y \leq y'$ it follows that

$$x + y - 1 \leq x' + y' - 1.$$

Since the function $t \mapsto \max\{0, t\}$ is monotone increasing, we obtain

$$\begin{aligned} T_{Ł}(x, y) &= \max\{0, x + y - 1\} \\ &\leq \max\{0, x' + y' - 1\} \\ &= T_{Ł}(x', y') \end{aligned}$$

4. We immediately have

$$\begin{aligned} T_{Ł}(x, y) &= \max\{0, x + y - 1\} \\ &= \max\{0, y + x - 1\} \\ &= T_{Ł}(y, x) \end{aligned}$$

5. For associativity, we compute:

$$\begin{aligned} T_{\mathrm{Luka}}(x, T_{\mathrm{Luka}}(y, z)) &= \max\{0, x+ \\ &\quad \max\{0, y + z - 1\} - 1\} \\ &= \max\{0, x + y + z - 2\} \end{aligned}$$

In general, we have

$$x+\max\{0, y+z-1\}-1 = \max\{x-1, x+y+z-2\},$$

and therefore

$$T_{\mathrm{Luka}}(x, T_{\mathrm{Luka}}(y, z))$$

$$= \max\{0, x + \max\{0, y + z - 1\} - 1\}$$
$$= \max\{0, x + y + z - 2\}$$

Since $x \leq 1$, we have $x - 1 \leq 0$, and hence

$$T_{\text{Luka}}(x, T_{\text{Luka}}(y, z)) = \max\{0, x + y + z - 2\}.$$

Analogously, we obtain

$$T_{Ł}(T_{Ł}(x, y), z) = \max\{0, \max\{0, x + y - 1\} + z - 1\}$$
$$= \max\{0, x + y + z - 2\}$$

Thus associativity has also been shown.

□

The three T-norms express different intuitions about how two statements interact. The minimum behaves in a particularly strict way: the conjunction can receive no value higher than the smaller of the two truth values. The product conjunction gives a smoother interpretation of the interaction and is often understood intuitively as the simultaneous support of both component formulas. The Łukasiewicz conjunction permits an even finer gradation, since the two truth values can, up to a certain point, "compensate" for one another.

Example 4.91. *Consider the propositional signature*

$$\Sigma_{\text{Device}} = \{\mathit{Fast}, \mathit{Quiet}, \mathit{Portable}\}$$

with the following informal meanings:

- *Fast: "The device operates quickly."*
- *Quiet: "The device is quiet."*
- *Portable: "The device is easy to transport."*

For a fuzzy set $F : \Sigma_{\text{Device}} \to [0, 1]$*, suppose that*

$$F(\mathit{Fast}) = 0.9, \qquad F(\mathit{Quiet}) = 0.8, \qquad F(\mathit{Portable}) = 0.6.$$

The truth value of the conjunction $\mathit{Fast} \wedge \mathit{Quiet}$ *now depends on the chosen T-norm:*

$$\begin{aligned} T_{\min}(0.9, 0.8) &= 0.8 \\ T_{\text{prod}}(0.9, 0.8) &= 0.72 \\ T_{\text{Ł}}(0.9, 0.8) &= 0.7 \end{aligned}$$

This illustrates that different T-norms can evaluate the same intuitive conjunction with different degrees of strictness.

4.3.3 S-Norms

After conjunction, we now turn to disjunction. In classical logic, a disjunction is true exactly when at least one of its component formulas is true. This idea should also be preserved in fuzzy logic. However, it is no longer sufficient here to distinguish only between "true" and "false". Instead, the truth degree of a disjunction should be computed from the truth degrees of its component formulas.

Here too, there is not just one reasonable possibility. Depending on the intended semantics, disjunction can be modeled more cautiously, more generously, or more sharply. For this reason, the behavior of fuzzy-logical disjunction is also specified axiomatically.

Definition 4.92 (S-Norm). *A function $S : [0,1]^2 \to [0,1]$ is called an* S-norm *if, for all $x, x', y, y', z \in [0,1]$, it satisfies the following properties:*

1. $S(x, 1) = 1$ *(absorbing element 1)*
2. $S(x, 0) = x$ *(neutral element 0)*
3. *If $x \leq x'$ and $y \leq y'$, then $S(x, y) \leq S(x', y')$ (monotonicity)*
4. $S(x, y) = S(y, x)$ *(commutativity)*
5. $S(x, S(y, z)) = S(S(x, y), z)$ *(associativity)*

These properties are precisely the counterpart of the conditions for T-norms. Property (1) means that if one component formula is completely true, then the entire disjunction should be completely true. Property (2) means that a completely false component formula does not change the truth value of the disjunction. Monotonicity, commutativity, and associativity correspond to the natural expectations for an "or" in a many-valued semantics.

As with T-norms, S-norms correctly generalize classical disjunction on the values 0 and 1.

Proposition 4.93. *For every S-norm* $S : [0,1]^2 \to [0,1]$, *we have*

$$S(0,0) = 0, \qquad S(1,0) = S(0,1) = S(1,1) = 1.$$

Proof. From the definition of an S-norm we obtain in particular:

$$S(x,0) = x \quad \text{for all } x \in [0,1],$$

$$S(x,1) = 1 \quad \text{for all } x \in [0,1],$$

and

$$S(x,y) = S(y,x) \quad \text{for all } x, y \in [0,1].$$

Substituting $x = 0$ into the neutral-element condition gives

$$S(0,0) = 0.$$

Substituting $x = 1$ gives

$$S(1,0) = 1.$$

By commutativity, we therefore obtain

$$S(0,1) = S(1,0) = 1.$$

Finally, the absorbing property with $x = 1$ yields

$$S(1,1) = 1.$$

This proves all the asserted identities. □

The following three S-norms are among the most important standard examples. Each of them corresponds to a natural way of interpreting disjunction in a fuzzy framework.

Definition 4.94. *Define functions* $S_{\text{max}}, S_{\text{prod}}, S_{\text{Luka}} : [0,1]^2 \to [0,1]$ *by*

$$\begin{aligned} S_{\text{max}}(x, y) &= \max\{x, y\} \\ S_{\text{prod}}(x, y) &= x + y - xy \\ S_{\text{Luka}}(x, y) &= \min\{x + y, 1\} \end{aligned}$$

for all $x, y \in [0,1]$. *The function* S_{max} *is called the* maximum disjunction, S_{prod} *the* probabilistic disjunction, *and* S_{Luka} *the* Łukasiewicz disjunction.

Proposition 4.95. *The functions* S_{max}, S_{prod}, *and* S_{Luka} *are S-norms.*

Proof. We verify the five properties from Definition 4.92 for each of the three functions.

1. The function S_{max}**.**

1. $S_{\text{max}}(x, 1) = \max\{x, 1\} = 1$.

2. $S_{\text{max}}(x, 0) = \max\{x, 0\} = x$, since $x \geq 0$.

3. If $x \leq x'$ and $y \leq y'$, then immediately

$$\max\{x, y\} \leq \max\{x', y'\}.$$

4. We have $\max\{x, y\} = \max\{y, x\}$.

5. For associativity, we compute

$$\begin{aligned} S_{\max}(x, S_{\max}(y, z)) &= \max\{x, \max\{y, z\}\} \\ &= \max\{x, y, z\} \\ &= \max\{\max\{x, y\}, z\} \\ &= S_{\max}(S_{\max}(x, y), z). \end{aligned}$$

2. The function S_{prod}. We have

$$S_{\text{prod}}(x, y) = x + y - xy = 1 - (1 - x)(1 - y).$$

This representation is especially useful.

1. $$S_{\text{prod}}(x, 1) = 1 - (1 - x)(1 - 1) = 1.$$

2. $$S_{\text{prod}}(x, 0) = 1 - (1 - x)(1 - 0) = 1 - (1 - x) = x.$$

3. Suppose that $x \leq x'$ and $y \leq y'$. Then

$$1 - x \geq 1 - x' \quad \text{and} \quad 1 - y \geq 1 - y'.$$

Since all factors are nonnegative, it follows that

$$(1 - x)(1 - y) \geq (1 - x')(1 - y').$$

Hence, after subtracting from 1, we obtain

$$1 - (1 - x)(1 - y) \leq 1 - (1 - x')(1 - y'),$$

that is,

$$S_{\text{prod}}(x, y) \leq S_{\text{prod}}(x', y').$$

4. The formula is symmetric in x and y, so

$$S_{\text{prod}}(x, y) = S_{\text{prod}}(y, x).$$

5. For associativity, we compute:

$$\begin{aligned} S_{\text{prod}}(x, S_{\text{prod}}(y, z)) &= 1 - (1 - x)(1 - S_{\text{prod}}(y, z)) \\ &= 1 - (1 - x)(1 - (y + z - yz)) \\ &= 1 - (1 - x)(1 - y)(1 - z). \end{aligned}$$

Analogously, we obtain

$$S_{\text{prod}}(S_{\text{prod}}(x, y), z) = 1 - (1 - x)(1 - y)(1 - z).$$

Thus S_{prod} is associative.

3. The function S_{Luka}.

1.

$$S_{\text{Luka}}(x, 1) = \min\{x + 1, 1\} = 1.$$

2.

$$S_{\text{Luka}}(x, 0) = \min\{x, 1\} = x.$$

3. If $x \leq x'$ and $y \leq y'$, then

$$x + y \leq x' + y'.$$

Since the function $t \mapsto \min\{t, 1\}$ is monotone increasing, we obtain

$$\min\{x + y, 1\} \leq \min\{x' + y', 1\}.$$

Hence

$$S_{\text{Luka}}(x, y) \leq S_{\text{Luka}}(x', y').$$

4. Commutativity follows immediately from commutativity of addition:

$$\begin{aligned} S_{\text{Luka}}(x, y) &= \min\{x + y, 1\} \\ &= \min\{y + x, 1\} \\ &= S_{\text{Luka}}(y, x) \end{aligned}$$

5. For associativity, it suffices to show that both bracketings yield the same value, namely $\min\{x + y + z, 1\}$.

   First consider

$$S_{\text{Luka}}(x, S_{\text{Luka}}(y, z)) = \min\{x + \min\{y + z, 1\}, 1\}.$$

   If $y + z \leq 1$, then $S_{\text{Luka}}(y, z) = y + z$, and therefore

$$S_{\text{Luka}}(x, S_{\text{Luka}}(y, z)) = \min\{x + y + z, 1\}.$$

   If $y + z \geq 1$, then $S_{\text{Luka}}(y, z) = 1$, and hence

$$\begin{aligned} S_{\text{Luka}}(x, S_{\text{Luka}}(y, z)) &= \min\{x + \min\{y + z, 1\}, 1\} \\ &= \min\{x + 1, 1\} \\ &= 1. \end{aligned}$$

   Since in this case also $x + y + z \geq 1$, this is again equal to $\min\{x + y + z, 1\}$.

   Thus, in all cases,

$$S_{\text{Luka}}(x, S_{\text{Luka}}(y, z)) = \min\{x + y + z, 1\}.$$

Analogously, one obtains

$$S_{\text{Luka}}(S_{\text{Luka}}(x, y), z) = \min\{x + y + z, 1\}.$$

This proves associativity as well.

□

The three S-norms differ in how permissively they treat disjunction. The maximum disjunction simply takes the larger truth value. The product disjunction allows both truth values to contribute simultaneously. The Łukasiewicz disjunction first adds the two values and then caps the result at 1. Thus, depending on the application, the same linguistic disjunction can be modeled in semantically different ways.

Example 4.96. *Consider the propositional signature*

$$\Sigma_{\text{Service}} = \{\mathit{Fast}, \mathit{Affordable}, \mathit{Reliable}, \mathit{Personal}\}$$

with the following informal meanings:

- *Fast: "The service is fast."*
- *Affordable: "The service is affordable."*
- *Reliable: "The service is reliable."*
- *Personal: "The service is personal."*

Suppose that a fuzzy set $F : \Sigma_{\text{Service}} \to [0, 1]$ is given by

$$F(\mathit{Fast}) = 0.7$$

$$
\begin{aligned}
F(\mathit{Affordable}) &= 0.4 \\
F(\mathit{Reliable}) &= 0.9 \\
F(\mathit{Personal}) &= 0.6
\end{aligned}
$$

Then the disjunction $\mathit{Fast} \vee \mathit{Affordable}$ *is evaluated as follows:*

$$
\begin{aligned}
S_{\text{max}}(0.7, 0.4) &= 0.7 \\
S_{\text{prod}}(0.7, 0.4) &= 0.7 + 0.4 - (0.7 \cdot 0.4) = 0.82 \\
S_{\text{Luka}}(0.7, 0.4) &= \min\{0.7 + 0.4, 1\} = 1
\end{aligned}
$$

This illustrates that different S-norms may interpret the same linguistic disjunction in different ways, ranging from a more cautious interpretation to a very generous one.

4.3.4 Negations

After considering in the two previous subsections how conjunctions and disjunctions can be evaluated in fuzzy logic, we now turn to negation. Here, too, the basic idea of classical logic is preserved: negation is meant, in a certain sense, to "reverse" the truth value of a formula. Since fuzzy logic no longer works only with the values 0 and 1, but with all numbers in the interval $[0, 1]$, this reversal has to be generalized accordingly.

Intuitively, the more true a formula is, the smaller the truth value of its negation should be. At the same time, negation should remain compatible with the classical boundary cases, mapping 1 to 0 and 0 to 1. We again collect these requirements axiomatically.

Definition 4.97 (Negation norm). *A function* $N : [0,1] \to [0,1]$ *is called a* negation norm *if, for all* $x, y \in [0,1]$, *the following properties hold:*

1. $N(1) = 0$ *and* $N(0) = 1$ *(complementarity)*
2. *if* $x \leq y$, *then* $N(x) \geq N(y)$ *(antitonicity)*

The first property ensures that negation agrees with classical logic at the endpoints. The second property expresses that larger truth values should become smaller truth values under negation. Thus, a negation should not merely exchange individual points; it should reverse the order in a systematic way.

In many applications, the standard negation is used.

Definition 4.98 (Standard negation). *Define* $N_0 : [0,1] \to [0,1]$ *by*

$$N_0(x) = 1 - x$$

for all $x \in [0,1]$.

Proposition 4.99. *The standard negation* $N_0 : [0,1] \to [0,1]$ *is a negation norm.*

Proof. We verify the two properties from the definition of a negation norm.

1. Complementarity. We have

$$N_0(1) = 1 - 1 = 0 \qquad \text{and} \qquad N_0(0) = 1 - 0 = 1.$$

2. Antitonicity. Let $x, y \in [0,1]$ with $x \leq y$. Subtracting both sides from 1 reverses the inequality, so

$$1 - x \geq 1 - y.$$

Hence

$$N_0(x) = 1 - x \geq 1 - y = N_0(y).$$

Moreover, for every $x \in [0,1]$, we also have $1 - x \in [0,1]$, so N_0 is indeed a function from $[0,1]$ to $[0,1]$. Therefore, N_0 is a negation norm. □

Example 4.100. *We consider the propositional signature*

$$\Sigma_{\text{Service}} = \{\textit{Fast}, \textit{Affordable}, \textit{Reliable}, \textit{Friendly}\}$$

with the following informal meanings:

- *Fast: "The service is fast."*
- *Affordable: "The service is affordable."*
- *Reliable: "The service is reliable."*
- *Friendly: "The service is friendly."*

Suppose that a fuzzy set $F : \Sigma_{\text{Service}} \to [0,1]$ *is given by*

$$\begin{aligned} F(\textit{Fast}) &= 0.7 \\ F(\textit{Affordable}) &= 0.4 \\ F(\textit{Reliable}) &= 0.9 \\ F(\textit{Friendly}) &= 0.8 \end{aligned}$$

Then, for negation, we obtain for instance

$$\llbracket \neg \mathit{Affordable} \rrbracket = N_0(0.4) = 0.6$$
$$\textit{and} \quad \llbracket \neg \mathit{Reliable} \rrbracket = N_0(0.9) = 0.1$$

Thus the standard negation does not merely reverse a truth degree in a formal sense; it reflects it around the midpoint of the interval $[0,1]$.

In classical propositional logic, conjunction and disjunction can be related to one another by means of negation. In fuzzy logic, this principle is generalized by a duality between T-norms and S-norms.

Definition 4.101 (N-Duality). *Let* $N : [0,1] \to [0,1]$ *be a negation norm, let* $T : [0,1]^2 \to [0,1]$ *be a T-norm, and let* $S : [0,1]^2 \to [0,1]$ *be an S-norm. Then* T *and* S *are called* N-dual *if*

$$T(x,y) = N(S(N(x), N(y)))$$
$$\textit{and} \quad S(x,y) = N(T(N(x), N(y)))$$

for all $x, y \in [0,1]$.

The underlying idea is the same as in the De Morgan laws: each connective can be reconstructed from the other by applying negation.

Theorem 4.102. *The following statements hold:*

1. T_{min} *and* S_{max} *are* N_0*-dual.*

2. T_{prod} *and* S_{prod} *are* N_0*-dual.*

3. T_{Luka} and S_{Luka} are N_0-dual.

Proof. We verify both equations of the duality in each case.

1. $T_{\min}$ and $S_{\max}$. For all $x, y \in [0,1]$, we have

$$\begin{aligned} N_0(S_{\max}(N_0(x), N_0(y))) &= 1 - \max\{1-x, 1-y\} \\ &= \min\{x, y\} \\ &= T_{\min}(x, y) \end{aligned}$$

Conversely, we obtain

$$\begin{aligned} N_0(T_{\min}(N_0(x), N_0(y))) &= 1 - \min\{1-x, 1-y\} \\ &= \max\{x, y\} \\ &= S_{\max}(x, y) \end{aligned}$$

2. T_{prod} and S_{prod}. First,

$$S_{\text{prod}}(N_0(x), N_0(y)) = (1-x) + (1-y) - (1-x)(1-y).$$

Expanding this expression gives

$$1 - xy.$$

Hence

$$N_0(S_{\text{prod}}(N_0(x), N_0(y))) = 1-(1-xy) = xy = T_{\text{prod}}(x, y).$$

For the converse direction, we compute

$$T_{\text{prod}}(N_0(x), N_0(y)) = (1-x)(1-y) = 1 - x - y + xy.$$

Therefore,

$$\begin{aligned} N_0(T_{\text{prod}}(N_0(x), N_0(y))) &= 1 - (1 - x) \cdot (1 - y) \\ &= 1 - (1 - x - y + xy) \\ &= x + y - xy \\ &= S_{\text{prod}}(x, y) \end{aligned}$$

3. T_{Luka} **and** S_{Luka}**.** We have

$$\begin{aligned} S_{\text{Luka}}(N_0(x), N_0(y)) &= \min\{(1 - x) + (1 - y), 1\} \\ &= \min\{2 - x - y, 1\} \end{aligned}$$

Thus

$$N_0(S_{\text{Luka}}(N_0(x), N_0(y))) = 1 - \min\{2 - x - y, 1\}.$$

If $x + y \leq 1$, then $2 - x - y \geq 1$, and therefore

$$1 - \min\{2 - x - y, 1\} = 1 - 1 = 0 = \max\{0, x + y - 1\}.$$

If $x + y \geq 1$, then $2 - x - y \leq 1$, and therefore

$$1 - \min\{2 - x - y, 1\} = 1 - (2 - x - y) = x + y - 1.$$

In both cases, we obtain

$$N_0(S_{\text{Luka}}(N_0(x), N_0(y))) = \max\{0, x+y-1\} = T_{\text{Luka}}(x, y).$$

For the converse direction,

$$T_{\text{Luka}}(N_0(x), N_0(y)) = \max\{0, (1 - x) + (1 - y) - 1\}$$

$$= \max\{0, 1 - x - y\}$$

Consequently,

$$\begin{aligned} N_0(T_{\text{Luka}}(N_0(x), N_0(y))) &= 1 - \max\{0, 1 - x - y\} \\ &= \min\{1, x + y\} \\ &= S_{\text{Luka}}(x, y) \end{aligned}$$

Thus all three pairs are N_0-dual. □

More generally, once a T-norm and a negation norm are fixed, the dual S-norm can be obtained by applying negation. Conversely, the corresponding T-norm can be recovered from an S-norm in the same way.

Theorem 4.103 (Duality of T-Norms and S-Norms). *Let $H : [0, 1]^2 \to [0, 1]$ be a function and define*

$$H'(x, y) = N_0(H(N_0(x), N_0(y))) = 1 - H(1 - x, 1 - y).$$

Then the following statements hold:

1. *If H is a T-norm, then H' is an S-norm.*

2. *If H is an S-norm, then H' is a T-norm.*

In both cases, H and H' are N_0-dual.

Proof. We first prove statement (1), and then statement (2).

(1) If H is a T-norm, then H' is an S-norm.

1. *Absorbing element* 1.

$$H'(x,1) = 1 - H(1-x,0) = 1 - 0 = 1,$$

since, as a T-norm, H has the property $H(a,0) = 0$.

2. *Neutral element* 0.

$$H'(x,0) = 1 - H(1-x,1) = 1 - (1-x) = x,$$

since $H(a,1) = a$.

3. *Monotonicity.* Let $x \leq x'$ and $y \leq y'$. Then

$$1 - x \geq 1 - x' \quad \text{and} \quad 1 - y \geq 1 - y'.$$

By the monotonicity of H, it follows that

$$H(1-x, 1-y) \geq H(1-x', 1-y').$$

After applying N_0, we obtain

$$\begin{aligned} H'(x,y) &= 1 - H(1-x, 1-y) \\ &\leq 1 - H(1-x', 1-y') \\ &= H'(x', y') \end{aligned}$$

4. *Commutativity.*

$$\begin{aligned} H'(x,y) &= 1 - H(1-x, 1-y) \\ &= 1 - H(1-y, 1-x) \\ &= H'(y,x) \end{aligned}$$

5. *Associativity.* We have

$$\begin{aligned} H'(x, &H'(y,z)) \\ &= 1 - H(1-x, 1-(1-H(1-y, 1-z))) \\ &= 1 - H(1-x, H(1-y, 1-z)) \end{aligned}$$

   Using the associativity of H, this yields

$$\begin{aligned} 1 - H(1-x, &H(1-y, 1-z)) \\ &= 1 - H(H(1-x, 1-y), 1-z) \\ &= H'(H'(x,y), z) \end{aligned}$$

   Thus H' is an S-norm.

(2) If H is an S-norm, then H' is a T-norm. We now use the defining properties of an S-norm.

1. *Absorbing element* 0.

$$H'(x,0) = 1 - H(1-x,1) = 1 - 1 = 0,$$

   since S-norms satisfy $H(a,1) = 1$.

2. *Neutral element* 1.

$$H'(x,1) = 1 - H(1-x,0) = 1 - (1-x) = x,$$

   since S-norms satisfy $H(a,0) = a$.

3. *Monotonicity.* Let $x \leq x'$ and $y \leq y'$. As before, we have

$$1 - x \geq 1 - x' \quad \text{and} \quad 1 - y \geq 1 - y'.$$

By the monotonicity of H, it follows that

$$H(1-x,1-y) \geq H(1-x',1-y').$$

Thus, after applying N_0, we obtain

$$H'(x,y) \leq H'(x',y').$$

4. *Commutativity.*

$$\begin{aligned} H'(x,y) &= 1 - H(1-x,1-y) \\ &= 1 - H(1-y,1-x) \\ &= H'(y,x) \end{aligned}$$

5. *Associativity.* As above, we compute:

$$\begin{aligned} &H'(x,H'(y,z)) \\ &\quad = 1 - H(1-x,1-(1-H(1-y,1-z))) \\ &\quad = 1 - H(1-x,H(1-y,1-z)) \\ &\quad = 1 - H(H(1-x,1-y),1-z) \\ &\quad = H'(H'(x,y),z) \end{aligned}$$

Thus H' is a T-norm.

In both cases, the duality is built directly into the definition of H'. □

4.3.5 Semantics and Consequence

In the preceding subsections we introduced the basic components of fuzzy logic: T-norms for conjunction, S-norms for disjunction, and negation functions for negation. Taken together, these three components determine how compound formulas are evaluated in a fuzzy logic. Thus, a particular choice of T-norm, S-norm, and negation determines a particular fuzzy-logical semantics.

Definition 4.104 (Fuzzy structure). *Let T be a T-norm, S an S-norm, and N a negation function. Then*

$$\Lambda = (T, S, N)$$

is called a fuzzy structure.

In practice, T and S are usually chosen so that they are dual to one another with respect to N. However, this is not a necessary condition for the notion of a fuzzy structure itself.

On the basis of a fuzzy structure, we can now define how formulas are evaluated. Each propositional variable is interpreted by a number in the interval $[0, 1]$, and compound formulas are evaluated recursively.

Definition 4.105 (Fuzzy truth function). *Let $\Lambda = (T, S, N)$ be a fuzzy structure, let Σ be a propositional signature, and let $F : \Sigma \to [0, 1]$ be a fuzzy set over Σ. The* fuzzy truth function

$$[\![\cdot]\!]_F^\Lambda : L(\Sigma) \to [0, 1]$$

is defined inductively by

$$[\![x]\!]_F^\Lambda = F(x)$$

for every propositional variable $x \in \Sigma$, *and by*

$$[\![\neg\varphi]\!]_F^\Lambda = N([\![\varphi]\!]_F^\Lambda),$$

$$[\![\varphi \wedge \psi]\!]_F^\Lambda = T([\![\varphi]\!]_F^\Lambda, [\![\psi]\!]_F^\Lambda),$$
$$[\![\varphi \vee \psi]\!]_F^\Lambda = S([\![\varphi]\!]_F^\Lambda, [\![\psi]\!]_F^\Lambda).$$

This definition is the fuzzy-logical counterpart of the classical truth function. The difference is that formulas no longer take only the values 0 and 1, but may take arbitrary degrees of truth between 0 and 1.

Example 4.106. *Consider the propositional signature*

$$\Sigma_{\text{Cafe}} = \{\mathit{Fast}, \mathit{Tasty}, \mathit{Affordable}, \mathit{Quiet}\}$$

with the following informal meanings:

- *Fast: "The service is fast."*
- *Tasty: "The food is tasty."*
- *Affordable: "The offer is affordable."*
- *Quiet: "The restaurant is quiet."*

Let $F : \Sigma_{\text{Cafe}} \to [0, 1]$ *be given by*

$$F(\mathit{Fast}) = 0.8$$

$$F(\mathit{Tasty}) = 0.9$$
$$F(\mathit{Affordable}) = 0.4$$
$$F(\mathit{Quiet}) = 0.7$$

For the fuzzy structure $\Lambda_0 = (T_{\min}, S_{\max}, N_0)$, *we obtain, for example,*

$$[\![\mathit{Fast} \wedge \mathit{Tasty}]\!]_F^{\Lambda_0} = \min\{0.8, 0.9\} = 0.8,$$

$$[\![\mathit{Affordable} \vee \mathit{Quiet}]\!]_F^{\Lambda_0} = \max\{0.4, 0.7\} = 0.7,$$

$$[\![\neg \mathit{Affordable}]\!]_F^{\Lambda_0} = 1 - 0.4 = 0.6.$$

With the same fuzzy set but a different fuzzy structure, the same formulas may receive different values. Hence the concrete semantics depends not only on the atomic truth degrees, but also on the chosen fuzzy structure.

4.3.6 Value Assignments and Fuzzy Consequence

In addition to evaluating individual formulas, we also want to express when a fuzzy set satisfies prescribed truth-degree assignments. For this purpose, we extend the syntax by expressions of the form $\varphi = x$.

Definition 4.107 (Value assignment). *Let* Σ *be a propositional signature, let* $\varphi \in L(\Sigma)$ *be a formula, and let* $x \in [0, 1]$. *Then*

$$\varphi = x$$

is called a value assignment *over* Σ. *The set of all value assignments over* Σ *is denoted by* $\mathcal{Z}(\Sigma)$.

The informal reading is straightforward: the formula φ is required to take exactly the truth degree x.

Definition 4.108 (Model). *Let Σ be a propositional signature, let $\Phi \subseteq \mathcal{Z}(\Sigma)$ be a set of value assignments, and let $\Lambda = (T, S, N)$ be a fuzzy structure. A fuzzy set $F : \Sigma \to [0,1]$ is called a* Λ-model *of Φ, written*

$$F \models_{\Sigma}^{\Lambda} \Phi,$$

if and only if, for every $(\varphi = x) \in \Phi$,

$$[\![\varphi]\!]_F^{\Lambda} = x.$$

The set of all Λ-models of Φ is denoted by $\mathrm{FMod}_{\Sigma}^{\Lambda}(\Phi)$.

Example 4.109. *Continuing the previous example, consider the set*

$$\begin{aligned}\Phi = \{\,&\mathit{Tasty} \wedge \mathit{Fast} = 0.8,\\ &\neg \mathit{Affordable} = 0.6,\\ &\mathit{Quiet} = 0.7\}\end{aligned}$$

Then, for $\Lambda_0 = (T_{\min}, S_{\max}, N_0)$, we have:

$$\begin{aligned}[\![\mathit{Tasty} \wedge \mathit{Fast}]\!]_F^{\Lambda_0} &= 0.8\\ [\![\neg \mathit{Affordable}]\!]_F^{\Lambda_0} &= 0.6\\ [\![\mathit{Quiet}]\!]_F^{\Lambda_0} &= 0.7\end{aligned}$$

Hence

$$F \models_{\Sigma_{\mathrm{Cafe}}}^{\Lambda_0} \Phi.$$

We can now define a corresponding consequence relation. The basic idea is the same as in classical logic: a set of constraints entails another constraint if every structure satisfying the former also satisfies the latter.

Definition 4.110 (Fuzzy consequence). *Let Σ be a propositional signature, let $\Phi, \Phi' \subseteq \mathcal{Z}(\Sigma)$, and let $\Lambda = (T, S, N)$ be a fuzzy structure. We define*

$$\Phi \models_{\Sigma}^{\Lambda} \Phi'$$

if and only if

$$\mathrm{FMod}_{\Sigma}^{\Lambda}(\Phi) \subseteq \mathrm{FMod}_{\Sigma}^{\Lambda}(\Phi').$$

Example 4.111. *Consider again Σ_{Cafe} and let*

$$\Phi = \{\neg \mathit{Affordable} = 0.6,\ \mathit{Fast} = 0.8\}.$$

Under $\Lambda_0 = (T_{\min}, S_{\max}, N_0)$ this first implies

$$\mathit{Affordable} = 0.4,$$

because $\neg \mathit{Affordable} = 0.6$ yields, by standard negation,

$$1 - [\![\mathit{Affordable}]\!]_F^{\Lambda_0} = 0.6 \quad \Rightarrow \quad [\![\mathit{Affordable}]\!]_F^{\Lambda_0} = 0.4.$$

Thus, for every fuzzy set F with $F \models_{\Sigma_{\mathrm{Cafe}}}^{\Lambda_0} \Phi$, we also have

$$[\![\mathit{Fast} \wedge \mathit{Affordable}]\!]_F^{\Lambda_0} = \min\{0.8, 0.4\} = 0.4.$$

Therefore

$$\Phi \models_{\Sigma_{\mathrm{Cafe}}}^{\Lambda_0} \mathit{Fast} \wedge \mathit{Affordable} = 0.4.$$

4.3.7 Compatibility with Classical Propositional Logic

Finally, fuzzy-logical consequence is compatible with classical propositional logic in the following sense. Classical interpretations can be regarded as special fuzzy sets whose truth degrees are restricted to the values 0 and 1.

Theorem 4.112. *Let Σ be a propositional signature, let $\Phi \subseteq L(\Sigma)$ be a set of propositional formulas with $\Phi \not\equiv \bot$, let $\psi \in L(\Sigma)$, and let $\Lambda = (T, S, N)$ be a fuzzy structure. Define*

$$\Phi' = \{\varphi = 1 \mid \varphi \in \Phi\}.$$

If

$$\Phi' \models^{\Lambda}_{\Sigma} \psi = 1$$

then also

$$\Phi \models_{\mathrm{PL}} \psi.$$

Proof. Let $\omega \in \mathrm{Mod}^{\mathrm{PL}}_{\Sigma}(\Phi)$ be a classical valuation satisfying Φ. We define the corresponding fuzzy set $F_\omega : \Sigma \to [0, 1]$ by

$$F_\omega(x) = \begin{cases} 1, & \text{if } \omega(x) = \text{true}, \\ 0, & \text{if } \omega(x) = \text{false}. \end{cases}$$

We now prove, by structural induction on $\varphi \in L(\Sigma)$, the following two claims:

1. if $[\![\varphi]\!]_\omega = \text{true}$, then $[\![\varphi]\!]^{\Lambda}_{F_\omega} = 1$;
2. if $[\![\varphi]\!]_\omega = \text{false}$, then $[\![\varphi]\!]^{\Lambda}_{F_\omega} = 0$.

Case 1: $\varphi = x$ for some $x \in \Sigma$. The claim is immediate, since

$$[\![x]\!]_\omega = \omega(x) \quad \text{and} \quad [\![x]\!]^\Lambda_{F_\omega} = F_\omega(x).$$

Thus the classical and fuzzy-logical evaluations agree on atomic formulas.

Case 2: $\varphi = \neg\chi$. If $[\![\neg\chi]\!]_\omega = \text{true}$, then $[\![\chi]\!]_\omega = \text{false}$. By the induction hypothesis,

$$[\![\chi]\!]^\Lambda_{F_\omega} = 0.$$

Hence

$$[\![\neg\chi]\!]^\Lambda_{F_\omega} = N(0) = 1,$$

because every negation function satisfies $N(0) = 1$.

Conversely, if $[\![\neg\chi]\!]_\omega = \text{false}$, then $[\![\chi]\!]_\omega = \text{true}$. Again by the induction hypothesis,

$$[\![\chi]\!]^\Lambda_{F_\omega} = 1,$$

and therefore

$$[\![\neg\chi]\!]^\Lambda_{F_\omega} = N(1) = 0.$$

Case 3: $\varphi = \chi \wedge \eta$. If $\chi \wedge \eta$ is classically true, then both χ and η are true. By the induction hypothesis,

$$[\![\chi]\!]^\Lambda_{F_\omega} = 1 \quad \text{and} \quad [\![\eta]\!]^\Lambda_{F_\omega} = 1.$$

Thus

$$[\![\chi \wedge \eta]\!]^\Lambda_{F_\omega} = T(1, 1) = 1.$$

If $\chi \wedge \eta$ is classically false, then at least one of the two formulas is false. Without loss of generality, assume that χ is false. By the induction hypothesis,

$$[\![\chi]\!]^{\Lambda}_{F_\omega} = 0,$$

and hence

$$[\![\chi \wedge \eta]\!]^{\Lambda}_{F_\omega} = T(0, [\![\eta]\!]^{\Lambda}_{F_\omega}) = 0,$$

because every T-norm satisfies $T(0, y) = 0$.

Case 4: $\varphi = \chi \vee \eta$**.** This follows analogously to the conjunction case, using the defining properties of an S-norm:

$$S(1, y) = 1 \qquad \text{and} \qquad S(0, 0) = 0.$$

This proves that classical truth values are preserved under the embedding F_ω. In particular, for every classical model ω of Φ we also have

$$F_\omega \models^{\Lambda}_{\Sigma} \Phi'.$$

Now assume that $\Phi' \models^{\Lambda}_{\Sigma} \psi = 1$. Then every model of Φ' is also a model of $\psi = 1$. Since, in particular, F_ω is a model of Φ', it follows that

$$[\![\psi]\!]^{\Lambda}_{F_\omega} = 1.$$

By the induction result just proved, ψ cannot be false under ω. Hence

$$\omega \models_{\mathrm{PL}} \psi.$$

Since ω was arbitrary, we obtain

$$\Phi \models_{\mathrm{PL}} \psi.$$

□

Failure of the Converse. The converse does not hold in general. A fuzzy-logical consequence may therefore be weaker than, or structurally different from, its classical counterpart. This shows that fuzzy logic extends classical logic, but does not simply replace it.

Example 4.113. *The converse of the theorem does not hold in general. To see this, consider the signature*

$$\Sigma = \{A, B, C\}$$

and the set of formulas

$$\Phi = \{A \vee B,\ \neg A \vee C,\ \neg B \vee C\}.$$

Classically, this entails

$$\Phi \models_{\mathrm{PL}} C.$$

Indeed, if $C = false$, then the three formulas would force $A \vee B$, $\neg A$, and $\neg B$, which is impossible.

Now consider the corresponding set of value constraints

$$\Phi' = \{A \vee B = 1,\ \neg A \vee C = 1,\ \neg B \vee C = 1\}$$

and the fuzzy structure

$$\Lambda_{\text{Luka}} = (T_{\text{Luka}}, S_{\text{Luka}}, N_0).$$

Let $F : \Sigma \to [0, 1]$ *be given by*

$$F(A) = F(B) = F(C) = 0.5.$$

Then we have

$$[\![A \vee B]\!]_F^{\Lambda_{\text{Luka}}} = S_{\text{Luka}}(0.5, 0.5) = \min\{0.5 + 0.5, 1\} = 1,$$

$$[\![\neg A \vee C]\!]_F^{\Lambda_{\text{Luka}}} = S_{\text{Luka}}(1-0.5, 0.5) = \min\{0.5+0.5, 1\} = 1,$$

$$[\![\neg B \vee C]\!]_F^{\Lambda_{\text{Luka}}} = S_{\text{Luka}}(1-0.5, 0.5) = \min\{0.5+0.5, 1\} = 1.$$

Hence

$$F \in \text{FMod}_\Sigma^{\Lambda_{\text{Luka}}}(\Phi').$$

On the other hand,

$$[\![C]\!]_F^{\Lambda_{\text{Luka}}} = F(C) = 0.5 \neq 1.$$

Therefore,

$$\Phi' \not\models_\Sigma^{\Lambda_{\text{Luka}}} C = 1,$$

although classically we already have $\Phi \models_{\text{PL}} C$*.*

4.3.8 Exercises with Solutions

Exercise 4.114. *In the following, we work with the propositional signature*

$$\Sigma = \{A, C, D, G, H\}.$$

For the evaluation, we use the fuzzy structures

$$\Lambda_0 = (T_{\min}, S_{\max}, N_0),$$
$$\Lambda_{\text{prod}} = (T_{\text{prod}}, S_{\text{prod}}, N_0),$$
$$\Lambda_{\text{Luka}} = (T_{\text{Luka}}, S_{\text{Luka}}, N_0).$$

Furthermore, let the following fuzzy sets be given:

$$F_1(A) = 0.1, \quad F_1(C) = 0.6, \quad F_1(D) = 0.2,$$
$$F_1(G) = 0.2, \quad F_1(H) = 1,$$

$$F_1'(A) = 0.4, \quad F_1'(C) = 0.4, \quad F_1'(D) = 0.5,$$
$$F_1'(G) = 0.5, \quad F_1'(H) = 1,$$

$$F_2(A) = 0.6, \quad F_2(C) = 0.3, \quad F_2(D) = 0.8,$$
$$F_2(G) = 0.9, \quad F_2(H) = 0.2,$$

$$F_3(A) = 0.5, \quad F_3(C) = 0.8, \quad F_3(D) = 0.6,$$
$$F_3(G) = 0.7, \quad F_3(H) = 0.8,$$

$$F_4(A) = 0.9, \quad F_4(C) = 0.2, \quad F_4(D) = 0.8,$$
$$F_4(G) = 0.9, \quad F_4(H) = 0.9.$$

Moreover, let

$$\Phi_4 = \{\, H \wedge \neg(G \vee A \vee G) = 0,\ C \wedge G = 0.1 \,\}.$$

Decide for each of the following statements whether it is true or false.

1. *We have* $F_1 \subseteq F_1'$.

2. *We have*

$$[\![\neg C \vee H \vee A]\!]_{F_2}^{\Lambda_0} = 0.7.$$

3. *We have*

$$[\![D \wedge C \wedge G]\!]_{F_3}^{\Lambda_{\text{prod}}} = 0.1.$$

4. *We have*

$$F_4 \models_{\Sigma}^{\Lambda_{\text{Luka}}} \Phi_4.$$

Solution 4.115. 1. ***False.***

For $F_1 \subseteq F_1'$ to hold, we would need, for every $x \in \Sigma$,

$$F_1(x) \leq F_1'(x).$$

However, this is not the case, since already for C we have

$$F_1(C) = 0.6 > 0.4 = F_1'(C).$$

Thus F_1 is not a fuzzy subset of F_1'.

2. ***True.***

We evaluate the formula under the fuzzy structure $\Lambda_0 = (T_{\min}, S_{\max}, N_0)$. First,

$$[\![\neg C]\!]_{F_2}^{\Lambda_0} = 1 - F_2(C) = 1 - 0.3 = 0.7.$$

It follows that

$$[\![\neg C \vee H]\!]_{F_2}^{\Lambda_0} = \max\{0.7, 0.2\} = 0.7,$$

and finally

$$[\![\neg C \vee H \vee A]\!]_{F_2}^{\Lambda_0} = \max\{0.7, 0.6\} = 0.7.$$

Hence the statement is true.

3. ***False.***

   Under product conjunction, associativity gives the following value for a threefold conjunction:

   $$[\![D \wedge C \wedge G]\!]_{F_3}^{\Lambda_{\text{prod}}} = F_3(D) \cdot F_3(C) \cdot F_3(G).$$

   Thus we obtain

   $$[\![D \wedge C \wedge G]\!]_{F_3}^{\Lambda_{\text{prod}}} = 0.6 \cdot 0.8 \cdot 0.7 = 0.336.$$

   The stated value 0.1 *is therefore false.*

4. ***True.***

   We check both value assignments from Φ_4.

   First consider the formula $H \wedge \neg(G \vee A \vee G) = 0$. *Under* $\Lambda_{\text{Luka}} = (T_{\text{Luka}}, S_{\text{Luka}}, N_0)$, *we have:*

   $$[\![G \vee A]\!]_{F_4}^{\Lambda_{\text{Luka}}} = \min\{0.9 + 0.9, 1\} = 1.$$

   Hence also

   $$[\![G \vee A \vee G]\!]_{F_4}^{\Lambda_{\text{Luka}}} = 1,$$

   and therefore

   $$[\![\neg(G \vee A \vee G)]\!]_{F_4}^{\Lambda_{\text{Luka}}} = 1 - 1 = 0.$$

Now it follows that

$$[\![H \wedge \neg(G \vee A \vee G)]\!]_{F_4}^{\Lambda_{\text{Luka}}} = \max\{0, 0.9 + 0 - 1\} = 0.$$

Thus the first value assignment is satisfied.

For the second formula, we obtain:

$$[\![C \wedge G]\!]_{F_4}^{\Lambda_{\text{Luka}}} = \max\{0, 0.2 + 0.9 - 1\} = 0.1.$$

Thus the second value assignment is satisfied as well.

Consequently, overall we have

$$F_4 \models_{\Sigma}^{\Lambda_{\text{Luka}}} \Phi_4.$$

Overall, these exercises show how differently the various fuzzy structures can evaluate the same formulas, and how important it is to specify the chosen semantics in each case.

4.4 Default Logic

Classical logics such as propositional logic, predicate logic, and many modal logics are highly suitable when a domain is to be described completely and without contradiction. In many everyday situations, however, our knowledge is incomplete, provisional, or shaped by exceptions. Human reasoning therefore does not rely only on strictly certain inferences, but also on plausible assumptions: what may be expected *normally* is initially inferred, yet it can be withdrawn again when new information becomes available.

This is precisely the point at which default logic comes into play. It makes it possible to treat rules of the form “Normally, ...” in a formal way. Such rules are not strict implications. A formula of the form $\varphi \to \psi$ states that ψ always follows from φ; a default rule, by contrast, expresses only that ψ *typically* follows from φ, provided that there is no information speaking against ψ. Default logic is therefore a suitable tool for reasoning with exceptions, reservations, and provisional conclusions.

The fundamental difference from the classical logics considered so far lies in non-monotonicity: new information may defeat earlier plausible conclusions. In the first subsection, we examine this property and show why it is indispensable for reasoning with everyday rules. We then introduce the syntax and semantics of Reiter’s default logic.

4.4.1 Non-monotonic Inferences

A characteristic feature of classical logical consequence is its monotonicity. In propositional logic, the following holds: if a formula ψ follows from a knowledge base Φ, then ψ still follows from every extension of Φ. Formally, this property is expressed as follows:

$$\text{If } \Phi \models_{\mathrm{PL}} \psi \text{ and } \Phi \subseteq \Phi', \text{ then also } \Phi' \models_{\mathrm{PL}} \psi.$$

This is intuitively very natural for strictly logical derivations: additional knowledge may make new conclusions possible, but it cannot destroy classical consequences that have already been secured.

Corollary 4.116. *Let* Σ *be a propositional signature, let* $\Phi \subseteq L(\Sigma)$*, and let* $\psi, \xi \in L(\Sigma)$*. If*

$$\Phi \models_{\text{PL}} \psi,$$

then also

$$\Phi \cup \{\xi\} \models_{\text{PL}} \psi.$$

Proof. By the definition of classical consequence, we have

$$\Phi \models_{\text{PL}} \psi \iff \text{Mod}_\Sigma(\Phi) \subseteq \text{Mod}_\Sigma(\psi).$$

For the extended knowledge base,

$$\text{Mod}_\Sigma(\Phi \cup \{\xi\}) = \text{Mod}_\Sigma(\Phi) \cap \text{Mod}_\Sigma(\xi) \subseteq \text{Mod}_\Sigma(\Phi).$$

Hence we also have

$$\text{Mod}_\Sigma(\Phi \cup \{\xi\}) \subseteq \text{Mod}_\Sigma(\psi),$$

and therefore

$$\Phi \cup \{\xi\} \models_{\text{PL}} \psi.$$

□

For reasoning with exceptions, however, this monotonicity is precisely too restrictive. Many everyday rules are not absolute, but valid only under normal circumstances. Consider, for instance, the signature

$$\Sigma_{\text{Wet}} = \{R, S, K\}$$

with the intended meanings

- R: "It is raining."
- S: "The street is wet."
- K: "The street has just been cleaned."

The rule "If it is raining, then the street is normally wet" cannot be modeled classically as a strict implication, because it allows for exceptions. The street might in fact remain dry despite the rain, for example because it is covered. A classical formula $R \to S$, however, would claim that rain always implies wetness, without exception.

The problem becomes especially clear when additional information is introduced. Suppose that at first we only know that it is raining. Then the conclusion "The street is wet" seems plausible. If we later learn, however, that the street has just been cleaned, we may want to withdraw this conclusion. This is exactly the kind of behavior that calls for a non-monotonic consequence relation.

Formally, we therefore aim for a relation $\mid\!\sim$ with the property

$$\Phi \cup \{R\} \mid\!\sim S \quad \text{but} \quad \Phi \cup \{R, K\} \not\mid\!\sim S.$$

In other words, new information may revise earlier plausible conclusions.

Example 4.117. *Let* Φ_{Wet} *be a knowledge base containing the statements*

$$\Phi_{\text{Wet}} = \{R\},$$

and suppose we also adopt the everyday rule "If it is raining, then the street is normally wet". Then the conclusion S is initially plausible. If the additional information K is added, namely that the street has just been cleaned, the conclusion S is no longer automatic. A classical consequence relation cannot represent this dynamic, whereas a non-monotonic relation can.

The central task of default logic is therefore to capture such plausible but revisable inferences formally. In the next step, we introduce default rules as a syntactic device for this purpose and then make precise how semantically consistent sets of conclusions can be constructed from them.

4.4.2 Syntax and Semantics of Reiter's Default Logic

Classical propositional logic works with strict consequences: if something follows from a set of knowledge, then this consequence remains valid even when additional knowledge is added. For many everyday forms of reasoning, this is too rigid. In such cases, we often want to use precisely *tentative* rules that may be withdrawn when new information becomes available. Default logic formalizes exactly such rules of the "normal case".

Definition 4.118 (Default Rule). *Let Σ be a propositional signature and let*

$$\varphi, \psi_1, \ldots, \psi_n, \chi \in L(\Sigma).$$

Then

$$\delta = \frac{\varphi : \psi_1, \ldots, \psi_n}{\chi}$$

is called a default rule *over* $L(\Sigma)$. *Here*

$$\text{pre}(\delta) = \varphi, \qquad \text{cons}(\delta) = \chi, \qquad \text{just}(\delta) = \{\psi_1, \ldots, \psi_n\}$$

are called the prerequisite, *the* consequent, *and the* justification *of the rule. We denote the set of all default rules over* $L(\Sigma)$ *by* $\text{Def}(L(\Sigma))$.

Remark 4.119. *A default rule is read as follows: "If the prerequisite is known and the justification remains consistent with the knowledge available so far, then the consequent may be tentatively assumed." Thus, a default rule is not a strict implication, but a rule for typical cases.*

Example 4.120 (Travel). *We consider the signature*

$$\Sigma_{\text{Travel}} = \{V, S, G\}$$

with

- V*: "The train is delayed."*
- S*: "The arrival will be postponed."*
- G*: "There is a strike."*

The rule

$$\delta_1 = \frac{V : S}{S}$$

expresses that, in the case of a delay, the arrival is normally expected to be postponed. The rule

$$\delta_2 = \frac{G : \neg S}{\neg S}$$

by contrast, models the opposite typical expectation in the case of a strike. Such rules are plausible, but not absolute; they may be overridden by additional information.

Definition 4.121 (Default Theory). *A default theory over $L(\Sigma)$ is a tuple*

$$T = (W, \Delta)$$

where $W \subseteq L(\Sigma)$ is a set of classical formulas and $\Delta \subseteq \mathrm{Def}(L(\Sigma))$ is a set of default rules. Here W represents the certain knowledge, while Δ contains the plausible additional rules.

Example 4.122 (A Small Default Theory). *For the signature $\Sigma_{\text{Travel}} = \{V, S, G\}$, we consider the default theory*

$$T_{\text{Travel}} = (W_{\text{Travel}}, \Delta_{\text{Travel}})$$

with

$$W_{\text{Travel}} = \{V\}, \qquad \Delta_{\text{Travel}} = \left\{ \frac{V : S}{S}, \frac{V : \neg S}{\neg S} \right\}.$$

The only certain information here is that the train is delayed. Two competing possibilities are now plausible: either the arrival will be postponed or it will not. This ambiguity is characteristic of default logic.

Definition 4.123 (Applicability). *Let Σ be a propositional signature and let $\Phi, \Phi' \subseteq L(\Sigma)$ be sets of formulas. A default rule*

$$\delta = \frac{\varphi : \psi_1, \ldots, \psi_n}{\chi}$$

is called applicable to Φ with respect to Φ' *if*

$$\varphi \in \Phi \qquad \text{and} \qquad \neg\psi_i \notin \Phi' \text{ for all } i = 1, \ldots, n.$$

If $\Phi' = \Phi$, we simply say that δ is applicable to Φ.

Remark 4.124. *The prerequisite φ determines* when *the rule is relevant at all. The justifications $\psi_1, \ldots, \psi_n$ determine* whether *the rule is still plausible in the current state of knowledge. It is precisely this second aspect that makes default rules non-monotonic.*

Example 4.125 (Applicability). *For the theory T_{Travel}, the following holds:*

- $\frac{V:S}{S}$ *is applicable to $\{V\}$, since V is present and $\neg S$ is not part of the state of knowledge.*
- $\frac{V:S}{S}$ *is not applicable to $\{V, \neg S\}$, because $\neg S$ defeats the justification S.*
- $\frac{V:\neg S}{\neg S}$ *is also applicable to $\{V\}$.*

Thus, already at the level of applicability, it becomes clear that competing defaults may lead to several possible conclusions.

Definition 4.126 (Operator and Extension). *Let $T = (W, \Delta)$ be a default theory over $L(\Sigma)$. We define the operator*

$$\hat{\Lambda}_T : 2^{L(\Sigma)} \to 2^{L(\Sigma)}$$

by

$$\hat{\Lambda}_T(\Phi) = Cn_\Sigma(W \cup \{\text{cons}(\delta) \mid \delta \in \Delta \textit{ is applicable to } \Phi\})$$

for every set $\Phi \subseteq L(\Sigma)$.

Furthermore, let $\Lambda_T(\Phi)$ be the smallest subset of Φ that is a fixed point of $\hat{\Lambda}_T$, that is,

$$\Lambda_T(\Phi) = \Phi' \quad \textit{with} \quad \Phi' \subseteq \Phi,\ \hat{\Lambda}_T(\Phi') = \Phi',$$

and assume that no proper subset of Φ' has the same property.

A set $E \subseteq L(\Sigma)$ is called an extension *of T if*

$$\Lambda_T(E) = E.$$

Remark 4.127. *An extension is a* stable plausible state of knowledge. *Here,* stable *means two things. First, an extension contains the certain initial knowledge W in full. Second, it remains unchanged when all classical consequences derivable from it are added or when the default rules already applicable in it are checked again. Thus, an extension is reached exactly when no new information can be obtained from the current knowledge, neither classically nor by further applications of default rules.*

The definition of an extension E therefore immediately yields the following three characteristic properties:

1. *E contains the certain knowledge:*

$$W \subseteq E.$$

2. *E is classically deductively closed:*

$$Cn_{\Sigma}(E) = E.$$

3. *E is closed under applications of defaults: If a default rule*

$$\delta \in \Delta$$

is applicable to E, then its consequence is already contained in E, that is,

$$\text{cons}(\delta) \in E.$$

Since, by item 2, E contains all classical consequences of itself, an extension is in general infinite. This is not a disadvantage, however: although E, as the set of all consequences, may be infinite, it can always be described by finitely many initial formulas, namely as the deductive closure of a finite set. In this sense, extensions are semantically comprehensive while still remaining syntactically manageable.

Theorem 4.128 (Characterization of Extensions). *Let Σ be a propositional signature and let $T = (W, \Delta)$ be a default theory over $L(\Sigma)$. A set $E \subseteq L(\Sigma)$ is an extension of T if and only if there is a sequence*

$$\delta_1, \ldots, \delta_n \in \Delta$$

such that

1.

$$E = Cn_\Sigma(W \cup \{\text{cons}(\delta_1), \ldots, \text{cons}(\delta_n)\}),$$

2. for all $i = 1, \ldots, n$, δ_i is applicable to

$$Cn_\Sigma(W \cup \{\text{cons}(\delta_1), \ldots, \text{cons}(\delta_{i-1})\})$$

with respect to E,

3. no default rule $\delta' \in \Delta \setminus \{\delta_1, \ldots, \delta_n\}$ is applicable to E.

Proof. We prove both directions.

(⇒) Let E be an extension of T. Then $E = \Lambda_T(E)$, and in particular

$$\begin{aligned} E &= \hat{\Lambda}_T(E) \\ &= Cn_\Sigma\left(W \cup \{\text{cons}(\delta) \mid \delta \in \Delta \text{ is applicable to } E\}\right). \end{aligned}$$

Thus E is classically generated by W together with the consequences of all default rules applicable to E.

Since propositional logic is compact, a finite subset of these consequences already suffices to generate the same deductive closure. Hence there are default rules

$$\delta_1, \ldots, \delta_n \in \Delta,$$

which represent exactly the rules applicable to E, such that

$$E = Cn_\Sigma(W \cup \{\text{cons}(\delta_1), \ldots, \text{cons}(\delta_n)\}).$$

We now order these rules in such a way that each next rule is already applicable after the preceding steps. This is possible because each of the selected consequences can occur in E only if its prerequisite is already derivable from W and earlier consequences. Otherwise, a rule would occur in E without sufficient support, contradicting the fact that E is a fixed point.

Thus the δ_i can be numbered so that, for every i, the following holds:

$$\delta_i \text{ is applicable to } \mathrm{Cn}_\Sigma(W \cup \{\mathrm{cons}(\delta_1), \ldots, \mathrm{cons}(\delta_{i-1})\}) \text{ with respect to } E$$

Moreover, by construction, no further rule outside this list is applicable to E. Hence (1)–(3) are satisfied.

(⇐) Conversely, suppose that (1)–(3) hold. We show that E is then an extension, that is, $\Lambda_T(E) = E$.

By (2), all $\delta_1, \ldots, \delta_n$ are stepwise applicable to E in the stated order. Therefore their consequences are included by the definition of the operator. By (3), there are no further default rules applicable to E. Hence

$$\hat{\Lambda}_T(E) = Cn_\Sigma(W \cup \{\mathrm{cons}(\delta_1), \ldots, \mathrm{cons}(\delta_n)\}) = E$$

by (1). Thus E is a fixed point of $\hat{\Lambda}_T$, and hence an extension. □

Example 4.129 (Two competing extensions). *We continue the theory T_{Reise} and additionally assume that* *the train is delayed, but that no further information is available:*

$$W = \{V\}, \qquad \Delta = \left\{ \frac{V : S}{S}, \frac{V : \neg S}{\neg S} \right\}.$$

Then there are two extensions:

$$E_1 = Cn_\Sigma(\{V, S\}), \qquad E_2 = Cn_\Sigma(\{V, \neg S\}).$$

For E_1, the first rule is applicable, whereas the second is not, since $\neg S$ blocks the justification S. Hence

$$\hat{\Lambda}_T(E_1) = Cn_\Sigma(\{V, S\}) = E_1.$$

Analogously, for E_2 the second rule is applicable, while the first is blocked. Therefore,

$$\hat{\Lambda}_T(E_2) = Cn_\Sigma(\{V, \neg S\}) = E_2.$$

Thus the theory has two stable but distinct plausible states of knowledge.

Remark 4.130. *Not every default theory has an extension. This is not a defect of the theory, but rather expresses the fact that some default rules may block one another so strongly that no stable state of knowledge arises. We shall examine this phenomenon more closely in what follows.*

Proposition 4.131 (Extensions are incomparable). *Let Σ be a propositional signature, let $T = (W, \Delta)$ be a default theory over $L(\Sigma)$, and let $E, E' \subseteq L(\Sigma)$ be extensions of T. If $E \subseteq E'$, then*

$$E = E'.$$

Proof. Since E and E' are extensions, we have

$$\Lambda_T(E) = E \qquad \text{and} \qquad \Lambda_T(E') = E'.$$

In particular, E is a fixed point of $\hat{\Lambda}_T$.

Now suppose that $E \subseteq E'$. By the definition of $\Lambda_T(E')$, $\Lambda_T(E')$ is the smallest subset of E' that is already a fixed point of $\hat{\Lambda}_T$. Since E is such a fixed point and at the same time $E \subseteq E'$, minimality yields

$$\Lambda_T(E') \subseteq E.$$

Since $\Lambda_T(E') = E'$, we therefore obtain $E' \subseteq E$. Together with $E \subseteq E'$, this implies

$$E = E'.$$

Thus two distinct extensions can never stand in a proper subset relation to one another. □

Proposition 4.132 (Unique extension under consistency of default assumptions). *Let Σ be a propositional signature and let $T = (W, \Delta)$ be a default theory over $L(\Sigma)$. If the set*

$$W \cup \left\{ \psi_1 \wedge \cdots \wedge \psi_n \wedge \chi \;\middle|\; \frac{\varphi : \psi_1, \ldots, \psi_n}{\chi} \in \Delta \right\}$$

is classically consistent, then T has exactly one extension.

Proof. We first prove the existence of an extension and then its uniqueness.

Step 1: Construction of a stable sequence. Define inductively

$$\begin{aligned} E_0 &:= W, \\ E_{i+1} &:= \hat{\Lambda}_T(E_i) \\ &= \mathrm{Cn}_\Sigma\Big(W \cup \{\mathrm{cons}(\delta) \mid \delta \in \Delta \text{ is applicable to } E_i\}\Big) \end{aligned}$$

for all $i \in \mathbb{N}$, and set

$$E := \bigcup_{i \in \mathbb{N}} E_i.$$

Step 2: Monotonicity of the construction. We show by induction that

$$E_i \subseteq E_{i+1}$$
$$\text{and every default rule applicable to } E_i$$
$$\text{is also applicable to } E_{i+1}.$$

For $i = 0$, this is precisely the point at which the consistency assumption is used: the possible default assumptions $\psi_1, \ldots, \psi_n$ are not contradicted by the certain knowledge W and the consequences already generated. Hence rules that are applicable once remain applicable even after their consequences have been added.

Assume that the assertion holds for i. Then

$$E_i \subseteq E_{i+1}$$

by the monotonicity of deductive closure. Moreover, adding consequences that are consistent with all default assumptions from the hypothesis cannot block any previously applicable rule. Thus the second assertion is preserved as well.

Step 3: E is a fixed point of $\hat{\Lambda}_T$. We now show that

$$E = \hat{\Lambda}_T(E).$$

(i) $E \subseteq \hat{\Lambda}_T(E)$: Let $\varphi \in E$. Then there is an i such that $\varphi \in E_i$. By construction,

$$E_i = Cn_\Sigma(W \cup \{\text{cons}(\delta) \mid \delta \in \Delta \text{ is applicable to } E_{i-1}\})\,.$$

By compactness of propositional logic, a finite subset of the consequences used is sufficient to derive φ. Each of these consequences comes from a default rule that is applicable to E_{i-1}, and by Step 2 it remains applicable to E. Thus φ is already derivable from

$$W \cup \{\text{cons}(\delta) \mid \delta \in \Delta \text{ is applicable to } E\}$$

is derivable. Hence $\varphi \in \hat{\Lambda}_T(E)$.

(ii) $\hat{\Lambda}_T(E) \subseteq E$: Conversely, let $\varphi \in \hat{\Lambda}_T(E)$. Then

$$\varphi \in Cn_\Sigma(W \cup \{\text{cons}(\delta) \mid \delta \in \Delta \text{ is applicable to } E\})\,.$$

Again, compactness provides a finite set of consequences from which φ already follows. For each of these consequences, choose a default rule that is applicable to E. Each such rule is already applicable from some stage E_j onward and remains applicable thereafter by Step 2.

Thus these consequences belong to some later E_k, and hence $\varphi \in E_k \subseteq E$.

Therefore $E = \hat{\Lambda}_T(E)$. Hence E is an extension of T.

Step 4: Uniqueness. Now let F be an arbitrary extension of T. We show by induction that $E_i \subseteq F$ for all $i \in \mathbb{N}$.

For $i = 0$, we have $E_0 = W \subseteq F$, because every extension contains the certain knowledge. Assume that $E_i \subseteq F$. Then every default rule applicable to E_i is also applicable to F by the consistency assumption; for F contains only formulas arising from W and admissible default consequences, and these are compatible with all default assumptions of the rules in Δ. Since F is an extension, it therefore also contains all corresponding consequences. Consequently, $E_{i+1} \subseteq F$.

It follows that

$$E = \bigcup_{i \in \mathbb{N}} E_i \subseteq F.$$

Since E and F are both extensions, the preceding proposition applied to $E \subseteq F$ already yields $E = F$. Thus T has exactly one extension. □

Proposition 4.133 (Inconsistency of an extension carries over to the theory). *Let Σ be a propositional signature and let $T = (W, \Delta)$ be a default theory over $L(\Sigma)$. If there is an extension E of T such that*

$$E \equiv \bot,$$

then the following hold:

1. $W \equiv \bot$,

2. every extension E' *of* T *also satisfies* $E' \equiv \bot$.

Proof. Let E be an inconsistent extension of T, that is, $E = \Lambda_T(E)$ and $E \equiv \bot$.

We first show that the certain knowledge W must already be inconsistent. Since E is deductively closed and inconsistent, for every formula $\psi \in L(\Sigma)$ we have

$$\psi \in E \quad \text{and} \quad \neg\psi \in E.$$

In particular, for every default rule

$$\delta = \frac{\varphi : \psi_1, \ldots, \psi_n}{\chi} \in \Delta$$

at least one of its justifications is refuted by E, since $\neg\psi_i \in E$ holds for all ψ_i. Thus no default rule is applicable to E. Hence the operator simplifies to

$$\hat{\Lambda}_T(E) = Cn_\Sigma(W).$$

Since E is an extension, we have $\hat{\Lambda}_T(E) = E$. It follows that

$$E = Cn_\Sigma(W).$$

Because $E \equiv \bot$ was assumed, we also have $Cn_\Sigma(W) \equiv \bot$, and hence already

$$W \equiv \bot.$$

This proves part (1).

For part (2), let E' be any extension of T. By part (1), W is inconsistent. Since every extension contains the certain knowledge, $W \subseteq E'$. Therefore E' is inconsistent as well: a consistent set cannot contain an inconsistent subset. Consequently,

$$E' \equiv \bot.$$

Both claims are therefore proved. □

4.4.3 Inference and Semi-Monotonicity

With the extensions of a default theory, we have already identified the possible "plausible states of knowledge" in the previous section. From these extensions we now obtain an inference relation by asking not merely what holds in *one* extension, but what holds in *all* of them. This cautious perspective is especially natural when several alternative plausible interpretations are possible.

Definition 4.134 (Default inference). *Let Σ be a propositional signature, let $T = (W, \Delta)$ be a default theory over $L(\Sigma)$, and let $\varphi \in L(\Sigma)$ be a propositional formula. We write*

$$T \mid\!\sim_{\mathrm{DL}} \varphi$$

if and only if φ is contained in every extension of T.

Remark 4.135. *This inference relation is* skeptical*: a formula is a default consequence only if it occurs in all admissible developments of the theory. It is therefore well suited to cautious reasoning under uncertainty.*

Example 4.136 (An unambiguous default case). *Consider the signature*

$$\Sigma_{\text{Coffee}} = \{K, W\}$$

with

- K*: "There is coffee."*
- W*: "The coffee is warm."*

and the default theory

$$T_{\text{Coffee}} = (\{K\}, \{\tfrac{K:W}{W}\}).$$

Here there is exactly one extension, namely

$$E = Cn_{\Sigma}(\{K, W\}).$$

Hence

$$T_{\text{Coffee}} \mid\!\sim_{\text{DL}} W.$$

If, by contrast, we add the competing default rule $\frac{K:\neg W}{\neg W}$*, then two extensions arise, and neither* W *nor* $\neg W$ *is then a skeptical default consequence.*

Example 4.137 (Non-monotonicity of inference). *Consider the signature*

$$\Sigma_{\text{TrafficLight}} = \{R, G, S\}$$

with

- R*: "The traffic light is red."*

- *G: "The traffic light is green."*
- *S: "Cars are allowed to drive."*

Let

$$T = \left(\{R\}, \left\{\frac{R : S}{S}\right\}\right).$$

Then T has exactly one extension, namely $Cn_{\Sigma}(\{R, S\})$, so

$$T \mathrel{|\!\!\sim}_{\mathrm{DL}} S.$$

Now extend the theory by the additional default rule

$$\frac{R : \neg S}{\neg S},$$

then two extensions arise, one with S and one with $\neg S$. Therefore, for the extended theory, it is no longer the case that

$$T' \mathrel{|\!\!\sim}_{\mathrm{DL}} S.$$

This shows that default reasoning is non-monotonic.

Remark 4.138. *When information is added to default theories, quite different changes may occur:*

1. *an extension may disappear,*
2. *the content of an extension may change,*
3. *or new extensions may arise.*

This instability is intentional, since it reflects the behavior of rules with exceptions.

Definition 4.139 (Semi-monotone Extension). *Let* Σ *be a propositional signature, and let*

$$T = (W, \Delta), \qquad T' = (W', \Delta')$$

be two default theories over $L(\Sigma)$ *such that*

$$W = W' \qquad \textit{and} \qquad \Delta \subseteq \Delta'.$$

If, for every extension E *of* T*, there exists an extension* E' *of* T' *such that*

$$E \subseteq E'$$

then T' *is called a* semi-monotone extension *of* T*.*

Remark 4.140. *Semi-monotonicity is therefore a weakened stability requirement: old plausible knowledge states need not be preserved unchanged, but they must not be lost entirely. Rather, every old state of knowledge should reappear within some new plausible state of knowledge.*

Definition 4.141 (Normal Default Rule). *A default rule*

$$\delta = \frac{\varphi : \psi_1, \ldots, \psi_n}{\chi}$$

is called normal *if*

$$\text{just}(\delta) = \{\text{cons}(\delta)\}.$$

That is, the only justification of a normal default rule is its own consequence. A default theory is called normal *if all of its default rules are normal.*

Example 4.142 (Normal and Non-normal Default Rules). *The rule*

$$\frac{B : F}{F}$$

is normal: the consequence F is at the same time the only justification.

By contrast, the rule

$$\frac{A : innocence}{acquittal}$$

is not normal, because the justification "innocence" differs from the consequence "acquittal".

Theorem 4.143 (Normal Default Theories Are Semi–monotone). *Let $T = (W, \Delta)$ be a normal default theory over $L(\Sigma)$, and let*

$$\delta = \frac{\varphi : \chi}{\chi}$$

be a normal default. Then

$$T' = (W, \Delta \cup \{\delta\})$$

is a semi-monotone extension of T.

Proof. Let E be an arbitrary extension of T. We show that there is an extension E' of T' such that

$$E \subseteq E'$$

holds.

We distinguish two cases.

Case 1: δ is not applicable to E.

Then δ is already blocked by E, so the new rule contributes nothing to the construction at E. Hence, with respect to the defaults applicable to E, it makes no difference whether we work with T or with T'. Thus

$$\hat{\Lambda}_{T'}(E) = \hat{\Lambda}_T(E) = E.$$

Therefore E itself is already an extension of T'. In particular, there exists an extension E' of T' with $E \subseteq E'$, namely $E' = E$.

Case 2: δ is applicable to E.

Then $\varphi \in E$ and $\neg\chi \notin E$. Since δ is normal, its justification is precisely χ itself. Thus χ is compatible with E, and adding χ does not destroy any classical information already secured in E.

We now use the characterization of extensions by finite generating sequences. Since E is an extension of T, there is a sequence of default rules

$$\delta_1, \ldots, \delta_n \in \Delta$$

such that

$$E = Cn_\Sigma(W \cup \{\text{cons}(\delta_1), \ldots, \text{cons}(\delta_n)\}),$$

and each δ_i is applicable at the corresponding stage.

Since δ is applicable to E and is normal, it can be appended to the end of this sequence. The only additional information is χ, and this contradicts neither E nor the justifications of the previous normal defaults. Hence the extended sequence

$$\delta_1, \ldots, \delta_n, \delta$$

is also an admissible generating sequence for T'.

Set

$$E' := Cn_\Sigma(W \cup \{\text{cons}(\delta_1), \ldots, \text{cons}(\delta_n), \chi\}).$$

Then E' is an extension of T', and by construction

$$E \subseteq E'.$$

Thus, in both cases, every extension of T is contained in an extension of T'. Therefore T' is semi-monotone over T. □

Remark 4.144. *The statement of the theorem is especially important from a didactic point of view: normal defaults form precisely the well-behaved class of default rules for which adding new rules does not abruptly destroy old plausible states of knowledge, but at most embeds them into larger plausible states of knowledge.*

4.4.4 Exercises and Solutions

Exercise 4.145. *We consider the propositional signature*

$$\Sigma = \{B, E, F, G\}.$$

We always read a default rule in the form

$$\frac{\varphi : \psi_1, \ldots, \psi_n}{\chi},$$

that is, with prerequisite φ*, justifications* $\psi_1, \ldots, \psi_n$*, and consequent* χ*.*

For the following exercises, let the two default theories be given by

$$T_3 = \left(\{B, \neg E, \neg G\}, \left\{ \begin{array}{l} \delta_1 = \dfrac{\top : \neg B, \neg E, G}{F}, \\ \delta_2 = \dfrac{G : E, G}{F}, \\ \delta_3 = \dfrac{\neg G : \neg E, \neg F}{\neg F}, \\ \delta_4 = \dfrac{\neg G : \neg E}{\neg F} \end{array} \right\} \right)$$

and

$$T_4 = \left(\{E, \neg G\}, \left\{ \begin{array}{l} \delta_1 = \dfrac{\top : \neg F, \neg G}{\neg F}, \\ \delta_2 = \dfrac{\neg F : G}{\neg B}, \\ \delta_3 = \dfrac{\top : \neg B, G}{\neg B}, \\ \delta_4 = \dfrac{\neg E : \neg B}{\neg F} \end{array} \right\} \right).$$

Decide for each of the following statements whether it is true or false.

1. *We have*

$$\text{pre}\left(\frac{\neg G : B, \neg F}{\neg F}\right) = \{B, \neg F\}.$$

2. *The rule*

$$\frac{\neg B : \neg E, G}{F}$$

is applicable to $\{\neg B, \neg E, G\}$.

3. $Cn_{\Sigma}(\{B, \neg E, \neg F\})$ *is an extension of* T_3.

4. *We have*

$$T_4 \mid\!\sim_{\text{DL}} \neg F.$$

Solution 4.146 (Solution to 1.). ***False.*** *In a default rule of the form*

$$\frac{\varphi : \psi_1, \ldots, \psi_n}{\chi}$$

φ *is the prerequisite, that is, the expression* before *the colon. Hence, in the given rule,*

$$\text{pre}\left(\frac{\neg G : B, \neg F}{\neg F}\right) = \neg G.$$

By contrast, the set $\{B, \neg F\}$ *consists of the justifications of the rule, so*

$$\text{just}\left(\frac{\neg G : B, \neg F}{\neg F}\right) = \{B, \neg F\}.$$

Solution 4.147 (Solution to 2.). ***True.*** *Let*

$$\delta = \frac{\neg B : \neg E, G}{F} \quad \textit{and} \quad \Phi = \{\neg B, \neg E, G\}.$$

First, we have

$$\text{pre}(\delta) = \neg B \in \Phi.$$

Moreover, for every justification $\psi \in \{\neg E, G\}$, *it must hold that* $\neg\psi \notin \Phi$.

For $\psi = \neg E$, *we have* $\neg\psi = E$, *and* $E \notin \Phi$. *For* $\psi = G$, *we have* $\neg\psi = \neg G$, *and* $\neg G \notin \Phi$.

Therefore, δ *is applicable to* Φ.

Solution 4.148 (Solution to 3.). ***False.*** *Every extension of a default theory must contain the certain knowledge. Hence, for every extension* E^* *of* T_3, *we have:*

$$\{B, \neg E, \neg G\} \subseteq E^*.$$

However, the set $Cn_\Sigma(\{B, \neg E, \neg F\})$ *does not contain* $\neg G$. *For this reason alone, it cannot be an extension of* T_3.

The important first quick check is this: before checking the default rules, every candidate set must at least contain the certain knowledge W. *If even one fact from* W *is missing, the candidate is immediately ruled out.*

Solution 4.149 (Solution to 4.). ***True.*** *We show that* T_4 *has exactly one extension and that* $\neg F$ *is contained in this extension.*

For the certain knowledge

$$W = \{E, \neg G\}$$

the default rule

$$\delta_1 = \frac{\top : \neg F, \neg G}{\neg F}$$

is applicable, because the prerequisite $\top$ *is always satisfied and the justifications* $\neg F$ *and* $\neg G$ *are not contradicted by* W*.*

The remaining rules are initially not applicable to W*:*

- $\delta_2 = \frac{\neg F : G}{\neg B}$ *is not applicable, since* G *is not contained in* W*.*
- $\delta_3 = \frac{\top : \neg B, G}{\neg B}$ *is likewise not applicable, since* G *is missing.*
- $\delta_4 = \frac{\neg E : \neg B}{\neg F}$ *is not applicable, since the prerequisite* $\neg E$ *is contradicted by* $E \in W$*.*

Thus we obtain the following stable extension:

$$E^* = Cn_\Sigma(\{E, \neg G, \neg F\}).$$

In E^**,* δ_1 *remains applicable, and none of the remaining rules is activated by the added information. Hence,*

$$\Lambda_{T_4}(E^*) = E^*,$$

and therefore E^* *is an extension of* T_4*.*

Since every extension of T_4 *must contain the certain knowledge* W*, and the only unblocked default rule already yields* $\neg F$*, there is no further extension. Consequently,*

$$T_4 \mid\!\sim_{\mathrm{DL}} \neg F.$$

4.5 Answer Set Programming

In the previous section, we encountered default logic as a highly expressive, but also comparatively abstract, formalism for nonmonotonic reasoning. For many applications in knowledge representation, however, it is useful to have a notation that is closer to logic programming and is therefore easier to implement and read. This is precisely where answer set programming comes in: it combines rules in a compact program syntax with a nonmonotonic semantics that also permits conclusions in the presence of exceptions and incomplete knowledge.

The basic idea is simple: a program describes which conclusions may be drawn under which conditions. In doing so, not only classical negation $\neg$ plays a role, but also default negation `not`. While $\neg B$ expresses that B is false, `not` B means that B is not provable. This distinction is crucial for the semantics developed below.

We begin with the syntax of extended logic programs. In the next subsection, we will then show how a first semantics can be obtained from such programs by means of minimal models. From this, the actual answer set semantics will subsequently be developed.

4.5.1 Extended Logic Programs

The syntax of answer set programming is based on rules of the form

$$H \leftarrow A_1, \ldots, A_n, \texttt{not}\ B_1, \ldots, \texttt{not}\ B_m, \qquad (4.1)$$

where $H, A_1, \ldots, A_n, B_1, \ldots, B_m \in \text{Lit}(\Sigma)$ are literals over a propositional signature Σ.

Definition 4.150 (Extended logic program). *Let Σ be a propositional signature. An* extended logic program *P over Σ is a finite set of rules of the form (4.1).*

For a rule

$$r: \quad H \leftarrow A_1, \ldots, A_n, \ \textit{\textbf{not}}\ B_1, \ldots, \textit{\textbf{not}}\ B_m$$

we call

$$\text{head}(r) = \{H\}$$

the head *of the rule, and*

$$\text{body}(r) = \{A_1, \ldots, A_n, \textit{\textbf{not}}\ B_1, \ldots, \textit{\textbf{not}}\ B_m\}$$

the body*. Furthermore, we define*

$$\begin{aligned} &\text{body}^+(r) = \{A_1, \ldots, A_n\} \\ \textit{and} \quad &\text{body}^-(r) = \{B_1, \ldots, B_m\} \end{aligned}$$

If $m = 0$, then r is called positive*. If, in addition, $n = m = 0$, then r is called a* fact*. In this case, instead of writing $H \leftarrow$, we simply write H.*

The reading of a rule is intuitive:

> If all positive prerequisites are satisfied and none of the negative prerequisites is known, then the head literal may be inferred.

In other words: the head is accepted whenever the positive conditions are satisfied and no counterevidence is found among the negative conditions.

Example 4.151. *Consider the propositional signature*

$$\Sigma_{\mathrm{Sys}} = \{\,Start, Alarm, Defect, Maintenance, Safe\,\}.$$

The program

$$P_{\mathrm{Sys}} = \left\{ \begin{array}{l} Alarm \leftarrow Defect,\ \mathtt{not}\ Maintenance; \\ Start \leftarrow Safe,\ \mathtt{not}\ Alarm; \\ Defect \leftarrow Maintenance; \\ Safe \end{array} \right\}$$

is an extended logic program over Σ_{Sys}.

Here the last rule is the fact $Safe$. *The rule*

$$Start \leftarrow Safe,\ \mathtt{not}\ Alarm$$

means that starting is permitted if the system is safe and no alarm is known. The rule

$$Alarm \leftarrow Defect,\ \mathtt{not}\ Maintenance$$

captures a typical exception pattern: a defect triggers an alarm as long as no maintenance situation is present.

In applications, rules are often formulated not only propositionally, but also with predicates, variables, and constants. In that case, a rule initially serves as a schema from which all *ground instances* are obtained by substituting constants.

Example 4.152. *Let*

$$\Sigma_{\text{Lab}} = \{Access/1, Employee/1, \\ Blocked/1, alice/0, bob/0\}$$

be a first-order signature. Consider the rule pattern

$$Access(x) \leftarrow Employee(x),\ \mathtt{not}\ Blocked(x).$$

Together with the facts

$$Employee(alice), \\ Employee(bob), \\ Blocked(bob)$$

grounding yields, among others, the rules

$$Access(alice) \leftarrow Employee(alice), \\ \mathtt{not}\ Blocked(alice);$$

$$Access(bob) \leftarrow Employee(bob), \\ \mathtt{not}\ Blocked(bob).$$

Intuitively, this means that Alice is granted access if she is an employee and no block is known; for Bob, this is precisely not the case because he is blocked.

The distinction between classical negation $\neg$ and default negation `not` is central. Classical negation is used when a statement is explicitly modeled as false. Default negation, by contrast, represents the absence of knowledge.

Example 4.153. *Let Rain and DryRoad be statements about the weather and the condition of the road. Then the two rules*

$$DryRoad \leftarrow \neg Rain \qquad \text{and} \qquad DryRoad \leftarrow \mathtt{not}\ Rain$$

are not to be understood in the same way.

The first rule says: the road is dry if we know that it is not raining. The second rule says: the road is dry if no rain has been detected.

This distinction is especially important in real knowledge bases: from the mere absence of information, one must not automatically infer that the opposite is true. Default negation models precisely this transition from lack of knowledge to a cautious conclusion.

For brevity, rules with an empty body are often written simply as facts. Thus, for example,

$$A$$

is only an abbreviation for

$$A \leftarrow .$$

Similarly, we usually write a program in set notation, using semicolons to separate the rules, for example

$$P = \{\, r_1;\, r_2;\, r_3 \,\}.$$

This avoids ambiguities that could otherwise arise from a notation using commas.

Example 4.154. *The program*

$$P_{\text{Warm}} = \left\{ \begin{array}{l} Heating \leftarrow Cold,\ \mathtt{not}\ Malfunction; \\ Cold \leftarrow WindowOpen; \\ Malfunction \leftarrow Maintenance; \\ WindowOpen \end{array} \right\}$$

contains one fact, one positive rule, and two rules with default negation. It is therefore a typical small ASP program: conclusions are drawn from an observed situation only when no contrary information is present.

Extended logic programs therefore form the syntactic basis of answer set programming. They make it possible to formulate rules concisely, model exceptions, and explicitly take missing knowledge into account. In the next step, we will make the actual semantics of these programs precise by means of minimal models and, later, answer sets.

4.5.2 Minimal Models

Before introducing the actual answer set semantics, we first consider a simpler special case: *positive* extended logic programs, that is, programs without default negation. In this case, the semantics can initially be described very naturally in terms of *closure* and *minimal* closed states. This preliminary step is important not only conceptually, but also technically, since it will later serve as a building block for answer set semantics.

Intuitively, a state describes which literals we take to be true. Such a state should model a program exactly when it respects all rules: whenever the body of a rule is satisfied, its head must also be contained in the state. Among all states with this property, we are then interested in the smallest one, because it contains precisely the literals that are genuinely forced by the program.

Definition 4.155 (State). *Let Σ be a propositional signature and let $S \subseteq \mathrm{Lit}(\Sigma)$. If S is consistent, that is, if there is no $A \in \Sigma$ such that*

$$A \in S \quad \textit{and} \quad \neg A \in S,$$

then S is called a state *over Σ.*

States are therefore consistent sets of literals. They correspond to ordinary truth assignments in propositional logic, except that here we work directly with literals rather than with formulas.

Example 4.156. *Consider the signature*

$$\Sigma_{\mathrm{Robot}} = \{\, Start, Charging, BatteryEmpty, Ready \,\}.$$

Then, for example,

$$\emptyset, \qquad \{Ready\}, \qquad \{Start, \neg BatteryEmpty\}$$

are states over Σ_{Robot}.

By contrast, the following are not states:

$$\{Start, \neg Start\} \qquad \textit{or} \qquad \{Ready, \neg Ready, Start\},$$

because both sets are inconsistent.

For positive programs, the central semantic notion is closure.

Definition 4.157 (Positive Program)**.** *Let Σ be a propositional signature. An extended logic program P over Σ is called* positive *if none of its rules contains default negation; that is, for every rule $r \in P$,*

$$\mathrm{body}^{-}(r) = \emptyset.$$

Equivalently, all rules have the form

$$H \leftarrow A_1, \ldots, A_n.$$

Definition 4.158 (Closure)**.** *Let Σ be a propositional signature, and let P be a positive extended logic program over Σ. A state $S \subseteq \mathrm{Lit}(\Sigma)$ is called* closed under P *if, for every rule*

$$r: \; H \leftarrow A_1, \ldots, A_n$$

in P,

$$\{A_1, \ldots, A_n\} \subseteq S \quad \Longrightarrow \quad H \in S.$$

The meaning is simple: everything that can be inferred by an applicable rule should actually be present in the state.

Example 4.159. *Let*

$$P_{\text{Team}} = \left\{ \begin{array}{l} \mathit{Authorized} \leftarrow \mathit{Member},\ \mathit{Active}; \\ \mathit{Active} \leftarrow \mathit{Contract}; \\ \mathit{Member} \end{array} \right\}$$

be a positive program. Then

$$S_1 = \emptyset$$

is not closed, because the fact $Member$ *is applicable but its head is not contained in the state.*

The state

$$S_2 = \{Member, Active, Authorized\}$$

is closed, by contrast, because the applicable rule

$$Authorized \leftarrow Member, Active$$

has its head in the state, and the fact $Member$ *is also satisfied.*

Also

$$S_3 = \{Member, Active, Authorized, Contract\}$$

is closed. However, it contains more than is necessary. Such superfluous literals will later be excluded by minimality.

We first record a small but important observation.

Proposition 4.160. *Let* Σ *be a propositional signature, and let* P *be a positive extended logic program. If* S *is closed under* P *and* $r \in P$ *is a fact, that is, a rule of the form* $H \leftarrow$*, then* $H \in S$.

Proof. Let r be a fact. Then $\text{body}(r) = \emptyset$. In particular,

$$\text{body}(r) \subseteq S.$$

Since S is closed under P, the head of every applicable rule must, by definition, also belong to the state. Hence

$$H \in S.$$

□

Closure alone is still too weak. A state may respect the rules of a program while nevertheless containing unnecessary literals. We therefore consider, among all closed states, those that are minimal with respect to inclusion.

Definition 4.161 (Minimal Model). *Let Σ be a propositional signature, and let P be a positive extended logic program over Σ. A state $S \subseteq \text{Lit}(\Sigma)$ is called a* minimal model *of P if the following conditions hold:*

1. *S is closed under P;*
2. *for every state S' that is closed under P, we have $S \subseteq S'$.*

A minimal model is therefore a closed state that has no proper smaller closed subset. It contains exactly what is forced by the program, and nothing beyond that.

Example 4.162. *Consider the program*

$$P_{\mathrm{M}} = \left\{ \begin{array}{l} Ready \leftarrow BatteryOk; \\ BatteryOk \leftarrow Charged; \\ Charged \end{array} \right\}$$

over the signature

$$\Sigma_{\mathrm{M}} = \{Ready, BatteryOk, Charged, Fault\}.$$

Then

$$M = \{Charged, BatteryOk, Ready\}$$

is closed: from the fact Charged we obtain BatteryOk, and from this we obtain Ready.

The state

$$M' = \{Charged, BatteryOk, Ready, Fault\}$$

is also closed, but it is not minimal. The additional literal Fault is not forced by any rule of the program.

In the positive case, the existence and uniqueness of the minimal model are particularly simple.

Theorem 4.163. *Let Σ be a propositional signature, and let P be a consistent positive extended logic program over Σ. Then P has exactly one minimal model.*

Proof. Let $\mathcal{C}$ be the set of all states that are closed under P. Since P is consistent, $\mathcal{C}$ is nonempty.

We first show that the intersection of all closed states is again closed. Define

$$M := \bigcap_{S \in \mathcal{C}} S.$$

Let

$$r : \ H \leftarrow A_1, \ldots, A_n$$

be a rule in P, and suppose that

$$\{A_1, \ldots, A_n\} \subseteq M.$$

Then each A_i belongs to every closed state $S \in \mathcal{C}$. Since every $S \in \mathcal{C}$ is closed, it follows for each such S that

$$H \in S.$$

Thus H belongs to all closed states and therefore to their intersection M. Hence M is closed.

Moreover, for every closed state $S \in \mathcal{C}$ we have

$$M \subseteq S,$$

because M, as the intersection of all closed states, is precisely the smallest set with this property. Therefore M is a minimal model.

Uniqueness follows immediately: if M_1 and M_2 were two minimal models, then in particular M_1 would be closed. By the minimality of M_2, we would have $M_2 \subseteq M_1$. Conversely, the minimality of M_1 would give $M_1 \subseteq M_2$. Hence $M_1 = M_2$.

Thus P has exactly one minimal model. □

This minimal model can also be described by means of an immediate derivation operator. To this end, we define an operator that assigns to a state all heads of those rules whose bodies are already satisfied.

Definition 4.164 (Characteristic Operator). *Let Σ be a propositional signature and let P be a positive extended logic program over Σ. The characteristic operator*

$$\Lambda_P : 2^{\mathrm{Lit}(\Sigma)} \to 2^{\mathrm{Lit}(\Sigma)}$$

is defined by

$$\begin{aligned}\Lambda_P(S) = \{H \mid (H \leftarrow A_1, \ldots, A_n) \in P \\ \textit{and } \{A_1, \ldots, A_n\} \subseteq S\}\end{aligned}$$

Furthermore, we define inductively

$$\Lambda_P^0(S) = S, \qquad \Lambda_P^{n+1}(S) = \Lambda_P(\Lambda_P^n(S)) \quad (n \geq 0).$$

Thus, the operator maps a state to the set of all immediately derivable heads.

Example 4.165. *For the program*

$$P_{\mathrm{M}} = \left\{ \begin{array}{l} \mathit{Ready} \leftarrow \mathit{BatteryOk}; \\ \mathit{BatteryOk} \leftarrow \mathit{Charged}; \\ \mathit{Charged} \end{array} \right\}$$

we have:

$$\Lambda_{P_{\mathrm{M}}}(\emptyset) = \{\mathit{Charged}\},$$

$$\Lambda_{P_{\mathrm{M}}}(\{\mathit{Charged}\}) = \{\mathit{Charged}, \mathit{BatteryOk}\},$$

$$\Lambda_{P_{\mathrm{M}}}(\{\mathit{Charged}, \mathit{BatteryOk}\}) = \{\mathit{Charged}, \\ \mathit{BatteryOk}, \mathit{Ready}\}$$

After this, nothing changes anymore. Precisely this stable set is the minimal model.

The connection between closure and the operator Λ_P is particularly close.

Theorem 4.166. *Let Σ be a propositional signature and let P be a positive extended logic program over Σ. A state $S \subseteq \mathrm{Lit}(\Sigma)$ is closed under P if and only if*

$$\Lambda_P(S) \subseteq S$$

holds.

Proof. We prove both directions.

"$\Rightarrow$" Assume that S is closed under P, and let $H \in \Lambda_P(S)$ be arbitrary. Then there is a rule

$$r: \; H \leftarrow A_1, \ldots, A_n$$

in P such that

$$\{A_1, \ldots, A_n\} \subseteq S.$$

Since S is closed, the applicability of r implies that $H \in S$ as well. Hence every element of $\Lambda_P(S)$ is contained in S, and therefore

$$\Lambda_P(S) \subseteq S.$$

"$\Leftarrow$" Conversely, assume that $\Lambda_P(S) \subseteq S$. Again, let

$$r: \; H \leftarrow A_1, \ldots, A_n$$

be a rule from P such that

$$\{A_1, \ldots, A_n\} \subseteq S.$$

Then, by the definition of Λ_P, we also have $H \in \Lambda_P(S)$. Since $\Lambda_P(S) \subseteq S$, it follows that $H \in S$. Thus S is closed under P. □

The minimal model is obtained from the operator Λ_P by repeated application starting from the empty state.

Theorem 4.167. *Let Σ be a propositional signature and let P be a consistent positive extended logic program over Σ. Then the sequence*

$$\emptyset \subseteq \Lambda_P(\emptyset) \subseteq \Lambda_P^2(\emptyset) \subseteq \cdots$$

stabilizes after finitely many steps, and its limit is the minimal model of P.

Proof. We first show that the sequence is monotonically increasing. In view of the preceding theorem, it suffices to show that

$$X \subseteq \Lambda_P(X)$$

for every state X that already contains all literals derived so far. Intuitively, this is clear: once a literal has become derivable, it remains derivable under further applications of the same positive rule set, since no default negation occurs and hence nothing can be "withdrawn."

More formally, let $X_0 = \emptyset$ and $X_{n+1} = \Lambda_P(X_n)$ for $n \geq 0$. Then

$$X_0 \subseteq X_1.$$

Assume that $X_n \subseteq X_{n+1}$. If a literal H already occurs in X_n as the head derived by some rule, then it remains contained in X_{n+1}. Hence

$$X_{n+1} \subseteq X_{n+2}.$$

Thus the sequence is monotonically increasing.

Since $\mathrm{Lit}(\Sigma)$ is finite, there can be only finitely many distinct states. A strictly increasing chain of subsets must therefore terminate after finitely many steps. Hence there exists some $k \geq 0$ such that

$$X_k = X_{k+1}.$$

With $M := X_k$, we therefore have

$$\Lambda_P(M) = M,$$

so M is a fixed point of Λ_P.

It remains to show that M is the minimal model. Since $\Lambda_P(M) = M$, the preceding theorem implies that M is closed. Now let S be any closed state. We prove by induction on n that

$$X_n \subseteq S$$

holds.

Base case: $X_0 = \emptyset \subseteq S$.

Induction step: Assume that $X_n \subseteq S$. Since S is closed and $X_{n+1} = \Lambda_P(X_n)$, the preceding theorem yields

$$X_{n+1} = \Lambda_P(X_n) \subseteq S.$$

Consequently, $X_n \subseteq S$ for all n, and in particular for the stabilization point $M = X_k$:

$$M \subseteq S.$$

Thus M is contained in every closed state and is therefore minimal. Together with the closedness already established, this shows that M is the minimal model of P. □

We thus obtain a very clear picture: for positive programs there is exactly one minimal model. It can be described either as the intersection of all closed states or as the fixed point of the derivation operator Λ_P. This characterization will later form the basis for the semantics of general programs with default negation, and hence for answer set programming in the proper sense.

4.5.3 Answer Sets

We now return to general extended logic programs, in particular to programs that contain default negation. For such programs, the notion of a minimal model alone is no longer sufficient, because the applicability of a rule may depend on the chosen interpretation: whether a rule can be applied is determined precisely by which literals are already known, and which are not, in the assumed view of the world.

The central idea of answer set programming is to simplify a program first *relative to a candidate state.* Under this assumption, the program is transformed into a positive program whose minimal model is then checked. A candidate is an answer set exactly when it coincides with the minimal model of its own reduct.

Definition 4.168 (Reduct). *Let* Σ *be a propositional signature, let* P *be an extended logic program over* Σ*, and let* $S \subseteq \mathrm{Lit}(\Sigma)$ *be a state. The* reduct *of* P *with respect to* S*, written* P^S*, is the positive program consisting of all rules*

$$H \leftarrow A_1, \ldots, A_n,\ \mathbf{not}\ B_1, \ldots, \mathbf{not}\ B_m$$

from P *for which*

$$\{B_1, \ldots, B_m\} \cap S = \emptyset$$

holds, with all occurrences of **not***-literals then deleted from the remaining rules. Formally:*

$$P^S = \left\{ H \leftarrow A_1, \ldots, A_n \;\middle|\; \begin{array}{l} H \leftarrow A_1, \ldots, A_n, \\ \mathbf{not}\ B_1, \ldots, \mathbf{not}\ B_m \in P, \\ \{B_1, \ldots, B_m\} \cap S = \emptyset \end{array} \right\}$$

The intuition is very simple:

- Rules whose negative preconditions are refuted by S are removed.
- In all remaining rules, the default negations are deleted.

Thus the reduct is a kind of "view" of the program under the assumption that S is already the correct description of the world.

Example 4.169. *Consider the signature*

$$\Sigma_{\text{Entscheid}} = \{Tea, Coffee\}$$

and the program

$$P_{\text{Drink}} = \left\{ \begin{array}{l} Tea \leftarrow \mathbf{not}\ Coffee; \\ Coffee \leftarrow \mathbf{not}\ Tea \end{array} \right\}.$$

For the state

$$S_1 = \{Tea\}$$

the reduct is

$$P_{\text{Drink}}^{S_1} = \{Tea\}.$$

For the state

$$S_2 = \{Coffee\}$$

we analogously obtain

$$P_{\text{Drink}}^{S_2} = \{Coffee\}.$$

By contrast, the state

$$S_3 = \{Tea, Coffee\}$$

yields

$$P_{\text{Drink}}^{S_3} = \emptyset,$$

because both rules are eliminated by the corresponding opposing assertion contained in the state.

Definition 4.170 (Answer Set). *Let* Σ *be a propositional signature and let* P *be an extended logic program over* Σ*. A state* $S \subseteq \text{Lit}(\Sigma)$ *is called an* answer set *of* P *if* S *is the minimal model of* P^S*.*

Thus, the semantics of ASP can be summarized as follows:

> A state is acceptable exactly when, after reducing the program with respect to that state, it is precisely the least closed interpretation of its own reduct.

Example 4.171. *For the program* P_{Drink} *from the previous example, we have*

$$P_{\text{Drink}}^{\{Tea\}} = \{Tea\},$$

and the minimal model of this reduct is indeed $\{Tea\}$*. Hence* $\{Tea\}$ *is an answer set.*

Similarly,

$$P_{\text{Drink}}^{\{Coffee\}} = \{Coffee\},$$

and the minimal model is $\{Coffee\}$. *Hence* $\{Coffee\}$ *is also an answer set.*

By contrast, the state $\{Tea, Coffee\}$ *is not an answer set, since its reduct is* $\emptyset$, *whose minimal model is* $\emptyset$ *rather than the state* $\{Tea, Coffee\}$.

A slightly richer example illustrates how answer sets behave for rules with variables and facts.

Example 4.172. *Let*

$$\Sigma_{\text{Access}} = \{Student/1, Permit/1, Blocked/1, alice/0, bob/0\}$$

and

$$P_{\text{Access}} = \begin{Bmatrix} Permit(x) \leftarrow \\ \quad Student(x),\ \mathbf{not}\ Blocked(x); \\ Student(alice); \\ Student(bob); \\ Blocked(bob) \end{Bmatrix}$$

After grounding, one obtains, among others, the rules

$$Permit(alice) \leftarrow Student(alice), \mathbf{not}\ \ Blocked(alice)$$

and $Permit(bob) \leftarrow Student(bob), \mathbf{not}\ \ Blocked(bob)$.

The state

$$S = \{Student(alice), Student(bob), \\ Blocked(bob), Permit(alice)\}$$

is an answer set: in the reduct P^S_{Access}, *the rule for alice is retained, the rule for bob is removed, and the minimal model of this reduct is exactly* S.

Thus, a program may have several answer sets, or it may have none at all. The former is typical of programs with genuine choices, whereas the latter is typical of inconsistent programs.

Example 4.173. *The program*

$$P_{\text{Choice}} = \left\{ \begin{array}{l} A \leftarrow \texttt{not}\ B; \\ B \leftarrow \texttt{not}\ A \end{array} \right\}$$

has exactly two answer sets, namely $\{A\}$ *and* $\{B\}$. *By contrast, the program*

$$P_{\text{Contradict}} = \{A; \neg A\}$$

has no answer set, because no consistent state can be the minimal model of its reduct.

Answer sets satisfy several basic properties. In particular, they are pairwise incomparable in the sense that no proper inclusion can occur between distinct answer sets.

Proposition 4.174. *Let* Σ *be a propositional signature and let* P *be an extended logic program. For answer sets* S_1 *and* S_2 *of* P*, the following statements hold:*

1. *If* $S_1 \subseteq S_2$*, then* $S_1 = S_2$.
2. *For every rule* $r \in P$ *with* $\text{body}(r) = \emptyset$*, we have* $\text{head}(r) \subseteq S_1$.

3. We have

$$S_1 \subseteq \bigcup_{r \in P} \text{head}(r).$$

Proof. We prove the three statements separately.

For 1. Assume that $S_1 \subseteq S_2$. By the definition of the reduct, we then have

$$P^{S_2} \subseteq P^{S_1}.$$

Indeed, the larger the candidate S_2 is, the more rules are eliminated; the reduct therefore becomes smaller.

Since S_1 is an answer set, S_1 is the minimal model of P^{S_1}. Because $P^{S_2} \subseteq P^{S_1}$, every model of P^{S_1} is also a model of P^{S_2}; in particular, S_1 is a model of P^{S_2}. Since S_2 is the minimal model of P^{S_2}, it follows that

$$S_2 \subseteq S_1.$$

Together with the assumption $S_1 \subseteq S_2$, this yields

$$S_1 = S_2.$$

For 2. Let $r \in P$ be a fact, that is, a rule of the form

$$H \leftarrow .$$

Then $r \in P^{S_1}$ for every state S_1, since facts contain no default negation and therefore are never removed. Since S_1 is the minimal model of P^{S_1}, it is in particular a closed model of P^{S_1}. Hence the head of the fact must belong to S_1:

$$H \in S_1.$$

For 3. Suppose that there were a literal $H \in S_1$ that does not occur in the head of any rule of P. Consider

$$S_1' := S_1 \setminus \{H\}.$$

We show that S_1' is still a model of P^{S_1}. To this end, let

$$r: \; K \leftarrow A_1, \ldots, A_n$$

be a rule from P^{S_1} such that

$$\{A_1, \ldots, A_n\} \subseteq S_1'.$$

Then also $\{A_1, \ldots, A_n\} \subseteq S_1$. Since S_1 is a model of P^{S_1}, it follows that $K \in S_1$.

Now K cannot be equal to H, because by assumption H occurs in no rule head of P and therefore in no rule head of P^{S_1}. Thus $K \neq H$, and hence

$$K \in S_1'.$$

Thus S_1' is also closed under P^{S_1}.

This contradicts the minimality of S_1 as the minimal model of P^{S_1}, since $S_1' \subsetneq S_1$. Hence no such H can exist, and therefore

$$S_1 \subseteq \bigcup_{r \in P} \text{head}(r).$$

□

The last two statements are useful because they strongly restrict the search for answer sets. In particular, it suffices to consider only candidates that contain all facts and do not contain more literals than can be generated by rule heads in the first place. This gives rise to a simple, though generally inefficient, brute-force approach.

Example 4.175. *For the program*

$$P_{\text{Drink}} = \left\{ \begin{array}{l} Tea \leftarrow \mathbf{not}\ Coffee; \\ Coffee \leftarrow \mathbf{not}\ Tea \end{array} \right\}$$

we only have to check the candidates

$$\{Tea\}, \quad \{Coffee\}$$

since other consistent sets either do not match the possible head literals or are excluded by the preceding proposition. Both candidates are indeed answer sets.

Finally, we define the consequence relation of answer set programming.

Definition 4.176 (ASP Consequence). *Let Σ be a propositional signature and let P be an extended logic program over Σ. A literal $L \in \mathrm{Lit}(\Sigma)$ is called an* ASP consequence *of P, written*

$$P \models_{\text{ASP}} L,$$

if and only if, for every answer set S of P, we have:

$$L \in S.$$

Thus, from a model-theoretic point of view, the semantics is very simple to read: what occurs in all answer sets is enforced by the program; what occurs only in some answer sets remains open.

Example 4.177. *For the program* P_{Drink}*, we have*

$$P_{\text{Drink}} \not\models_{\text{ASP}} neither\, Tea\; nor\; Coffee,$$

because there are two answer sets, namely $\{Tea\}$ *and* $\{Coffee\}$.

By contrast, for the program

$$P_{\text{Access}}$$

from the example above, we have

$$\begin{aligned} P_{\text{Access}} &\models_{\text{ASP}} Student(alice) \quad \textit{and} \\ P_{\text{Access}} &\models_{\text{ASP}} Blocked(bob). \end{aligned}$$

because these literals occur in every answer set.

This captures the basic idea of answer set programming: a program is not read simply through arbitrary models, but through exactly those models that remain stable after reduction. This stability makes ASP a particularly suitable formalism for knowledge representation with exceptions, alternatives, and nonmonotonic reasoning.

4.5.4 Exercises and Solutions

Exercise 4.178. *Consider the propositional signature*

$$\Sigma = \{A, B, C, D\}$$

and the extended logic program

$$P = \{B \leftarrow \neg D; \quad C; \quad A; \quad \neg D \leftarrow \neg B\}.$$

Decide, for each of the following statements, whether it is true or false.

1. $\leftarrow \neg B, \texttt{not}\ D$ *is a rule over* Σ.
2. *We have* $\text{head}(D \leftarrow C) = \{C\}$.
3. $\{A, \neg C, C\}$ *is a state in* P.
4. $\{A, C\}$ *is a minimal model of* P.

Solution 4.179. *We check the statements one by one.*

1. ***False.*** *According to our definition, every rule has the form*

   $$H \leftarrow A_1, \ldots, A_n, \texttt{not}\ B_1, \ldots, \texttt{not}\ B_m,$$

   that is, with a non-empty *head* H. *The expression*

   $$\leftarrow \neg B, \texttt{not}\ D$$

   has no head and is therefore not a valid rule over Σ *in this formalism. In some variants of ASP, such expressions are called* constraints, *but such rules have not been allowed here.*

2. ***False.*** *In the rule*

$$D \leftarrow C$$

the head of the rule is the literal D. Hence

$$\text{head}(D \leftarrow C) = \{D\},$$

not $\{C\}$. The literal C occurs in the body, not in the head.

3. ***False.*** *A state must be consistent, meaning that for no atom may both the atom and its classical negation be contained in it. The set*

$$\{A, \neg C, C\}$$

is not consistent, because it contains both C and $\neg C$. Thus this set is not a state.

4. ***True.*** *We show that $S = \{A, C\}$ is a minimal model of P.*

   First, S is closed under P:

   - *The facts A and C are contained in S.*
   - *The rule $B \leftarrow \neg D$ is not applicable in S, because $\neg D \notin S$.*
   - *The rule $\neg D \leftarrow \neg B$ is likewise not applicable, because $\neg B \notin S$.*

Thus no rule derives anything additional, and S is closed.

Now consider minimality: Because of the facts, every closed state must contain at least A and C. Hence, for every closed state S',

$$\{A, C\} \subseteq S'.$$

Therefore, $S = \{A, C\}$ is the smallest closed state and hence a minimal model of P.

Exercise 4.180. *Consider the propositional signature*

$$\Sigma = \{A, B, C, D, E\}$$

and the extended logic program

$$P = \left\{ \begin{array}{l} B \leftarrow \neg D; \\ D; \\ \neg C \leftarrow \texttt{not}\ \neg B; \\ \neg A \leftarrow \texttt{not}\ \neg E, \texttt{not}\ B; \\ \neg B \leftarrow D, \texttt{not}\ \neg C \end{array} \right\}.$$

Also consider the expression

$$P_2 = \left\{ \begin{array}{l} \neg C \leftarrow \texttt{not}\ \neg B; \\ \neg A; \\ \neg B \leftarrow D \end{array} \right\}.$$

Decide, for each of the following statements, whether it is true or false.

1. $C \leftarrow \texttt{not}\ \neg E, \texttt{not}\ \neg D$ *is positive.*

2. *We have* $P^{S_2} = P_2$ *with* $S_2 = \{\neg A, D\}$.

3. $\{\neg A, \neg C, D\}$ *is an answer set of* P.

4. *We have* $P \models_{\text{ASP}} \neg C$.

Solution 4.181. *We consider the statements one by one.*

1. ***False.*** *A program is* positive *precisely when it contains no default negation. The rule*

$$C \leftarrow \texttt{not}\ \neg E, \texttt{not}\ \neg D$$

contains two occurrences of **not**. *Hence it is not positive. The fact that classical negation occurs inside the negated expressions does not change this: what matters here is the symbol* **not** *in the body of the rule.*

2. ***False.*** *We compute the reduct of* P *with respect to*

$$S_2 = \{\neg A, D\}.$$

We proceed rule by rule:

- *The rule* $B \leftarrow \neg D$ *contains no default negation and therefore remains unchanged.*
- *The fact* D *also remains unchanged.*

- *In the rule*

$$\neg C \leftarrow \textbf{\textit{not}}\ \neg B$$

$\neg B \notin S_2$, so the rule remains, and after removing ***not*** *it becomes the fact*

$$\neg C.$$

- *In the rule*

$$\neg A \leftarrow \textbf{\textit{not}}\ \neg E, \textbf{\textit{not}}\ B$$

neither $\neg E$ nor B belongs to S_2. Hence this rule also remains and, after removing ***not****, becomes the fact*

$$\neg A.$$

- *In the rule*

$$\neg B \leftarrow D, \textbf{\textit{not}}\ \neg C$$

we do have $D \in S_2$, but $\neg C \notin S_2$. Therefore the rule remains and becomes

$$\neg B \leftarrow D.$$

Thus

$$P^{S_2} = \left\{ \begin{array}{l} B \leftarrow \neg D; \\ D; \\ \neg C; \\ \neg A; \\ \neg B \leftarrow D \end{array} \right\}.$$

This is therefore not *equal to* P_2. *In particular, when forming the reduct, the remaining rules of* P *were not simply omitted; moreover, in the rule* $\neg C \leftarrow \textbf{not}\ \neg B$, *the negative part has to be replaced by the fact* $\neg C$.

3. **True.** *We show that*

$$S = \{\neg A, \neg C, D\}$$

is an answer set of P. *To do so, we determine the reduct* P^S.

- *The rule* $B \leftarrow \neg D$ *remains unchanged.*
- *The fact* D *remains unchanged.*
- *In* $\neg C \leftarrow \textbf{not}\ \neg B$ *we have* $\neg B \notin S$, *so this becomes the fact* $\neg C$.
- *In* $\neg A \leftarrow \textbf{not}\ \neg E, \textbf{not}\ B$, *neither* $\neg E$ *nor* B *belongs to* S, *so this becomes the fact* $\neg A$.
- *The rule* $\neg B \leftarrow D, \textbf{not}\ \neg C$ *is removed because* $\neg C \in S$.

Hence

$$P^S = \left\{ \begin{array}{l} B \leftarrow \neg D; \\ D; \\ \neg C; \\ \neg A \end{array} \right\}.$$

We now determine the minimal model of this positive program. From the facts we immediately obtain D,

$\neg C$, and $\neg A$. The rule $B \leftarrow \neg D$ is not applicable, since $\neg D$ is not in the model. Thus the minimal model is exactly

$$\{\neg A, \neg C, D\} = S.$$

Therefore S is indeed an answer set of P.

4. **False.** *The statement*

$$P \models_{\text{ASP}} \neg C$$

means that $\neg C$ must occur in all *answer sets of P.*

However, we have just seen that P has at least the answer set

$$\{\neg A, \neg C, D\}$$

. There is, however, also the answer set

$$\{\neg A, \neg B, D\}.$$

For this state, the rule $\neg C \leftarrow \texttt{not}\ \neg B$ is removed in the reduct, while the rule $\neg B \leftarrow D, \texttt{not}\ \neg C$ becomes

$$\neg B \leftarrow D$$

together with the facts D and $\neg A$. The minimal model is then exactly $\{\neg A, \neg B, D\}$.

Since this answer set contains no *$\neg C$, $\neg C$ is not contained in all answer sets. Hence*

$$P \not\models_{\text{ASP}} \neg C.$$

Additional remark. *The two answer sets of P are therefore*

$$\{\neg A, \neg C, D\} \qquad \textit{and} \qquad \{\neg A, \neg B, D\}.$$

This is exactly why $\neg C$ is not an ASP consequence, even though it occurs in one of the two answer sets.

Chapter 5

Concluding Reflections

5.1 Summary

This textbook has guided the reader step by step through central areas of logic and has shown how varied logical reasoning, formal language, and semantic interpretation can be. The starting point, in Chapter 0, was a concise overview of the basic aims and methods of logic. Building on this foundation, Chapter 1 introduced propositional logic as the first major foundational system. There, we first clarified syntax and semantics and introduced notions such as formula, interpretation, model, and logical consequence. The focus then shifted to structured reasoning: we examined equivalences, normal forms, and important transformations before turning to calculi, especially resolution, as well as to initial algorithmic procedures for satisfiability checking and automated reasoning.

Chapter 2 extended this perspective to predicate logic, where statements are no longer made merely about truth values, but about objects and the relations between them. After introducing syntax and semantics, we studied fundamental metatheoretical results such as the substitution lemma, the coincidence lemma, and Herbrand's theorem. Here, too, normal forms and resolution procedures played a central role, allowing logical consequences to be reduced systematically to questions of unsatisfiability. At the same time, it became clear that many questions concerning first-order logical consequence are not decidable by algorithms in general. Nevertheless, methods such as first-order resolution provide powerful tools in many practical applications. In addition, we encountered the description logic ALC as an important building block for ontologies. A particular highlight of the chapter was Gödel's incompleteness results, which impressively demonstrate that formal systems have principled limitations. Finally, the discussion of second-order predicate logic once again illustrated how strongly expressive power and complexity influence one another in logical systems.

Chapter 3 dealt with logics involving modalities. The aim here was to consider statements not only as true or false, but also with respect to their scope of validity, for example in terms of possibility, necessity, time, or knowledge. General modal logic was given a semantic interpretation by means of Kripke structures and was complemented by a tableau calculus. On this basis, temporal logics such as Linear Temporal Logic and Computation Tree Logic were introduced as systems that formally model different conceptions of time. We also considered epistemic and dynamic extensions, which make it possible to express the knowledge of several agents as well as changes in states of knowledge.

Chapter 4 finally introduced selected non-classical logics. Intuitionistic logic showed that one can reject the law of excluded middle and still develop a coherent logical system. Paraconsistent logic made clear that even inconsistent bodies of knowledge can be handled meaningfully without making all reasoning trivial. Fuzzy logic generalized the classical true/false scheme to continuous truth values and thereby opened a way to represent gradual and vague statements. With default logic and answer set programming, we then encountered forms of nonmonotonic reasoning in which new information can revise earlier conclusions. These logics are especially important for realistic applications because they can deal with exceptions, uncertainty, and limited knowledge.

Taken as a whole, this book has shown that logic is far more than a collection of formal rules. Logic is a precise tool for analyzing language, knowledge, inference, and computability. It combines mathematical rigor with a high degree of practical relevance and thus forms a foundation for many areas of computer science, philosophy, linguistics, and artificial intelligence. Readers who have understood the topics treated in this book possess a solid basis for further exploring both classical and modern logical systems.

5.2 Further Directions and Applications

This book, "Formal Logic", has necessarily presented only a selected part of the vast and still highly active landscape of logic. Yet this limitation is also one of its strengths: the systems discussed here provide a solid foundation on which many more advanced theories and applications in mathematics, computer science, philosophy, and linguistics can be built.

The practical importance of first-order predicate logic is especially clear. In formal verification, it is used to formulate properties of programs and hardware systems precisely and then check them automatically. For example, one can show that certain forms of access are possible only under specified conditions, that loop invariants are preserved, or that a system never reaches undesirable states. In knowledge representation, predicate logic provides a natural formal language for recording facts and relations within a domain; on this basis, conclusions can then be drawn, for instance in expert systems or semantic knowledge graphs. It also plays a central role in database theory, since many queries can be formulated logically and understood as logical operations over relations. In this way, logic becomes a bridge between mathematical precision and practical data processing.

Modal logic, too, has become important far beyond its classical setting. In computer science, it allows systems with changing states to be modeled, such as communication protocols, concurrent processes, or embedded systems. Temporal logics are used especially in model checking, where one investigates whether a system is correct not only in its present state, but also across all possible future developments with respect to certain safety or liveness properties. In distributed systems, for example, one can express that a message will eventually arrive, that a certain state always remains preserved, or that a reaction to an event occurs within an appropriate amount of time. The modal perspective is therefore valuable because it captures not merely isolated statements, but entire state spaces and their transitions.

Combined and extended logics form a particularly active area of research. Hybrid logics connect modal and predicative means of expression and make it possible not only to speak about possibility and necessity, but also to name individual states explicitly. This allows highly fine-grained statements about systems to be formulated, for example in the modeling of agents, states of knowledge, or navigable structures. Such extensions show that logical systems are not rigid; rather, they can be adapted deliberately to new questions without losing their formal rigor.

Non-classical logics open up further perspectives. Intuitionistic logic plays a central role in constructive mathematics, because here a proof is not meant merely to secure the truth of a statement, but can also provide a constructive procedure. In type theory, this close connection between proofs and constructions is especially fruitful; it forms a basis for modern proof systems and for large parts of theoretical computer science. Paraconsistent logics, in turn, are useful when bodies of data contain contradictions, but these contradictions should not make every arbitrary statement derivable. This is relevant, for example, in heterogeneous knowledge sources, faulty data, or merged databases, where inconsistencies may be the rule rather than the exception.

In artificial intelligence, default logic and answer set programming are important tools for nonmonotonic reasoning. They model situations in which one initially makes reasonable assumptions, but must revise them when new information becomes available. In real application domains, precisely this kind of behavior is typical: a system must not operate only with complete knowledge, but must also be able to handle exceptions, uncertainty, and incomplete information. Answer set programming is therefore particularly useful for planning problems, diagnostic tasks, and configuration problems, because it can structure complex search spaces in a way that is both logical and efficient.

Fuzzy logic extends the classical model of truth by introducing gradual degrees of truth, making it attractive wherever statements are not simply completely true or false, but hold to different degrees. Typical applications can be found in control engineering, for example in automatic control systems for washing machines, air-conditioning units, or driver-assistance systems, where vague terms such as "warm", "strong", or "near" must be processed formally. Fuzzy logic can also support decision-making by combining several imprecise criteria and deriving robust recommendations for action. Its strength lies precisely in translating concepts shaped by everyday language into a formally controllable model.

Another forward-looking field is the automation of logical reasoning. Theorem provers and SMT solvers are now used in many areas, for instance in software verification, the analysis of hardware designs, and the checking of mathematical proofs. A typical example is the search for memory errors or unreachable states in a program: instead of examining the code only by hand, one translates certain properties into logical formulas and lets a solver analyze them. Such tools show that logic is not merely a theoretical discipline, but contributes directly to the reliability of digital systems.

In addition, logic is increasingly intertwined with machine learning in neuro-symbolic artificial intelligence. In this field, data-driven learning methods are combined with symbolic structures that can be interpreted logically. This is especially interesting because neural networks are powerful at recognizing patterns, but often remain difficult to explain, whereas logical systems offer transparency, traceability, and formal guarantees. The combination of both approaches promises systems that can learn flexibly while still remaining justifiable and verifiable.

Finally, the limits of logic themselves reveal its particular depth. Undecidability, incompleteness, and the boundaries of formal provability are not mere marginal phenomena, but central insights into formal systems. The study of Gödel's incompleteness theorems or Turing's halting problem shows that there are questions which, in principle, cannot be solved by a single algorithmic procedure. Precisely this insight sharpens our understanding of what logic can achieve – and what it cannot.

Overall, logic remains a highly relevant and productive discipline. It continues to develop both in theoretical foundational research and in its many applications. Readers who wish to go further will find natural connections to areas such as formal methods, knowledge representation, automated theorem proving, model checking, and philosophical logic. Logic is therefore not only an instrument of precise thought, but also a key to understanding complex systems in science and technology.

5.3 Recommended Further Reading

Readers who would like to continue studying logic after this book will find the following works to be meaningful and fruitful complements. They not only deepen individual aspects of logic, but also show how closely logical thinking is connected with computer science, mathematics, philosophy, and theoretical reflection.

A particularly natural continuation is the book "Logic: Foundations, the P-versus-NP Problem, and Information-Theoretic Perspectives". It connects directly with the topics treated here, but goes one step further by relating classical foundations of logic to questions from information theory and computational complexity. For readers interested in the limits of formal systems, this work offers a valuable deepening of perspective. It shows that logic is not only an instrument for analyzing correct inferences, but also a way of approaching fundamental questions about what can be efficiently recognized, proved, or computed at all.

Also recommended is "Object-Oriented Programming in Java". At first glance, this is a programming book, yet here too there is a close connection with logic. Programming requires precise and structured thinking, which overlaps with logical work in many respects: conditions must be formulated unambiguously, processes must be organized clearly, and structures must be modeled cleanly. Those who learn to program also train the kind of analytical thinking that is highly useful for understanding formal systems. Conversely, a solid grasp of logical structures can help in designing programs more systematically and in recognizing sources of error more reliably.

The book "Theory of Computation" forms an especially important bridge to logic. It treats formal languages, automata, computability, Turing machines, reductions, and undecidability—precisely those topics that are closely connected with the logical foundations of computer science. Many proof techniques developed in logic reappear there in modified form. Readers who want to understand why some problems are solvable in principle while others are not, how formal models work, and why undecidability is not a marginal phenomenon but a basic feature of formal systems will benefit from this book in a particularly direct way.

Building on this, "Complexity Theory" leads even deeper into the theoretical limits and possibilities of computation. While logic asks what can be expressed and proved at all, complexity theory investigates how costly such problems are. This makes clear that not only solvability, but also practical efficiency, plays a central role. Concepts such as P, NP, $PSPACE$, and NP-completeness are therefore significant for logical thinking as well, because they show that there is often a substantial gap between theoretical possibility and practical feasibility.

The book "Algorithms and Data Structures" complements this perspective in a very concrete way. It presents the tools by which problems can be handled efficiently in the first place. Logic and algorithmics stand in a close relationship here: logic provides the precise language of thought, while algorithms provide the controlled implementation of that thought in procedures. Readers who already know the logical foundations will be especially well prepared to see how abstract considerations can be translated into concrete computational and decision procedures. At the same time, this work provides an ideal preparation for more advanced topics such as formal verification, automated theorem proving, and questions from complexity theory.

The book "Linear Algebra and Analytic Geometry" is also an important complement. Although its subject matter initially appears closer to mathematics than to logic, the connection is stronger than it may seem at first. Linear algebra trains the ability to work with abstract structures, proofs, definitions, and precise chains of reasoning—exactly the skills that are also essential for logical thought. Moreover, linear structures play a central role in many areas of computer science, including coding theory, cryptography, data analysis, and numerical methods. A reader who understands both logic and linear algebra therefore gains an especially strong foundation for further mathematical and computational topics.

Finally, the book "Philosophy" is recommended. It complements formal logic by providing the broader philosophical perspective in which questions of truth, justification, rationality, and language are reflected upon not only technically, but also conceptually and practically. Logic is a core field of philosophy, yet it gains its full significance only in dialogue with epistemology, philosophy of language, philosophy of science, and ethics. This book shows how philosophical thinking sharpens our judgment and how closely argumentative clarity, conceptual precision, and reflective action belong together. Readers who want to understand logic not merely as a formal procedure, but as part of a broader theory of thought, will find this a particularly worthwhile complement.

Taken together, these works form a coherent intellectual network: logic provides the formal foundation, computer science its algorithmic and theoretical application, mathematics its structural deepening, and philosophy its conceptual and methodological horizon. Their special value lies precisely in this connection. Anyone who wants to understand logic deeply benefits not only from studying logic itself, but also from looking at the disciplines in which it continues to operate, prove its strength, and raise new questions.

Acknowledgements

I sincerely thank you for your interest in this book and for engaging with the topics it presents. It is a great pleasure to share this work with readers who are interested in precise reasoning, formal methods, and the possibilities opened up by logical systems.

If, while reading, you notice any ambiguities, errors, or suggestions for improvement, I would be very grateful to hear from you. Comments and corrections help to refine this work further and make it even more useful for future readers.

Thank you very much for your attention and interest.

Bibliography

[1] Herbert B. Enderton. *A Mathematical Introduction to Logic*, 2nd ed., Academic Press, 2000.

[2] Elliott Mendelson. *Introduction to Mathematical Logic*, 5th ed., Chapman & Hall/CRC, 2009.

[3] George S. Boolos, John P. Burgess, Richard C. Jeffrey. *Computability and Logic*, 5th ed., Cambridge University Press, 2007.

[4] Wolfgang Rautenberg. *A Concise Introduction to Mathematical Logic*, 4th ed., Springer, 2010.

[5] Kurt Gödel. *On Formally Undecidable Propositions of Principia Mathematica and Related Systems I.* Monatshefte für Mathematik und Physik 38(1), 1931, pp. 173–198.

[6] Ernest Nagel, James R. Newman. *Gödel's Proof.* New York University Press, 1958.

[7] Patrick Blackburn, Maarten de Rijke, Yde Venema. *Modal Logic.* Cambridge University Press, 2001.

[8] E. Allen Emerson. *Temporal and Modal Logic.* In: J. van Leeuwen (Ed.), *Handbook of Theoretical Computer Science*, Vol. B, Elsevier, 1990, pp. 995–1072.

[9] Amir Pnueli. *The Temporal Logic of Programs.* In: *Proceedings of the 18th IEEE Symposium on Foundations of Computer Science (FOCS)*, 1977, pp. 46–57.

[10] Zohar Manna, Amir Pnueli. *The Temporal Logic of Reactive and Concurrent Systems: Specification.* Springer, 1992.

[11] A. S. Troelstra, D. van Dalen. *Constructivism in Mathematics, Vol. 1: Basic Concepts.* North-Holland, 1988.

[12] Arend Heyting. *Intuitionism: An Introduction.* North-Holland, 1966.

[13] Jaakko Hintikka. *Knowledge and Belief.* Cornell University Press, 1962.

[14] R. Fagin, J. Y. Halpern, Y. Moses, M. Vardi. *Reasoning About Knowledge.* MIT Press, 1995.

[15] Graham Priest. *An Introduction to Non-Classical Logic: From If to Is*, 2nd ed., Cambridge University Press, 2008.

[16] Graham Priest, Richard Routley, Jean Norman (Eds.). *Paraconsistent Logic: Essays on the Inconsistent.* Philosophia Verlag, 1989.

[17] Lotfi A. Zadeh. *Fuzzy Sets. Information and Control* 8, 1965, pp. 338–353.

[18] George J. Klir, Bo Yuan. *Fuzzy Sets and Fuzzy Logic: Theory and Applications.* Prentice Hall, 1995.

[19] Raymond Reiter. *A Logic for Default Reasoning. Artificial Intelligence* 13(1-2), 1980, pp. 81–132.

[20] Michael Gelfond, Vladimir Lifschitz. *The Stable Model Semantics for Logic Programming.* In: *Proc. 5th Intl. Conf. on Logic Programming*, 1988, pp. 107–136.

Copyright Page

Bibliographic information from the German National Library:
The German National Library lists this publication in the German National Bibliography; detailed bibliographic data can be accessed online at `dnb.dnb.de`.

Publisher:

BoD · Books on Demand GmbH,
Überseering 33,
22297 Hamburg,
bod@bod.de

Printing:
Libri Plureos GmbH,
Friedensallee 273,
22763 Hamburg

ISBN: 978-3-6951-6210-9